**142**
# Structure and Bonding

**Series Editor: D. M. P. Mingos**



For further volumes:
http://www.springer.com/series/430

# Structure and Bonding

Series Editor: D. M. P. Mingos

Recently Published and Forthcoming Volumes

**Molecular Electronic Structures of Transition Metal Complexes I**
Volume Editors: D. M. P. Mingos, Peter Day, Jens Peder Dahl
Vol. 142, 2012

**Fuel Cells and Hydrogen Storage**
Volume Editors: Andrew Bocarsly, D. M. P. Mingos
Vol. 141, 2011

**Zintl Ions**
Principles and Recent Developments
Volume Editor: Thomas F. Fässler
Vol. 140, 2011

**Zintl Phases**
Principles and Recent Developments
Volume Editor: Thomas F. Fässler
Vol. 139, 2011

**Inorganic 3D Structures**
Volume Editor: Angel Vegas
Vol. 138, 2011

**Molecular Catalysis of Rare-Earth Elements**
Volume Editor: Peter W. Roesky
Vol. 137, 2010

**Metal-Metal Bonding**
Volume Editor: Gerard Parkin
Vol. 136, 2010

**Functional Phthalocyanine Molecular Materials**
Volume Editor: Jianzhuang Jiang
Vol. 135, 2010

**Data Mining in Crystallography**
Volume Editors: Hofmann, D. W. M., Kuleshova, L. N.
Vol. 134, 2010

**Controlled Assembly and Modification of Inorganic Systems**
Volume Editor: Wu, X.- T.
Vol. 133, 2009

**Molecular Networks**
Volume Editor: Hosseini, M. W.
Vol. 132, 2009

**Molecular Thermodynamics of Complex Systems**
Volume Editors: Lu, X., Hu, Y.
Vol. 131, 2009

**Contemporary Metal Boron Chemistry I**
Volume Editors: Marder, T. B., Lin, Z.
Vol. 130, 2008

**Recognition of Anions**
Volume Editor: Vilar, R.
Vol. 129, 2008

**Liquid Crystalline Functional Assemblies and Their Supramolecular Structures**
Volume Editor: Kato, T.
Vol. 128, 2008

**Organometallic and Coordination Chemistry of the Actinides**
Volume Editor: Albrecht-Schmitt, T. E.
Vol. 127, 2008

**Halogen Bonding**
Fundamentals and Applications
Volume Editors: Metrangolo, P., Resnati, G.
Vol. 126, 2008

**High Energy Density Materials**
Volume Editor: Klapötke, T. H.
Vol. 125, 2007

**Ferro- and Antiferroelectricity**
Volume Editors: Dalal, N. S., Bussmann-Holder, A.
Vol. 124, 2007

**Photofunctional Transition Metal Complexes**
Volume Editor: V. W. W. Yam
Vol. 123, 2007

# Molecular Electronic Structures of Transition Metal Complexes I

Volume Editors:
D. Michael P. Mingos · Peter Day · Jens Peder Dahl

With contributions by

J.P. Dahl · C.D. Flint · H.B. Gray · N.E. Henriksen ·
R.K. Hocking · K.M. Lancaster · K.B. Møller · J.H. Palmer ·
E.I. Solomon · W.C. Trogler · J.R. Winkler

Springer

*Editors*
D. Michael P. Mingos
Inorganic Chemistry Laboratory
Oxford University
South Parks Road
Oxford OX1 3QR, UK
michael.mingos@st-edmund-hall.oxford.ac.uk

Peter Day
The Royal Institution of Great Britain
Albermarle Street 21
W1X 4BS London
United Kingdom
profpday@msn.com

Jens Peder Dahl
Department of Chemistry
Technical University of Denmark
DTU 207, 2800 Lyngby
Denmark
dahlhome@webspeed.dk

ISSN 0081-5993 e-ISSN 1616-8550
ISBN 978-3-642-27369-8 e-ISBN 978-3-642-27370-4
DOI 10.1007/978-3-642-27370-4
Springer Heidelberg Dordrecht London New York

Library of Congress Control Number: 2011944826



Printed on acid-free paper

Springer is part of Springer Science+Business Media (www.springer.com)

# Series Editor

Prof. D. Michael P. Mingos

Inorganic Chemistry Laboratory
Oxford University
South Parks Road
Oxford OX1 3QR, UK
*michael.mingos@st-edmund-hall.oxford.ac.uk*

# Volume Editors

D. Michael P. Mingos

Inorganic Chemistry Laboratory
Oxford University
South Parks Road
Oxford OX1 3QR, UK
*michael.mingos@st-edmund-hall.oxford.ac.uk*

Peter Day

The Royal Institution of Great Britain
Albermarle Street 21
W1X 4BS London
United Kingdom
*profpday@msn.com*

Jens Peder Dahl

Department of Chemistry
Technical University of Denmark
DTU 207, 2800 Lyngby
Denmark
*dahlhome@webspeed.dk*

# Editorial Board

# Structure and Bonding
# Also Available Electronically

*Structure and Bonding* is included in Springer's eBook package *Chemistry and Materials Science*. If a library does not opt for the whole package the book series may be bought on a subscription basis. Also, all back volumes are available electronically.

For all customers who have a standing order to the print version of *Structure and Bonding*, we offer the electronic version via SpringerLink free of charge.

If you do not have access, you can still view the table of contents of each volume and the abstract of each article by going to the SpringerLink homepage, clicking on "Chemistry and Materials Science," under Subject Collection, then "Book Series," under Content Type and finally by selecting *Structure and Bonding*.

You will find information about the

- Editorial Board
- Aims and Scope
- Instructions for Authors
- Sample Contribution

at springer.com using the search function by typing in *Structure and Bonding*.

Color figures are published in full color in the electronic version on SpringerLink.

## Aims and Scope

The series *Structure and Bonding* publishes critical reviews on topics of research concerned with chemical structure and bonding. The scope of the series spans the entire Periodic Table and addresses structure and bonding issues associated with all of the elements. It also focuses attention on new and developing areas of modern structural and theoretical chemistry such as nanostructures, molecular electronics, designed molecular solids, surfaces, metal clusters and supramolecular structures. Physical and spectroscopic techniques used to determine, examine and model structures fall within the purview of *Structure and Bonding* to the extent that the focus is on the scientific results obtained and not on specialist information concerning the techniques themselves. Issues associated with the development of bonding models and generalizations that illuminate the reactivity pathways and rates of chemical processes are also relevant.

The individual volumes in the series are thematic. The goal of each volume is to give the reader, whether at a university or in industry, a comprehensive overview of an area where new insights are emerging that are of interest to a larger scientific audience. Thus each review within the volume critically surveys one aspect of that topic and places it within the context of the volume as a whole. The most significant developments of the last 5 to 10 years should be presented using selected examples to illustrate the principles discussed. A description of the physical basis of the experimental techniques that have been used to provide the primary data may also be appropriate, if it has not been covered in detail elsewhere. The coverage need not be exhaustive in data, but should rather be conceptual, concentrating on the new principles being developed that will allow the reader, who is not a specialist in the area covered, to understand the data presented. Discussion of possible future research directions in the area is welcomed.

Review articles for the individual volumes are invited by the volume editors.

In references *Structure and Bonding* is abbreviated *Struct Bond* and is cited as a journal.

Impact Factor in 2010: 4.659; Section "Chemistry, Inorganic & Nuclear":
Rank 4 of 43; Section "Chemistry, Physical": Rank 25 of 127

*This volume is dedicated to*
*Carl Johan Ballhausen*

# Preface

When I heard in August 2010 that Carl Ballhausen had died at the age of 84, I was very sad because he had played such an important part in the development of my interests in coordination and theoretical inorganic chemistry in the 1960s. Indeed lectures given by Carl, Al Cotton, Dick Fenske, and Harry Gray at a NATO Summer School held at Selsdon Park Hotel near London in 1967 did much to stimulate my interest in theoretical inorganic chemistry. Later, when I embarked on my first semiempirical molecular orbital calculations to understand why nitrosyl complexes adopted linear and bent geometries, I was greatly assisted by "Molecular Orbital Theory," which Carl coauthored with Harry Gray (W.A. Benjamin New York, 1965). My sense of loss must have been small when compared with those of the many graduate students and postdoctoral fellows who had been supervised by him. His many international collaborators would also have missed his profound theoretical insights and friendly, jovial manner. Therefore, I was pleased when Peter Day after consulting with Jens Peder Dahl proposed a volume of *Structure and Bonding* dedicated to Carl's life and his seminal contributions to quantum chemistry. The large number of Carl's co-workers who have gone on to occupy senior academic positions made it relatively easy to propose a list of authors for the volume. Their response to invitations to contribute has been so enthusiastic that it proved necessary to expand the project into two volumes. The first one deals with the application of Carl's Ligand Field Theory to spectroscopic and chemical properties of transition metal complexes. The second volume concentrates on more theoretical topics and reviews the development of modern ab initio theoretical methods and discusses the broader implications of fundamental theoretical concepts.

Jens Peder Dahl was closely associated with Carl's research group for many years, and his introductory chapter summarizes his scientific achievements and in particular his contributions to the inorganic renaissance which unfolded from 1950. Harry Gray recounts his very fruitful collaboration with Carl which resulted in a general molecular orbital description of the metal-oxo bond, and he and Jay Winkler review more recent experimental and spectroscopic data on metal-oxo complexes of the later transition metals. Colin Flint also provides a brief review of his work with Carl in 1967–1968, and then describes his subsequent research on the assignments of the vibronic sidebands in the emission spectra of chromium ammine complexes. He also comments on the Jahn–Teller effect in the emissive state.

Joshua Palmer discusses the transition metal complexes of the noninnocent, electron-rich corrole macrocycle. This includes a detailed summary of investigations to determine the physical oxidation states of formally copper(III), iron(IV), and cobalt(IV) corroles. The electronic structures and reactivities of other metallocorroles are also discussed, and comparisons made between corrole and porphyrin complexes. William Trogler reviews thin film conductivity sensors that use metal phthalocyanine semiconductors which offer properties distinct from those of organic thin film transistor sensors. Phthalocyanines (Pc) are planar ligands that are closely related to porphyrins (Por), as both can be deprotonated to form dianions that yield a wide variety of square planar metal(II) chelate complexes MPc and MPor. Porphyrins are crucial redox active metalloenzymes that play key roles in metabolism, oxygen storage, photosynthesis, and other biological processes. Phthalocyanines are also important dye and paint pigments, as well as having electronic device applications in CD-recordable media, xerography, photovoltaics, and organic light-emitting devices. Kyle Lancaster reviews the concept of outer-sphere coordination (OSC) in the context of bioinorganic chemistry. A distinction is made between electronic and structural OSC, which arises from the interaction of the protein matrix with inner sphere ligands. Electronic OSC entails the electronic interaction between the polypeptide and inner-sphere ligands. These effects principally arise from hydrogen-bonding interactions, though through-space dipolar interactions are also encountered.

Rosalie Hocking and Ed Solomon in the next chapter note that Carl Ballhausen taught them how to interpret ground and low lying excited state spectral features in terms of ligand field and covalency effects. With the availability and development of synchrotron radiation, this has been extended to the X-ray region where ligand field and covalency effects have an impact on the metal K and L-edges. The L-edge intensity and its distribution over the multiplets are particularly powerful in experimentally determining the bonding in highly covalent systems, where ligand spectral features can obscure the use of more traditional methods. Klaus Møller and Niels Henriksen in their review discuss how time-resolved X-ray diffraction may lead to important insights into the dynamics of the chemical bond. The focus of the review is the underlying theoretical concepts which are required in order to extract the dynamics of the chemical bond from the time-resolved X-ray diffraction data.

In the first chapter of the second volume, Tom Ziegler reviews the development of electronic structure theories for transition metal complexes from the 1950s to the present day. Jan Linderberg also focuses on the emergence of quantitative means for the study of the electronic properties of complexes and molecules. The development, refinement, and application of the orbital picture have elucidated the electric and magnetic features of molecules and used for the interpretation of electronic transitions, electron spin resonance parameters, rotatory dispersion, nuclear quadrupole couplings, and geometric bonding patterns. In a chapter which addresses some fundamental issues, John and James Avery discuss the theory of Sturmians and generalized Sturmians. They show that when generalized Sturmians are used as basis functions, calculations on the spectra and physical properties of few electron atoms can be performed with great ease and good accuracy. Brian Sutcliffe considers the implications of Carl Ballhausen's research on vibrational spectra, and it is suggested that his use of the Born–Oppenheimer approximation is capable of some refinement and extension in the light of later developments. A consideration of the potential energy surface in the context of a full Coulomb Schrödinger Hamiltonian in which translational and rotational motions are explicitly considered would seem to require a reformulation of the Born–Oppenheimer approach. The resulting potential surface for vibrational motion allows for the rotational motion and the nuclear permutational symmetry of the molecule. Tony McCaffery considers the role played by angular momentum in chemical physics, an interest which was first aroused by a Carl Ballhausen lecture early in the author's scientific career. Later came a deeper understanding of the fundamental nature of angular momentum and the power of its formal algebraic expression. Spectroscopy using light of precisely defined energy and (z-component of) angular momentum represents a unique experimental probe with the potential to reveal the underlying physics of chemical processes. Experiments using circularly polarized emission of gas phase molecules led to new insights in the field of molecular collision dynamics. Further work, and that of others, has suggested an alternative formulation of the mechanics of bimolecular collisions

Frank Neese and his coauthors summarize recent developments in AI methods for strongly correlated electronic systems, and they discuss their implementation in highly efficient quantum chemistry programs which allow one to calculate – from first principles – the spectroscopic and magnetic properties of transition metal complexes with open d- and f-shells. For a long time, this field was the domain of ligand field theory (LFT), subject to various assumptions and approximations which are solely justified by the success of using this theory for the interpretation of experimental data. Yet the chemical significance of the ligand field parameters, while being under intense debate, remains unclear as far as the roots of LFT in its relation to rigorous quantum chemistry are concerned. In the present review, the authors attempt to answer the question how well ligand field performs from the point of view of state-of-the-art first principle calculations and how to connect the two areas. In his chapter, Steve Berry notes that the Gibbs Phase Rule relating the number of degrees of freedom $f$ of a system to the number of components $c$ and the number of coexisting phases $p$ is a central, universally used relation. However, for small systems, notably atomic clusters the Phase Rule shows the coexistence of two or more phases in thermodynamic equilibrium over *bands* of temperature and pressure (with no other forces acting on them). He demonstrates that it is consistent with the laws of thermodynamics and even allows one to estimate the upper size limit of any particular system for which such apparent violation could be observed.

These two volumes provide a fine endorsement of the way in which Carl's contributions have influenced so many scientists and led them explore important new areas of transition metal chemistry, quantum theory, and spectroscopy. The authors have successfully illuminated the fundamental and broad question which guided much of Carl's research – "What is a molecule and in particular a transition metal complex?". As an interesting postscript, Carl Ballhausen and Christian Klixbull Jørgenson were both in University of Copenhagen in the early 1950s and did their doctoral theses with Jannik Bjerrum at about the same time. At an early stage they collaborated on a series of papers titled "Studies of Absorption Spectra," but their relationship was not always an easy one. Jorgenson moved from Copenhagen at the age of 29 and did research at Cyanamid and the University of Geneva in Switzerland. He died in 2001 and a pair of commemorative volumes were published as part of the *Structure and Bonding* Series in 2004 (volumes 106 and 107). The four volumes will remind us and future generations the important role played by Danish coordination and theoretical chemists in the renaissance of inorganic chemistry.

Oxford
August 2011

D. Michael P. Mingos

# Contents

Struct Bond (2012) 142: 1–16
DOI: 10.1007/430_2011_48

Published online: 14 September 2011

# Carl Johan Ballhausen (1926–2010)

**Jens Peder Dahl**

**Abstract** In this introductory article, I summarize Carl Johan Ballhausen's scientific achievements. Ballhausen was an outstanding contributor to the inorganic renaissance, which unfolded over a period of several years from 1950 forward. I have endeavoured to let the biography be carried along by a parallel account of the quantum chemical developments.

**Keywords** Biography · Complexes · Crystal field · Inorganic renaissance · Ligand field

## Contents



J.P. Dahl (✉)
Department of Chemistry, Technical University of Denmark, DTU 207, 2800 Lyngby, Denmark
e-mail: dahlhome@webspeed.dk

## 1 Introduction

Carl Johan Ballhausen was born on April 4, 1926 in Vesterbro, Copenhagen, the son of Supreme Court Attorney Carl Ballhausen and Vibeke Ballhausen, b. Stein. The family moved to Baunegaardsvej 20 in Gentofte, where Carl Johan grew up. He received his formal education at Gentofte Statsskole where he passed the studentereksamen in 1945. In school, Ballhausen had taken an interest in science and conducted chemistry and physics experiments at home together with his friend Aage Winther. After his studentereksamen, he started to study chemistry at the University of Copenhagen, with Professor Jannik Bjerrum as his supervisor.

However, a serious traffic accident tied him to the bed from 1949 to 1952. He made the best of this situation to become acquainted with quantum mechanics, by self-tuition and discussions with Aage Winther. His scientific lifework was to become one of strengthening the use of quantum mechanics in chemistry.

He got his mag.scient. degree in 1954 and his dr.phil. degree in 1958.[1] In the years 1959–1996, Ballhausen was professor of physical chemistry at the University of Copenhagen.

In the following, I give an account of Ballhausen's career and achievements. I concentrate on his scientific work, but Ballhausen was also a man of letters, a manager and a collector. As he might have put it himself, with one of his characteristic phrases: "In my Father's house are many mansions". I was Ballhausen's first Ph.D. student, and we have solved several problems together. This makes it difficult for me to draw an entirely objective picture of the man. What I will do is that I will take a blackboard, a *tabula rasa*, and split it into three sub-boards. On the first of these, I will briefly recall the level of quantum chemistry about 1950, and also have a look at the status of chemistry in Denmark and elsewhere at that time. Then we will be prepared to look at the young Ballhausen in the proper settings and fill in the central sub-board. On the third sub-board, we will look at Ballhausen as a Danish professor and an international figure, and also touch upon his alternative activities. Finally, I invite the reader to follow one of Ballhausen's suggestions, "Step back from the blackboard", to see the integrated picture.

[1] The mag.scient. degree was essentially equivalent to a Ph.D. degree. The master's degree from the University of Copenhagen was called cand.mag.

## 2 The Rise of Quantum Chemistry

The exact foundation of modern quantum mechanics was laid in 1925–1926, by Werner Heisenberg, Max Born and Pascual Jordan, and by the Austrian physicist Erwin Schrödinger. The latter's formulation in terms of wave functions has, in particular, proved well suited for the description of atoms, molecules and solids, and their interaction with light. Thus, the Schrödinger equation supplies the basis for our understanding of the chemical bond and for the description of, say, atomic and molecular spectra.

During the first 25 years, 1926–1950, the Schrödinger equation was applied to many atoms and molecules, and this resulted in several quantum mechanical methods or *theories,* which were applicable to atoms and molecules. They are the *quantum chemical* theories. Here, we will mention the three major theories of the time and indicate their most conspicuous features.

### 2.1 Valence Bond Theory

This theory starts from Heitler and London's 1927-description of the electron-pair bond in the hydrogen molecule $H_2$. The wave function for the electron-pair bond is the covalent function $a(1)b(2) + b(1)a(2)$, where $a$ and $b$ are 1s atomic orbitals on the two atomic centres. Spin is included by multiplying with the spin singlet function $\sqrt{\frac{1}{2}}(\alpha(1)\beta(2) - \beta(1)\,\alpha(2))$. The description may be improved by adding small contributions from the ionic functions $a(1)a(2)$ and $b(1)b(2)$. For a heteropolar bond, the description is similar, but with larger weight for one of the ionic functions. For a polyvalent atom, for instance carbon, one obtains directed covalent bonds by hybridizing the atomic orbitals before constructing the electron-pair bonds. This was first suggested by John C. Slater and Linus Pauling in 1931. As an example, let us consider the methane molecule $CH_4$ for which we have four bonds, directed toward the corners of a regular tetrahedron. We may imagine such a molecule being formed in two steps from four hydrogen atoms in their $^2S$ ground states, electron configuration $(1s)^1$, and a carbon $^3P$ ground-state atom with its $(1s)^2(2s)^2(2p)^2$ electron configuration. First, we promote a $2s$ electron to a $2p$ electron and form a set of four orthogonal and equivalent $sp^3$ hybrid orbitals on the carbon atom. The carbon atom is now in a so-called valence state, a concept introduced by J.H. van Vleck in 1933. From there, we can construct a covalent electron-pair function from each hybrid orbital and a $1s$-orbital on a hydrogen atom.

Performing a good valence bond (VB) calculation is obviously a quite complicated procedure. It involves a determination of the energy of valence states and a calculation of dissociation energies via evaluation of very many difficult integrals. In addition, there are many molecules for which a single VB structure will not do. The electronic state must then be described by a linear combination of VB-functions, with coefficients to be determined. This corresponds to what Pauling called resonance between structures. For good reasons, work based on the VB theory before 1950 was largely qualitative and descriptive.

## 2.2 Molecular Orbital Theory

Molecular orbital (MO) theory starts from Burrau's 1926-description of the hydrogen molecular ion $H_{2^+}$ and the 1927-discussions by E.U. Condon and by F. Hund of the $H_2$ molecule. Dominant contributions during the next few years were due to R.S. Mulliken, J.E. Lennard-Jones, and others. In MO theory, the electrons are assigned to molecular orbitals that in principle extend over the whole molecule. The molecular orbitals may be determined in a self-consistent manner in a similar way as atomic orbitals are determined for atoms. About 1950 it was a well-defined theory, but quantitative work could only be completed for small molecules. However, it became popular to introduce heavy approximations and neglect many complicated integrals, with others being assigned semi-empirical values. Particularly popular was the Hückel molecular orbital method for conjugated hydrocarbons, introduced by E. Hückel in 1931.

## 2.3 Crystal-Field Theory

Crystal-field (CF) theory before 1950 was not so much a theory of valence. It was primarily a theory of the magnetic properties of transition-metal and rare-earth salts. In these salts, each metal ion is located at the centre of a complex, consisting of the metal ion and a set of atoms, molecules or radicals, so-called ligands. The number of ligands is called the coordination number. Typically, this number will be six, with the ligands sitting at the corners of a more or less regular octahedron. The metal ion may normally be considered to be a dipositive or tripositive ion. The metal ions in a salt are sufficiently well separated that Heisenberg exchange may be neglected. Thus, it is the electronic structure of a single metal ion that determines the magnetic susceptibility of the salt. The electronic structure of an ion is that of a perturbed free ion. The effect of the perturbation is first and foremost to reduce the degeneracy defined by the orbitals in the free atom. As shown by Hans A. Bethe in 1929, the pattern of the splitting in a field of given symmetry is primarily determined by group theory. For an ion in the first transition series, the pertinent orbitals are the five $3d$ orbitals $\left[3d_{z^2},\ 3d_{x^2-y^2},\ 3d_{yz},\ 3d_{zx},\ 3d_{xy}\right]$. They span an irreducible representation of the rotation group in three dimensions. In an octahedral field, this five-dimensional representation splits into two, namely, a three-dimensional $t_{2g}$ representation and a two-dimensional $e_g$ representation. The energy difference between the corresponding orbitals is called $10Dq$. It is an empirical parameter whose value depends on the ion and the strength of the perturbing octahedral field.

Apart from Bethe's monumental contribution, and also contributions by Kramers, it was largely van Vleck and his research group that developed crystal-field theory.

They were also responsible for extending its use, for instance to the study of electronic transitions within the incomplete inner shells of a transition-metal or rare-earth complex.

The name crystal-field theory tends to indicate that a complex must be immersed in a crystal, and hence that the theory is part of solid-state physics. When van Vleck was awarded the Nobel prize, in 1977, it was in fact the Nobel prize in physics, and he shared it with two other solid-state physicists, namely, Philip W. Anderson and Sir Nevill F. Mott. One may, however, also consider a complex as an isolated molecular unit. The crystal field will then be restricted to a field from the ligands of the complex. The formalism remains, however, unchanged, so one keeps the name crystal-field theory. But the theory is now more open for chemical applications. Quote van Vleck [37].

> For applications to molecular spectra, we are interested merely in the free molecule $XY_6$ as such, but in connection with magnetism, it is imperative to include an external field, since here the cluster $XY_6$ is embedded in the crystal.

## 3 Theoretical Inorganic Chemistry Around 1950

The quantum-chemical theories that we have mentioned did not attract much attention among Danish chemists at the time. An exception was R.W. Asmussen who measured the magnetic susceptibility of a large number of transition-metal complexes and collected them in his dissertation from 1944, *Magnetokemiske undersøgelser over uorganiske kompleksforbindelser*. He adapted Pauling's model of a transition-metal complex, according to which the binding between the central ion and the ligands involves suitably hybridized orbitals, built from *s*, *p*, and *d* atomic orbitals. The geometry around the central ion determines the hybridization and hence the number of electrons left over to contribute to the magnetic susceptibility. By reversing this argument, the measured susceptibility should allow us to make assertions about the geometry of a complex. Asmussen concluded that such a method would take us a long way, but there were several cases where the magnetic data were consistent with more than one geometry. Pauling's model lacked the flexibility to deal properly with such cases.

Asmussen did his doctor's work at the Department of Chemistry A [Kemisk Laboratorium A (KLA)] of the Technical University of Denmark (DTU). Shortly after the defense of his dissertation, he became professor at Kemisk Laboratorium B (KLB), DTU.

### 3.1 Kemisk Laboratorium A

At KLA, there was an old tradition for studying the chemistry of complex compounds, dating back to Sophus M. Jørgensen who was the director of KLA from 1866 to 1908, and professor at the University of Copenhagen from 1887. Jørgensen synthesized a large number of complexes and rationalized their structures by means of chain models. This was at the same time as Alfred Werner in Switzerland formulated his coordination theory in which such compounds would be truly three-dimensional structures. This led to a bitter controversy between the two men, lasting from 1893 to 1899 when Jørgensen withdrew from the discussion.[2] The controversy affected the scientific life at KLA far into the twentieth century.

In 1948, Jannik Bjerrum became the director of KLA and professor at the University of Copenhagen. His dissertation, from 1937, was entitled *Metal Ammine Formation in Aqueous Solutions*. Bjerrum was a prominent complex chemist, but he never became comfortable with quantum mechanics. He did not make a secret of this, however.

## *3.2 The Inorganic Renaissance*

About 1950, it was common practice to divide the field of chemistry into three subfields, organic chemistry, inorganic chemistry, and physical chemistry. Organic chemistry was conceived of as being fairly systematic. Physical chemistry was distinguished by its logical structure. Inorganic chemistry, however, was by many regarded as messy and essentially a collection of poorly related concepts and methods. There was an apparent need for a renewal, for an *inorganic renaissance!* Renewal was expected to come about through a greater use of physical methods, both experimental and theoretical. Inorganic chemistry might then be defined as *the integrated study of the formation, composition, structure and reactions of the chemical elements and compounds, excepting most of those of carbon*. This definition was, in particular, favoured by the Australian chemist R.S. Nyholm.

The main scientific activity at KLA around 1950 was in the field of inorganic chemistry. It now turned out that two young co-workers at the department were to play internationally prominent roles in the carrying out of the inorganic renaissance. The two men were Ballhausen and Axel Christian Klixbüll Jørgensen. After his hospitalization, Ballhausen had continued his chemistry studies with Bjerrum and became mag.scient. in 1954. In the same year, he became a scientific assistant at KLA. Klixbüll Jørgensen, on the other hand, got his master's degree in 1954 and also became a scientific assistant at KLA. The two men had very different personalities and only interacted scientifically for a short while. Klixbüll Jørgensen died in 2001, and this journal devoted Vol. 106 (2004) to his memory.

In the present volume of Structure and Bonding, we focus upon Ballhausen's scientific achievements. In accordance with his own conception, they may be viewed as a coherent contribution to the inorganic renaissance. In particular, he advanced the study of coordination compounds by theoretical, pedagogic, and

[2] In 1913, Werner was honoured with the Nobel Prize for his work.

experimental means. The aim was to understand structure and properties by combining the results of theoretical calculations with experimental results. Following the technological developments during and after the second world war, experimental research equipment at the universities was being renewed, during the 1950s and beyond. Electromechanical calculators had improved, and the modern computer was on its way.

## 4 The Early Years

Ballhausen's own conception of being a contributor to the inorganic renaissance is revealed in the preface of his dissertation.

> For many years, the conjugated hydrocarbons were the only "large" molecules that were treated by quantum mechanics. This is no longer the case. On the contrary, inorganic complex compounds have during the last few years turned out to be treatable in an elegant way by means of the crystal-field theory. It is accordingly my hope that this little book will help to take the "inorganic renaissance" further. (Translated from Danish by the present author.)

The title of Ballhausen's dissertation was "Elektrontilstande i Komplekser af 1. overgangsgruppe. En studie i Krystalfelt Teori" which translates into "Electronic states in complexes of the first transition series. A study in crystal field theory." It builds on the scientific work of the early years, to which we now turn.

As a preliminary, let us recall that the first transition series includes the elements Sc (scandium), Ti (titanium), V (vanadium), Cr (chromium), Mn (manganese), Fe (iron), Co (cobalt), Ni (nickel), Cu (copper), and Zn (zink), with the atomic numbers 21–30. The ground electron configuration of a dipositive transition element is that of an argon core $1s^2 2s^2 2p^6 3s^2 3p^6$ plus a partly filled $3d$ shell,

$$\begin{array}{cccccccccc} Sc^{2+} & Ti^{2+} & V^{2+} & Cr^{2+} & Mn^{2+} & Fe^{2+} & Co^{2+} & Ni^{2+} & Cu^{2+} & Zn^{2+} \\ 3d^1 & 3d^2 & 3d^3 & 3d^4 & 3d^5 & 3d^6 & 3d^7 & 3d^8 & 3d^9 & 3d^{10} \end{array}$$

For each ion, the electron configuration gives rise to one or more terms for the free ion. For the $Cu^{2+}$ ion, we get a $^2D$ term. For $Ni^{2+}$, we get five terms, $^3F$, $^1D$, $^3P$, $^1G$, and $^1S$, in the order of increasing energy. The separation into terms is due to the electron–electron repulsion. A further separation into levels is caused by spin-orbit coupling. This is all well known from the quantum mechanics of free atoms. Let us now consider a complex consisting of a transition metal ion and its associated ligands.

Within CF theory the electronic states of such a complex are determined from the free ion states by means of perturbation or variation theory. As discussed in Sect. 2.3, the degeneracy of the five $d$ orbitals is at least partly destroyed, and this causes the free ion terms to split into levels with less degeneracy. Thus, the $^2D$ term of a $Cu^{2+}$ ion splits into two levels with different energies in a cubic field. The designations of these levels, namely, $^2T_{2g}$ and $^2E_g$, are determined by group theory. If the symmetry is lower than cubic more energy levels will occur. For instance, there will be four levels in a tetragonal field.

In order to calculate the crystal-field levels and their energies, the crystal field must of course be known. Until 1950, it would most often be considered sufficient to know the qualitative form of the field. But, as pointed out in Sect. 2.3, one is free to consider the complex as an isolated molecular unit. The theory may therefore be made a quantitative one by inserting the actual form of the electrostatic potential from the ligands. In the 1950s, several molecular crystal-field calculations of this type were performed. Admittedly, simplifications were introduced like representing the potential from a ligand by the potential from a point charge or an electric dipole. The analytical form of the potential could then be written down, and the energy levels could be determined numerically. This was, in particular, illustrated by Ilse and Hartmann in 1951, in a study of Ti(III) complexes.

## 4.1 Copper(II) Complexes

Ilse and Hartmann's work caught the attention of Bjerrum and his students, in particular Ballhausen and Klixbüll Jørgensen. They opened a joint research project, which may be characterized by the title of the first publication: *Studies on Absorption Spectra. Results of Calculations on the Spectra and Configuration of Copper (II) ions*. Bjerrum had a comprehensive knowledge of the coordination chemistry of copper. Many complexes of the $Cu^{2+}$ ion were known, and so were many of their absorption spectra. Others were measured by Jørgensen. The challenge was now to account for the relation between geometry and colour in these compounds. Ballhausen met this challenge by performing several crystal-field calculations by methods related to those used by Hartmann and Ilse. The ligands were, in particular, replaced by electric dipoles whose strengths and locations were treated as semi-empirical parameters to be determined from spectroscopic data and common chemical sense. The calculations were published separately [2], but the results put to use in the article.

In general, transition-metal complexes show weak absorption bands in the visible region. They are due to electronic transitions in incomplete inner shells and may be discussed by crystal-field theory, as mentioned in Sect. 2.3. In principle, each transition from the ground level to the higher ones gives rise to an absorption band. Of the models considered, the regular tetrahedron and the regular octahedron give two energy levels and hence only one band, whilst the tetragonal configuration should theoretically give three bands.

For a given transition, the measured wave number $\nu$ at the band maximum, is related to the excitation energy $\Delta E$ by the usual expression $\Delta E{=}hc\nu$. If, therefore, $E_0$, $E_1$ ,$E_2$ are the energies of the ground state and two excited states, respectively, and $\nu_1$ and $\nu_2$ are the wave numbers at the two band maxima, then $\nu_2/\nu_1{=}(E_2 - E_0)/(E_1 - E_0)$. Writes Ballhausen: "In this relation we have a connection between the absorption maxima and the values of the dipoles at the different sites." The relation between geometry and colour that the formula gives was exploited for a number of $Cu^{2+}$ complexes, leading to a consistent set of geometries. It was, for instance, concluded that none of the complexes could be tetrahedral and that octahedral complexes like $Cu(H_2O)_6^{2+}$ deviated slightly from a regular octahedral constitution. Thus, $Cu(H_2O)_6^{2+}$ itself was found to have tetragonal symmetry. The configuration found was that of a deformed octahedron, in which the copper ion has four neighbours arranged in a square and two neighbours at a slightly larger distance and placed on the fourfold axis. The deformation of the octahedron is consistent with the so-called Jahn–Teller theorem which, in a qualitative formulation, states that a non-linear molecule in a degenerate state – in the present case a $^2E_g$ state – will undergo a deformation so that it loses as much symmetry as possible.

The intensities of crystal-field transitions were also commented upon in the copper(II) project. In the simplest theoretical analysis, such intensities are zero, because both the ground state and the excited states are even under inversion. The intensity originates in the coupling between electronic and nuclear motions. The study of such couplings, and hence of intensities and of the Jahn–Teller theorem, remained one of Ballhausen's major interests throughout the years.

## 4.2 *Other Transition Metal Complexes*

After the $Cu^{2+}$ complexes, focus was put on other complexes, for instance $Ni^{2+}$ complexes which have a $3d^8$ electron configuration. The calculational angle of attack was the same as before, but the calculations were more demanding, because the crystal field in principle mixes all states deriving from a given electron configuration. In contrast to the configuration of the $Cu^{2+}$ ion, which only gives a single state, $^2D$, the $3d^8$ configuration of the $Ni^{2+}$ ion gives rise to the states $^3F$, $^1D$, $^3P$, $^1G$, and $^1S$, as already mentioned in the previous subsection. Forgetting about spin-orbit coupling, the triplet states do not mix with the singlet states, so we have two independent crystal-field problems, one for a paramagnetic ion and another for a diamagnetic ion. Concentrating on the paramagnetic ion, it turns out to be characterized by four states in an octahedral field. Thus, three absorption bands are predicted. Calculations of the band maxima in the dipole model were performed and the geometries of several complexes discussed.

## 4.3 *Harvard, Bell Labs, Chicago*

Having earned his mag.scient. degree, Ballhausen left for the United States to spend the years 1955–1956 at Harvard University as a postdoctoral fellow with the prominent theoretical chemist Professor W.E. Moffitt. Together, the two men wrote a 30-page review article in Annual Review of Physical Chemistry [6], on crystal-field theory and its origin, basis, and applications. It was an article that Ballhausen always was fond to refer to. After his stay at Harvard, he spent some time at KLA, but then went back to the U.S. as a visiting scientist (1957–1958) at Bell Telephone Laboratories. Here, he had a close interaction with A. D. Liehr. Together, they made complete crystal-field calculations, including spin-orbit coupling, for $Ni^{2+}$ and $V^{3+}$ in a cubic field. Furthermore, they performed detailed work on the calculation of intensities in inorganic complexes and on the Jahn–Teller theorem applied to complexes. Finally, they discussed the electronic structure of some metal-aromatic complexes, using a molecular orbital description with crystal-field approximations [7].

In 1958, Ballhausen became assistant professor at the University of Chicago. In the same year, on June 19, he defended his dissertation at the University of Copenhagen (95 pages, written in Danish). It was based on the papers [2, 3] and [6–9]. Accordingly, it discussed both Ballhausen's very early work at KLA and work carried out in the USA. With the exception of [7], the papers all discussed work performed in the framework of crystal-field theory. The dynamic interaction between metal ion and ligand electrons was as always neglected. The next step would be to replace crystal-field theory with ligand field theory.

## 5 Professor at Fysisk Kemisk Institut

In 1959, Ballhausen was appointed professor of chemistry at the University of Copenhagen and director of its Institute of Physical Chemistry [Fysisk Kemisk Institut (FKI)]. The institute was located on Blegdamsvej 19, next to the Niels Bohr Institute. Ballhausen's predecessor was the reaction kineticist J.A. Christiansen. He had employed Arne E. Nielsen, who worked on chemical nucleation, and Thor A. Bak, who worked on theoretical chemical kinetics and the theory of irreversible thermodynamics. They both contributed to the positive atmosphere at the institute. Thor Bak had, in particular, a very stimulating effect on students. Yet, another important staff member was Jørgen Koefoed, who taught physical chemistry at DTU and became professor there in 1962.

In the early 1960s FKI, together with the other chemistry departments of the university, moved to the newly built H.C. Ørsted Institute where it was revived as Chemistry Department III with – now professor – Thor Bak as director, and Chemistry Department IV with Ballhausen as director. In the following, I shall use FKI as an unbroken designation for the institute on Blegdamsvej and Chemistry Department IV.

At FKI Ballhausen built an international centre from which a large number of scientific papers were published in the fields of molecular quantum mechanics, spectroscopy and structural chemistry. Ballhausen was himself the author of more than 100 scientific papers and 5 books, including his dissertation.

## 5.1 Ligand Field Theory

In the research environment which Ballhausen created at FKI, crystal-field theory played a modest role. Focus was from the very beginning put on molecular orbital theory. In the MO theory, the electronic structure of a molecule or a complex is described by a set of molecular orbitals that extend over the whole region of the molecule or complex. Relative to crystal-field theory, MO theory is a ligand field theory. The ligands are treated on the same footing as the otherwise exclusive metal ion. As to the excited states of a complex, there will be new types besides those that may be described more or less well by crystal-field theory. They will be states essentially localized on the ligands and so-called charge transfer states. The various states and the molecular orbitals may be characterized by their symmetries in a similar way as in crystal-field theory.

By referring to MO theory as a ligand field theory, it is important to note that the name ligand field theory also is used for other, supposedly more approximate theories. Ballhausen has given an unsurpassed inverse definition [27, 30],

> Crystal-field theory can be defined as an electronic theory of inorganic complexes, which neglects the overlap of electrons associated with the metal, with electrons associated with the ligands. The corresponding theory with overlap not neglected is termed ligand field theory.

The electronic theory that was used at FKI from 1959 and onward was a progressive ligand field theory, which developed in accordance with the theoretical and experimental possibilities.

## 5.2 Molecular Orbital Calculations

As already mentioned, I was Ballhausen's first Ph.D. student. I performed a self-consistent MO calculation on the metallo-organic molecule ferrocene, $Fe(C_5H_5)_2$. The calculation [14], which was done on an electro-mechanical calculator, was based on the assumption that the wave function was a single Slater determinant (a Hartree–Fock function). It was also a semi-quantitative calculation, since a number of integrals had to be treated approximatively. But the calculation did lead to a set of molecular orbitals by means of which the electron density in the molecule could be determined, and the nature of excited states could be discussed. During the 1960s, similar calculations were carried out for other, selected complexes, with the co-operation of students and visitors, and by the use of computers. These calculations were useful at the time, but the actual electronic structure of inorganic complexes has turned out to have a number of facets, which are not well reflected in the results of simple Hartree–Fock calculations. A great improvement is obtained by the use of many-determinant wave functions (configuration interaction) as shown and discussed by Helge Johansen and Sten Rettrup [38].

In the years 1960–1961, Harry B. Gray was a postdoctoral fellow at FKI. He and Ballhausen developed an alternative MO theory for inorganic complexes. It was a semi-empirical and chemically oriented MO theory, with a lot of information about free-atom states built into it. This information was stored in the so-called valence state ionization energies (VSIEs), which could be determined from the electronic spectra of free atoms and ions. The VSIEs depend on the atomic charges, and this allowed the MO calculations to become self-consistent. The molecular orbitals were calculated by diagonalizing an energy matrix in which a non-diagonal element $H_{ij}$ was expressed in terms of the VSIE-dependent diagonal elements $H_{ii}$ and $H_{jj}$, the relation being the so-called Wolfsberg–Helmholz relation, $H_{ij} = F\ S_{ij}M_{ij}$. Here, $F$ is a constant with a value close to 2, $S_{ij}$ is an overlap integral, and $M_{ij}$ is either the arithmetic or the geometric mean of $H_{ii}$ and $H_{jj}$. This MO method is referred to as the extended Wolfsberg–Helmholz method, where the adjective extended is supposed to remind the reader that this is a self-consistent theory.

## 5.3 Integrating Theory and Experiment

When applied in combination with experimental investigations, the developed theory turned out to give a reliable picture of a complex and its properties. This was convincingly demonstrated in the article *The electronic structure of the vanadyl ion* [15]. It was based on the solid compound $VOSO_4 \cdot 5H_2O$ whose crystal structure was known and showed that vanadium was hexacoordinated. MO calculations were performed for the octahedral ion $VO(H_2O)_5^{2+}$. The resulting molecular orbitals were then used to suggest energies and symmetries of the lowest excited states of $VOSO_4 \cdot 5H_2O$. An absorption spectrum of this compound dissolved in water was available, but because the molecular units are randomly oriented in a solution, this spectrum contained no information about the symmetries of the observed electronic transitions. To get this kind of information, it would be necessary to work with a crystal and polarized light. Hence, a crystal spectrophotometer was built, with the assistance of Arne E. Nielsen, and a complete absorption spectrum was obtained. A reflection spectrum was also recorded, and the magnetic properties of the crystal were measured too. Integrating the theoretical and experimental results subsequently led to a complete understanding of the vanadyl ion.

Ballhausen often expressed the opinion that chemistry is an experimental science. He was of course aware that this was a meaningless statement. But wherever it was possible he would integrate a theoretical calculation with experimental measurements. In accordance with this attitude, he filled FKI with spectroscopic equipment and instruments for structural analysis, including X-ray diffractometers. Spectroscopically, measurements could be performed with linearly and circularly polarized light, with pulsating magnetic fields, temperatures down to 1.7K and an applied uniaxial stress. A number of students, post-docs, and visitors contributed – with staff member Ib Trabjerg as anchorman – to the use of the equipment in connection with their research projects. Typically, these projects would integrate theory and experiment.

As already mentioned, Ballhausen was much interested in the coupling between electronic and nuclear motions. Spin-orbit coupling and magnetic interactions also had his interest. These couplings express themselves in the fine structure of spectra and in the temperature dependence of the fine structure. The form of the fine structure is too intricate to be detailed here. But as the attached list of publications shows, it is treated in several of his articles.

The vanadyl article was followed by an article on the MO description of square planar complexes [17], as for instance $Ni(CN)_5^{2-}$. The two articles set a standard in theoretical inorganic chemistry. The vanadyl work was published in volume 1 of the newly started journal *Inorganic Chemistry*. Now, fifty years later, volume 50 has been published. The Editor looks back and states, *inter alia*:

> In theoretical chemistry, simple molecular orbital theory (Hiickel and semi-em-pirical Wolfsberg–Helmholz) was just being applied to metal complexes such as the vanadyl ion and mononuclear metal carbonyls when Inorganic Chemistry started, but none of the power, sophistication, and ease of use of today's density functional theory (DFT) existed. Today, DFT calculations are found in a large fraction of our submissions, often when the paper is predominantly experimental in nature.

This statement reflects that the theoretical inorganic renaissance was in fact carried through. At the beginning, one had to fight approximations and limitations to generate an electronic structure. Today, fifty years later, we have far more powerful and reliable tools at our disposal. Also, the development of such methods was furthered at FKI. I have already mentioned Helge Johansen and Sten Rettrup's work. With reference to the editor's focus on DFT theory, it should be mentioned that Tom Ziegler – a pioneer in the application of density functional theory to for instance inorganic complexes – was a student of Ballhausen.

## 5.4 Scientific Monographs

At the same time as Ballhausen cultivated the scientific activity at FKI did he also found time to write his influential books. Apart from the dissertation, the first of these was *Introduction to Ligand Field Theory* [16]. It is probably the publication for which he is best known to the broad audience of chemists. It is a classic in the world of inorganic chemistry. It contains chapters on atomic theory and group theory, crystal-field theory, molecular orbitals, spin-orbit coupling, and vibronic interactions. Last, but not least, it also contains an actual discussion of the properties of a number of inorganic complexes. In the preface, he writes, *inter alia*:

> Since I am a chemist writing for chemists, my emphasis and notation will probably appear clumsy to the physicists primarily responsible for the theory. For this I make no excuse. Elegant derivations and condensed notation are in my opinion not desirable in an introduction to a field.

Ballhausen had the ability to make difficult material available to the many by illustrating his presentation with worked-out examples rather than generalizing theorems.

The two books with Harry Gray [19, 32] are good pedagogical introductions to molecular orbital theory with emphasis on orbital symmetry and computational procedures. They are based on the lectures of the authors and the development of the extended Wolfsberg–Helmholz method.

Ballhausen's latest book [30], *Molecular Electronic Structures of Transition Metal Complexes* appeared in 1979, 25 years after his first article. It can be seen as his answer to the question "What is a molecule – in particular a transition metal complex?" He starts with his conclusion from a series of articles on the chemical bond [31], "Chemistry is one huge manifestation of quantum mechanics". He then introduces the Born–Oppenheimer approximation as the basis for applying electronic and nuclear coordinates, and lets the picture of a molecule unfold itself with the concepts of electronic states, potential surfaces, transitions, vibronic couplings, etc. The presentation is traditional, but contains many refinings in the discussion of a molecule's ground state as well as its excited states. The world of transition metal complexes is favoured through the choice of examples.

### 5.5 Exactly Solvable Models

It follows implicitly from the preceding pages that most of Ballhausen's work was based on the theory of stationary states. But during his later years, he became quite interested in non-stationary time-dependent states as well, in particular through models that could be solved exactly [34–36]. Such work gave him a special intellectual satisfaction.

## 6 The World Outside FKI

Ballhausen was a trusted and respected man. Internationally, he had strong contacts, and many foreign post-docs and visitors have spent some time at FKI, just as Ballhausen in several periods was a visiting professor at foreign universities. He was also a frequent lecturer at international summer schools, including professor C. A. Coulson's summer schools in quantum chemistry in Oxford. In 1992, he was elected a foreign member of The National Academy of Sciences, USA.

A number of Ballhausen's works were published in the Scandinavian journal *Acta Chemica Scandinavica*. This journal was founded in 1947 in co-operation between the chemical societies in Denmark, Norway, Sweden, and Finland. Ballhausen was a member of the Danish editorial committee from 1960, in the period 1963–1973 as its chairman. His work for the journal contributed much to its recognition as an international journal of good quality.

Ballhausen became a member of the Royal Danish Academy of Sciences and Letters in 1965. One function of the Academy which is exceptional is that it elects the members of the board of the Carlsberg Foundation. In 1972–1996, Ballhausen was a member of this board, and of the boards of the Carlsberg Laboratory and the Foundation's Institute of Biology.

In private, Ballhausen had strong literary interests which he shared with his wife Ingrid B., b. Vesterdal. During his later years, he built a remarkable collection of Danish books from the 1500s and the 1600s, and among booklovers, he was a highly esteemed bibliographer.

## 7 Publications

A complete list of publications is available on the website *Early Ideas in the History of Quantum Chemistry*, set up by Dr. Udo Anders at http://www.quantum-chemistry-history.com. Dr. Anders has supplemented his article about Ballhausen with photos, talks, and copies of some important publications.

When Ballhausen somewhat reluctantly agreed to have his list of publications presented on the web, he also made a list of what he considered his own favourites. This is essentially the list presented below minus the last two entries. References to this list are by numbers in square brackets.

## References


1. Bjerrum J, Ballhausen CJ, Jørgensen CK (1954) Results of calculations on the spectra and configuration of copper(II) ions. Acta Chem Scand 8:1275
2. Ballhausen CJ (1954) Theory of copper (II)-spectra. Dan Mat Fys Medd. 29, No.4
3. Ballhausen CJ (1955) The spectra of nickel (II) complexes. Dan Mat Fys Medd. 29, No.8
4. Ballhausen CJ, Jørgensen CK (1955) The spectra of cobalt (II) complexes. Acta Chem Scand 9:397
5. Ballhausen CJ, Jørgensen CK (1955) d-Electrons in crystal fields of different symmetries. Dan Mat Fys Medd. 29, No.14
6. Moffitt WE, Ballhausen CJ (1956) Review of the crystal field theory of inorganic complexes, Ann Rev Phys Chem. 7. Reprinted in Symmetry in Chemical Theory, Dowden, Hutchinson & Ross Inc., 1973
7. Liehr AD, Ballhausen CJ (1957) Electronic structure of metal aromatic complexes. Acta Chem Scand 11:207
8. Liehr AD, Ballhausen CJ (1957) Intensities in inorganic complexes. Phys Rev 106:1161
9. Ballhausen CJ, Moffitt WE (1956) On the dichroism of certain Co (III) complexes. J Inorg Nucl Chem 3:178
10. Ballhausen CJ (1958) Elektrontilstande i Komplekser af 1. Overgangsgruppe. Disputats for den filosofiske Doktorgrad, Københavns Universitet. 95 sider. Translated into English. Bell Telephone Laboratories 1958
11. Liehr AD, Ballhausen CJ (1958) Inherent configurational instability of octrahedral complexes in $E_g$ electronic states. Ann of Physics 3:304

12. Ballhausen CJ, Liehr AD (1958) Intensities of inorganic complexes, II tetrahedral complexes. Mol Spectroscopy, 2:342. Ibid. 4 (1960) 190. Reprinted in Symmetry in Chemical Theory. Dowden, Hutchinson & Ross Inc. 1973
13. Liehr AD, Ballhausen CJ (1959) Complete theory of Ni(II) and V(III) in cubic crystalline fields. Ann of Phys 6:134. Reprinted in Symmetry in Chemical Theory, Dowden, Hutchinson & Ross Inc. 1973
14. Dahl JP, Ballhausen CJ (1961) The electronic structure of ferrocene. Mat Fys Medd. Dan Vid Selsk 33 No.5
15. Ballhausen CJ, Gray HB (1962) The electronic structure of the vanadyl ion. Inorganic Chem 1:111
16. Ballhausen CJ (1962) Introduction to ligand field theory, McGraw-Hill, New York and London. 298 pages. Translated into Russian 1965 and into Japanese 1967
17. Gray HB, Ballhausen CJ (1963) A molecular orbital theory for square planar metal complexes. J Am Chem Soc 85:260
18. Ballhausen CJ (1963) A vibrational analysis of the $Mn0_4^-$ bands. Theoret Chim Acta 1:285
19. Ballhausen CJ, Harry B Gray (1964) Molecular orbital theory, W.A. Benjamin, Inc 273 pages
20. Ballhausen CJ, Bjerrum N, Dingle R, Eriks K, Hare CR (1965) Polarized spectra of tetracyano nickelate crystals. Inorg Chem 4:514
21. Hansen AE, Ballhausen CJ (1965) Electronic structure of copper acetate monohydrate based on a coupled chromophore model. Trans Far Soc 61:631
22. Ballhausen CJ (1965) Jahn-Teller configurational instability in square-planar complexes. Theor Chim Acta 3:368
23. Ballhausen CJ, de Heer J (1965) Ligand-field treatment of the Jahn-Teller distortion in $VCl_4$. J Chem Phys 43:4304
24. Dahl JP, Ballhausen CJ (1968) Molecular orbital theories of inorganic complexes. Adv Quantum Chem 4:170
25. Ballhausen CJ, Trabjerg I (1972) Confirmation of the Red band system in $MnO_4^-$ as $^1T_1 \leftarrow {}^1A_1$ from the effect of pressure on single crystal absorption spectra. Mol Phys 24(4):689–693
26. Kral M, Moscowitz A, Ballhausen CJ (1973) A calculation of the rotatory strengths of the electron-transfer transitions of the tris-(1,10-phenanthroline) iron(II) ion. Theor Chim Acta 30:339–346
27. Ballhausen CJ (1971) Crystal and ligand field theory. Int J Quant Chem 5:373
28. Ballhausen CJ, Dahl JP (1974) The ligand field as pseudo-potential. Theor Chim Acta 34:169–174
29. Solomon EI, Ballhausen CJ (1975) Identification of the structures of the $^3T_{1g}(I) \leftarrow {}^3A_{2g}$ band in the $Ni(H_2O)_6^{++}$ complex. Mol Phys 29(1):279–299
30. Ballhausen CJ (1979) Molecular electronic structures of transition metal complexes. McGraw-Hill, New York, 207 p
31. Ballhausen CJ (1979) Quantum mechanics and chemical bonding in inorganic complexes. J Chem Ed 56, 3 parts:215, 294, 357
32. Ballhausen CJ, Gray HB (1980) Molecular electronic structures. An introduction. Benjamin/ Cummings, Reading, MA, 133 p
33. Ballhausen CJ, Avery J (1981) Line intensities in an $A \rightarrow E\ (x)\ \varepsilon$ Jahn–Teller electronic transition. Mol Phys 43:1199–1203
34. Ballhausen CJ (1992) Time-dependent raising and lowering operators for the ageing, forced and damped oscillator in quantum mechanics. Chem Phys Lett 192:49–54
35. Ballhausen CJ (1994) Oscillating Huang-Rhys factor in the time-relaxed harmonic oscillator. J Phys Chem 98:3270–3271
36. Ballhausen CJ (1995) Time-dependent raising and lowering operators applied to the forced morse oscillator. J Phys Chem 99:2530–2533
37. van Vleck JH (1938) J Chem Phys 7:72
38. Johansen H, Rettrup S (1983) Chem Phys 74:77

Struct Bond (2012) 142: 17–28
DOI: 10.1007/430_2011_55

Published online: 6 October 2011

# Electronic Structures of Oxo-Metal Ions

**Jay R. Winkler and Harry B. Gray**

**Abstract** The dianionic oxo ligand occupies a very special place in coordination chemistry, owing to its ability to donate π electrons to stabilize high oxidation states of metals. The ligand field theory of multiple bonding in oxo-metal ions, which was formulated in Copenhagen 50 years ago, predicts that there must be an "oxo wall" between Fe–Ru–Os and Co–Rh–Ir in the periodic table. In this tribute to Carl Ballhausen, we review this early work as well as new developments in the field. In particular, we discuss the electronic structures of beyond-the-wall (groups 9 and 10) complexes containing metals multiply bonded to O- and N-donor ligands.

**Keywords** Ferryl · Ligand field theory · Oxo wall · Vanadyl

## Contents



J.R. Winkler and H.B. Gray (✉)
Beckman Institute, California Institute of Technology, Pasadena, CA 91125, USA
e-mail: winklerj@caltech.edu; hbgray@caltech.edu

# 1 Introduction

Transition metal ions in aqueous solutions typically are coordinated by multiple water molecules. Binding to a Lewis acidic metal center increases the Brønsted acidity of the water ligand such that, depending on the pH of the medium, one or two protons can be lost producing, respectively, hydroxo and oxo ligands. In some cases, the acidity of a coordinated water is so great that the oxo ligand cannot be protonated even in concentrated acid solutions. The short MO bond distances in these oxo complexes (1.6–1.7 Å) are indicative of metal to ligand multiple bonding [1].

Oxo-metal complexes are so pervasive that a unique nomenclature evolved to characterize the structural unit. Koppel and Goldmann in 1903 suggested that aqueous $VO^{2+}$ be known as vanadyl, in analogy to the uranyl ion, and replacing what they viewed as a less consistent term, hypovanadate [2]. Over the ensuing years, oxo complexes of several different metals became known as "-yl" ions. A 1957 IUPAC report on inorganic nomenclature recognized that certain radicals[1] containing oxygen or other chalcogens have special names ending in "-yl", and the Inorganic Nomenclature Commission in 1957 gave provisional approval to retention of the terms vanadyl ($VO^{2+}$), chromyl ($CrO_2{}^{2+}$), uranyl ($UO_2{}^{2+}$), neptunyl ($NpO_2{}^{2+}$), plutonyl ($PuO_2{}^{2+}$), and americyl ($AmO_2{}^{2+}$) [3]. But by 1970, only chromyl and the actinyls were still viewed favorably by IUPAC [4]. The preferred lexical alternative was oxo-metal, wherein $O^{2-}$ is considered as a ligand bound to a metal center. The unfortunate confluence of trivial and systematic names for these compounds leads to abominations such as the resilient "oxoferryl", a term used all too often to describe complexes containing the $FeO^{2+}$ group.

The electronic structures of oxo-metal complexes were placed on a firm footing in 1962 by Carl Ballhausen (B) and one of us (G) based on experimental and theoretical investigations of the vanadyl ion [5].

# 2 The B&G Bonding Scheme

## *2.1 The Vanadyl Ion*

Vanadyl complexes are characterized by their rich blue color. Furman and Garner reported one of the earliest spectra of aqueous vanadyl: a single, slightly asymmetric absorption band maximizing at 750 nm ($\varepsilon \sim 17\ M^{-1}\ cm^{-1}$) is the lone feature in the visible region [6]. Measurements reported a year later on complexes of the vanadyl ion with thiocyanate revealed that the asymmetry is due to a second

[1] IUPAC defined a radical as a group of atoms which occurs repeatedly in a number of different compounds.

absorption feature: a shoulder at ~650 nm in vanadyl sulfate; a resolved 560 nm maximum in the $SCN^-$ complex [7].

Jørgensen first used crystal field (CF) theory to explain the absorption spectrum of the aqueous vanadyl ion, its complexes with donor ligands (edta, oxalate, acetylacetonate, and tartrate), as well as that of molybdenyl chloride ($MoOCl_5{}^{2-}$) [8]. Importantly, the spectra of the edta and tartrate complexes exhibit a third weak band maximizing in the 25,000–30,000 $cm^{-1}$ range ($\varepsilon < 50\ M^{-1}\ cm^{-1}$). He recognized that the octahedral CF-splitting parameter for $V^{4+}$ should be ~25,000 $cm^{-1}$: the bands at 13,100 and 16,000 $cm^{-1}$ in the $VO^{2+}$ spectrum are at much lower energies than expected for transitions derived from $t_{2g}{\rightarrow}e_g$ parentage. Jørgensen suggested that the 25,000–30,000 $cm^{-1}$ bands in $VO(edta)^{2-}$ and VO $(tart)^{2-}$ are in better agreement with this transition. His analysis treated the $d^1$ $VO^{2+}$ complexes as tetragonally Jahn–Teller distorted cubic complexes, analogous to $d^9$ $Cu^{2+}$ ions but with axial compression rather than elongation. The limiting Jahn–Teller distortion in vanadyl complexes would produce the diatomic $VO^{2+}$ ion [8].

In 1962, B&G developed a molecular orbital energy level scheme to describe the absorption and electron paramagnetic resonance spectra of the vanadyl ion [5]. The $VO^{2+}$ molecular orbital splitting pattern (Fig. 1) is reminiscent of the CF analysis put forth by Jørgensen, but it is important to emphasize that a pure CF model without oxo-metal π bonding cannot provide an adequate description of the vanadyl electronic structure. The B&G MO model provides remarkably accurate predictions of the energy, intensity, and polarization of the absorption features in the vanadyl spectrum, as well as the $g$-values extracted from EPR spectra. The key feature of the model is the substantial destabilization of the degenerate $d_{xz,yz}$ orbital pair, owing to π-bonding with the oxo-ligand. The lowest energy absorption feature (~13,000 $cm^{-1}$) was attributed to the ${}^2B_2(xy) \rightarrow {}^2E(xz, yz)$ excitation, corresponding to promotion of an electron from a nonbonding to the VO π-antibonding orbital. That the energy of the ${}^2B_2(xy) \rightarrow {}^2E(xz, yz)$ transition in $VO^{2+}$ is nearly as great as that of the $t_{2g}{\rightarrow}e_g$ excitation in $Ti(OH_2)_6{}^{3+}$ [9, 10] emphasizes the importance of VO π-bonding. The shoulder at ~16,000 $cm^{-1}$ arises from the ${}^2B_2(xy) \rightarrow {}^2B_1(x^2 - y^2)$ transition and provides a direct estimate of the octahedral ligand field splitting parameter (10Dq). For comparison, 10Dq values for the $V^{2+}$ [10, 11] and $V^{3+}$ [10, 12–14] hexaaqua ions are estimated to be ~12,000 and 18,800 $cm^{-1}$, respectively. The absorption spectrum reveals that the strength of the V-$OH_{2(equatorial)}$ interaction in the vanadyl ion lies somewhere between that of $V^{2+}$ and $V^{3+}$.

Particularly noteworthy is the finding that the $VO^{2+}$ absorption spectrum is unaffected by changes in pH from −0.5 to 1.5 [15]. The extremely low pKa of $VO^{2+}$ is a direct consequence of strong VO π-bonding: two oxygen $2p$ orbitals are used for π-bonds to V(IV) and only a nonbonding $sp_\sigma$ hybrid orbital remains for interaction with a proton. The energy of the hybrid orbital, with its substantial $2s$ character, is poorly suited to bonding with $H^+$.

It is curious that the aqueous vanadyl ion is $VO(OH_2)_5{}^{2+}$ rather than $V(OH)_2(OH_2)_4{}^{2+}$. Indeed, Jørgensen considered these and other isomers in his

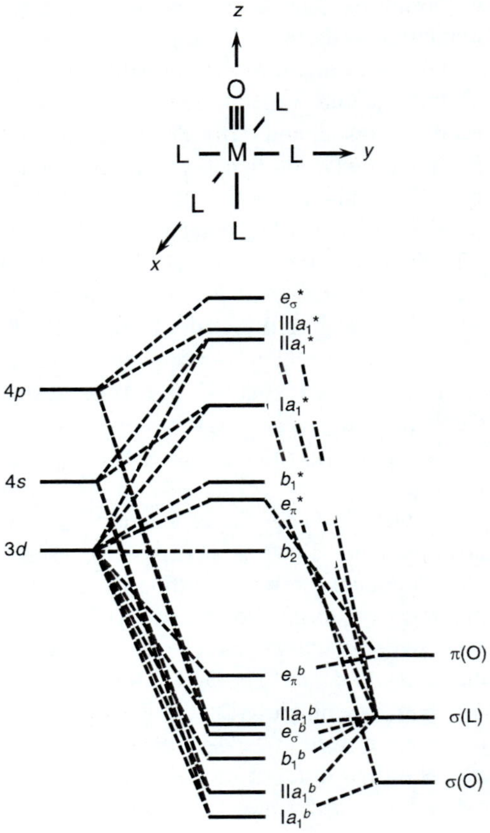


**Fig. 1** B&G molecular orbital model of the electronic structures of tetragonal oxo-metal complexes

analysis of the vanadyl spectrum [8]. Taube attributed the formation of -yl ions to the fact that the Lewis acidic metal center exerts a selective polarization of the surrounding water molecules, leading to double deprotonation of one water ligand rather than single deprotonation of two waters [16]. He argued that -yl ion formation is a consequence of the polarizability of $O^{2-}$, which is substantially greater than that of $HO^-$. The equivalent statement in the B&G molecular orbital language is that $O^{2-}$ is a far stronger $\pi$-donor than $HO^-$ such that the net stabilization gained by forming two $\pi$-bonds to one $O^{2-}$ ligand is greater than that resulting from one $\pi$-bond to each of two $HO^-$ ligands.

A minor controversy surrounding the B&G model developed in late 1963 with the publication by Selbin and coworkers of the spectrum of $VO(acac)_2$ dissolved in organic glasses at 77 K [17]. At cryogenic temperature the absorption bands narrowed, lost some intensity, and the lowest energy feature resolved into three distinct maxima. The authors attributed the four bands observed between 12,000 and 18,000 $cm^{-1}$ to the four distinct ligand field excitations expected for a $d^1$ complex with $C_{2v}$ symmetry. Workers in Copenhagen addressed this issue in 1968 with X-ray and optical spectroscopic measurements on crystalline $VO(SO_4)\cdot 5H_2O$ [18]. An improved X-ray crystal structure determination and low temperature

(20 K) single-crystal absorption spectra provided convincing evidence in support of the original B&G assignments. It is likely that the additional peaks observed by Selbin in the 77 K spectrum of $VO(acac)_2$ arise from vibrational fine structure in the VO stretching mode (*vide infra*). Such a progression is indicative of a distortion in the VO bond, consistent with the B&G $n \rightarrow \pi^*(VO)$ assignment.

## 2.2 The Chromyl and Molybdenyl Ions

Three months after the B&G analysis of the vanadyl electronic structure appeared in *Inorganic Chemistry* [5], Curt Hare (H) and one of us (G) published an interpretation of chromyl ($CrOCl_5{}^{2-}$) and molybdenyl ($MoOCl_5{}^{2-}$) spectra and electronic structures [19]. The assignments parallel those for the vanadyl ion: the lowest energy feature ($CrO^{3+}$, 12,900 $cm^{-1}$; $MoO^{3+}$, 13,800) is attributed to the ${}^2B_2(xy) \rightarrow {}^2E(xz, yz)$ transition; the next higher energy band ($CrO^{3+}$, 23,500 $cm^{-1}$; $MoO^{3+}$, 23,000) is ${}^2B_2(xy) \rightarrow {}^2B_1(x^2 - y^2)$. It is interesting to note that although the lowest energy bands in $CrO^{3+}$ and $MoO^{3+}$ are about the same position as the corresponding feature in the $VO^{2+}$ spectrum, the second band is some 5,000–7,000 $cm^{-1}$ higher energy in the group 6 ions. The energy of the second band provides a direct estimate of 10Dq. This ligand field splitting parameter for the molybdenyl ion is nearly identical with that of $MoCl_6{}^-$ [20, 21], and about 4,000 $cm^{-1}$ greater than that of $MoCl_6{}^{3-}$ [10], as expected for the higher oxidation state metal.

Garner and coworkers cast doubt on the ${}^2B_2(xy) \rightarrow {}^2B_1(x^2 - y^2)$ assignment for the 23,000 $cm^{-1}$ band in $MoOCl_5{}^{2-}$ [22]. Calculations placed the ${}^2B_2(xy) \rightarrow {}^2E(xz, yz)$ and ${}^2B_2(xy) \rightarrow {}^2B_1(x^2 - y^2)$ transitions at 15,600 and 23,000 $cm^{-1}$, but indicated that absorption corresponding to the latter transition would be obscured by $\pi Cl \rightarrow Mo$ charge-transfer features. We addressed this controversy with measurements of the spectra of $MoO(HSO_4)_4{}^-$ and $MoO(H_2PO_4)_4{}^-$: both ions exhibit weak ($\varepsilon < 30\ M^{-1}\ cm^{-1}$) absorptions at 14,000 and 26,000 $cm^{-1}$ that are assigned to ${}^2B_2(xy) \rightarrow {}^2E(xz, yz)$ and ${}^2B_2(xy) \rightarrow {}^2B_1(x^2 - y^2)$ transitions, respectively [23]. With the exception of a 2,000–3,000 $cm^{-1}$ blue shift in the second band, consistent with the stronger ligand field expected for O-donating equatorial ligands, the spectra match that of $MoOCl_5{}^{2-}$ quite closely. It is unlikely, then, that the 23,000 $cm^{-1}$ band in the latter complex is due to $Cl \rightarrow Mo$ charge transfer. Indeed, the spectra of $MoO(HSO_4)_4{}^-$ and $MoO(H_2PO_4)_4{}^-$ place a lower limit (35,000 $cm^{-1}$) on the energy of $\pi O \rightarrow Mo$ charge transfer in molybdenyl ions. It follows that bands observed at 28,000 and 32,500 $cm^{-1}$ in $MoOCl_5{}^{2-}$ likely arise from $\pi Cl \rightarrow Mo$ charge transfer.

Low-temperature (5 K) single-crystal polarized absorption spectra of $(Ph_4As)$ $[MoOCl_4]$ exhibit rich vibrational fine structure in both ${}^2B_2(xy) \rightarrow {}^2E(xz, yz)$ and ${}^2B_2(xy) \rightarrow {}^2B_1(x^2 - y^2)$ absorption systems [23]. The ${}^2B_2 \rightarrow {}^2E$ band is characterized by progressions in 900 and 165 $cm^{-1}$: the high-frequency progression corresponds to the MoO stretching mode in the ${}^2E$ excited state; and the 100 $cm^{-1}$

reduction in frequency relative to the ground state is consistent with population of a MoO $\pi^*$ orbital. Franck–Condon analysis of the absorption profile is consistent with a 0.09(1) Å distortion of the MoO bond in the $^2E$ state. Vibrational progressions in MO stretching modes, sometimes apparent even at room temperature in fluid solutions, are diagnostic of $(xy)\rightarrow(xz,yz)$ excitations in tetragonal oxo-metal complexes. The lower frequency vibrational progression was originally attributed to the symmetric OMoCl umbrella bending mode, but later work on related complexes suggests that the $b_1$(OMoCl) bend is a better assignment [24]. Vibrational fine structure in the $^2B_2(xy) \rightarrow {}^2B_1(x^2 - y^2)$ system corresponds to the symmetric MoCl stretch and Franck–Condon analysis suggests a 0.07(1) Å distortion along each MoCl bond in the $^2B_1$ excited state.

A final point of interest is the finding that molybdenyl ions are luminescent [25–27]. Crystalline samples of $(Ph_4As)[MoOX_4]$ (X=Cl, Br) exhibit emission maxima at ~900 nm, and at cryogenic temperatures a progression in the MoO stretching mode is resolved [27]. The luminescence lifetime of $(Ph_4As)[MoOCl_4]$ is 160 ns at room temperature, increasing to 1.4 μs at 5 K, but no luminescence from this molybdenyl ion could be detected in fluid solutions or frozen glasses. Using the Strickler–Berg approximation [28] to estimate the radiative decay rate constant for the $^2B_2(xy) \rightarrow {}^2E(xz, yz)$ transition, a room temperature quantum yield of ~0.01 can be extracted for crystalline $(Ph_4As)[MoOCl_4]$.

## 2.3 *The Oxo-Metal Triple Bond*

G&H were the first to represent the interaction between the molybdenum and oxygen atoms in $MoO^{3+}$ as a triple bond (M≡O) [19]. This assignment of bond order follows directly from the molecular orbital model of the electronic structure: two electrons in a MO $\sigma$-bonding orbital; and four electrons in a doubly degenerate pair of MoO $\pi$-bonding orbitals. The lone electron in the $b_2(xy)$ orbital is nonbonding with respect to the MO interaction. The triple bond (M≡O) is the proper formulation of electronic structures in $d^0$, $d^1$, and $d^2$ (low spin) tetragonal monooxo-metal complexes. It is distressing to see the MO interaction in these cases represented as a double bond, likely by those who think (incorrectly) that it is analogous to the double bond in organic carbonyl complexes. But, just as carbon monoxide possesses a triple bond, so do $VO^{2+}$, $CrO^{3+}$, $MoO^{3+}$, and $WO^{3+}$. Metal nitrido complexes also feature strong MN multiple bonding: indeed, $d^0$, $d^1$, and $d^2$ (low spin) tetragonal nitrido-metal complexes all contain MN triple bonds (M≡N).

## 2.4 *Lower Bond Orders*

Population of $e(xz,yz)$ $\pi^*$ orbitals in tetragonal oxo-metal complexes reduces the MO bond order. Although triple bonding is assured in $d^0$ and $d^1$ complexes, the bond order in complexes with two or three *d*-electrons depends critically on the spin state (Fig. 2). Tetragonal monooxo-complexes with $d^2$ configurations have MO bond orders of 3 and 2.5 in the low-spin ($^1A_1$) and high-spin ($^3E$) states, respectively. For $d^3$ configurations, the bond orders are 2.5 ($^2E$) and 2 ($^4B_1$). The relative energies of high- and low-spin states depend on a balance between the $e(xz,yz) - b_2(xy)$ energy gap ($\Delta_\pi$) and the electron–electron repulsion energies. This relationship can be represented graphically with modified Tanabe–Sugano (TS) diagrams [29, 30] (Fig. 3). Owing to differences in MO bond lengths in the high- and low-spin states, there is a "forbidden region" of the TS diagram (Fig. 3, gray area) that excludes a range of $\Delta_\pi$ values for ground states. Taking typical values for Racah parameters ($B = 500\ \mathrm{cm}^{-1}$, $C/B = 4$), we find that the $^3E$-$^1A_1$ spin crossover occurs with $\Delta_\pi \sim 9{,}500\ \mathrm{cm}^{-1}$ in $d^2$ oxo-metal complexes, but formation of a low-spin ground state requires $\Delta_\pi > 13{,}000\ \mathrm{cm}^{-1}$. This value of $\Delta_\pi$ is about the same as that found in the $d^1$ -yl ions, suggesting that $d^2$ oxo and nitrido-metal complexes are likely to have low-spin ($^1A_1$) ground states, but that high-spin complexes might be thermally accessible. It is important to remember that the TS diagrams represent vertical energy differences [31]; the energy required for thermal population of a high-spin state in $d^2$ complexes will be less than the vertical energy difference by several thousand wavenumbers. A larger $e(xz,yz) - b_2(xy)$ energy gap is required to produce low-spin $d^3$ oxo-metal complexes; the $^2E$-$^4B_1$ spin crossover occurs at $\Delta_\pi \sim 13{,}000\ \mathrm{cm}^{-1}$, but $\Delta_\pi > 17{,}000\ \mathrm{cm}^{-1}$ is required for a low-spin ground state.

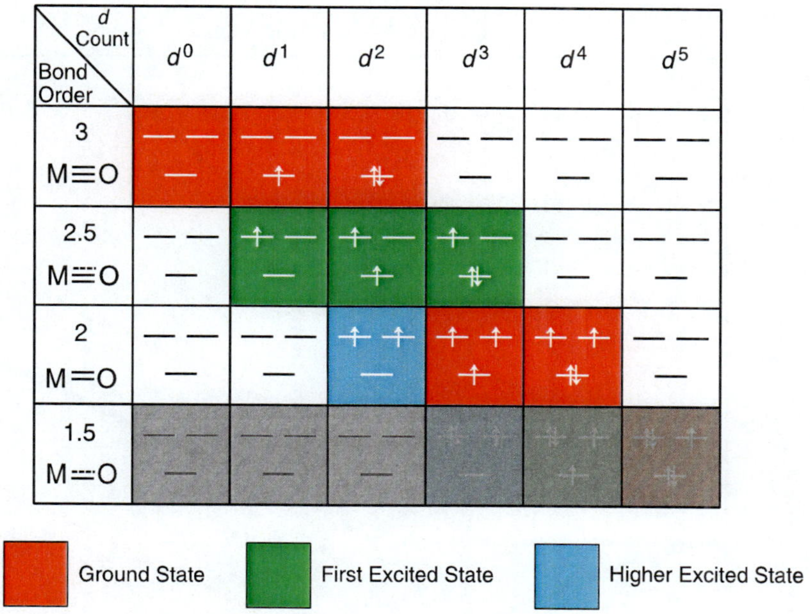


**Fig. 2** Correlation of MO bond order with *d*-electron count in oxo-metal complexes. The *gray shaded region* corresponds to a π bond order of 0.5; the oxo ligands in these complexes are expected to be extremely basic

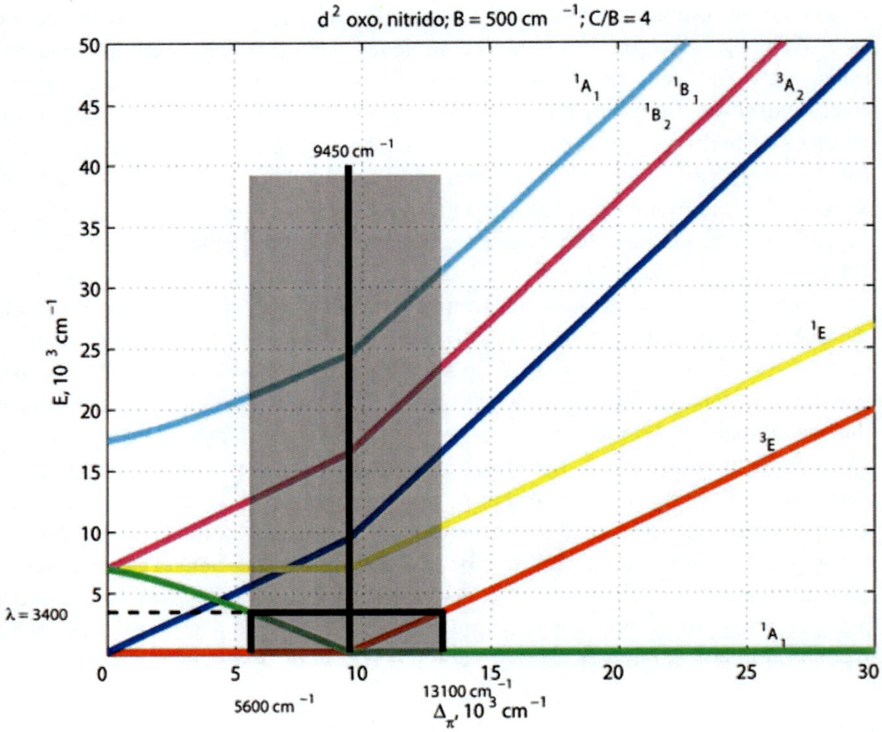
d$^2$ oxo, nitrido; B = 500 cm$^{-1}$; C/B = 4
E, 10$^3$ cm$^{-1}$
$\Delta_\pi$, 10$^3$ cm$^{-1}$
9450 cm$^{-1}$
5600 cm$^{-1}$
13100 cm$^{-1}$
λ = 3400
$^1A_1$
$^1B_1$
$^1B_2$
$^3A_2$
$^1E$
$^3E$
$^1A_1$

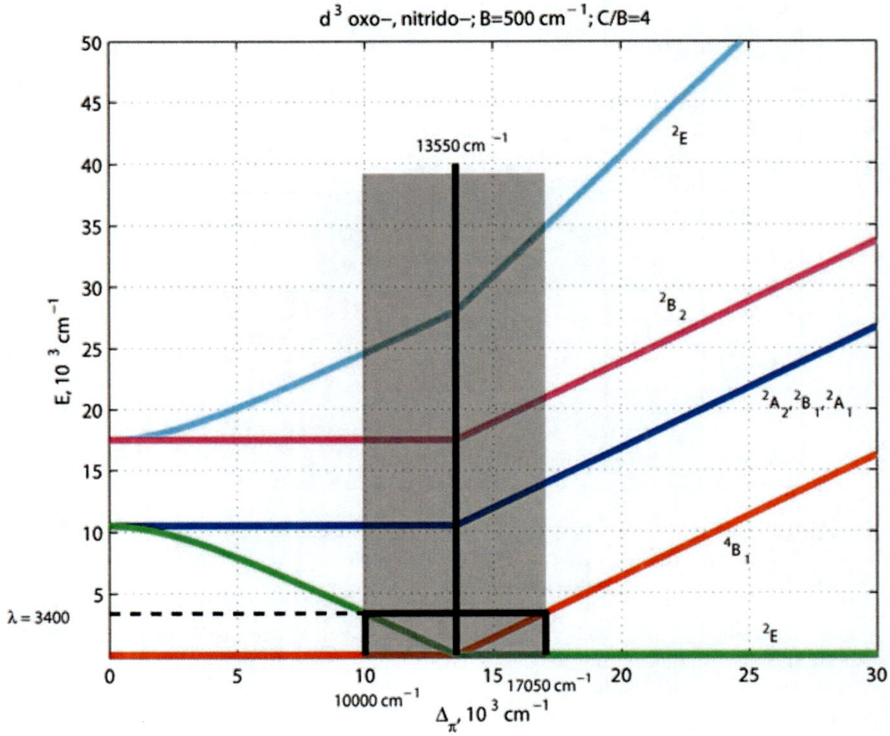
d$^3$ oxo–, nitrido–; B=500 cm$^{-1}$; C/B=4
E, 10$^3$ cm$^{-1}$
$\Delta_\pi$, 10$^3$ cm$^{-1}$
13550 cm$^{-1}$
10000 cm$^{-1}$
17050 cm$^{-1}$
λ = 3400
$^2E$
$^2B_2$
$^2A_2$, $^2B_1$, $^2A_1$
$^4B_1$
$^2E$

Substantially stronger ME π-bonding (E=O, N) is required to produce low-spin ground states in $d^3$ than in $d^2$ complexes. Nitrido ligands are stronger π donors than oxo ligands; ligand field theory predicts that the ground state in $d^3$ tetragonal oxo-metal complexes will be $^4B_1$ (high spin), whereas nitrido complexes of this electronic configuration are more likely to have $^2E$ (low-spin) ground states. Rounding out the possible oxo-metal configurations, most tetragonal $d^4$ oxo-metal complexes have a $^3A_2$ ground state and MO bond order of 2, whereas $d^5$ complexes have a bond order of 1.5. In some ferryl ($d^4$) complexes with very weak equatorial ligand fields, the $x^2-y^2$ orbital drops in energy, producing an $S = 2$ ground state (*e.g.*, $^5(Fe{=}O)^{2+}$), but retaining an MO bond order of 2 [32–34].

In tetragonal *trans*-dioxometal complexes with 0, 1, or 2 *d*-electrons, 4 bonds (two σ and two π) will be divided between two MO interactions. The average MO bond order is two, such that a double bond representation is not incorrect, although a structure with one full bond and two half bonds is a more accurate depiction of the interaction.

## 3 The Oxo Wall

The B&G bonding model defines electronic structural criteria for the existence of oxo-metal complexes. Complexes with tetragonal symmetry can have no more than 5 *d*-electrons and still retain some MO multiple bonding. In the absence of π-bonding to the metal, the oxo will be extremely basic and unstable with respect to protonation or attack by electrophiles. Achieving the low *d*-electron counts necessary for MO multiple bonding becomes increasingly difficult for metals on the right half of the transition series. Eventually, the metal oxidation states required for oxo formation become so great that the complexes are unstable with respect to elimination of $H_2O_2$ or $O_2$. The combined requirements of low *d*-electron counts and limiting metal oxidation states erect an "oxo wall" between groups 8 and 9 in the periodic table (Fig. 4). To the right of this wall, tetragonal oxo complexes are not likely to be found. Hence, while iron, ruthenium, and osmium form -yl complexes, the same cannot be said for cobalt, rhodium, and iridium.

**Fig. 3** TS-type diagrams correlating the energies of electronic states in $d^2$ and $d^3$ oxo-metal complexes with the strength of the $\Delta_\pi$ ligand field splitting parameter. *Gray areas* correspond to forbidden $\Delta_\pi$ zones for ground-state complexes

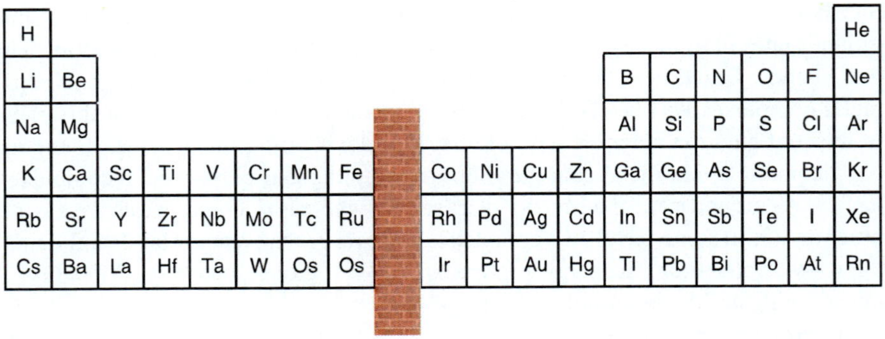


**Fig. 4** Periodic table of the elements with groups 8 and 9 separated by the oxo wall

## 4 Beyond the Oxo Wall

The oxo wall is a construct built on the concepts of oxo-metal π bonding in tetragonal complexes. To find complexes beyond the wall, reduced coordination numbers and geometry changes are required to liberate orbitals to host nonbonding (or weakly antibonding) electrons. Multiply bonded oxo-metal (and related) complexes of this type are not violations of the oxo wall but, rather, are entirely consistent with the metal-ligand π-bonding principles embodied in the original B&G model.

The well-known Wilkinson $d^4$ Ir(V) complex [35], Ir(O)(mesityl)$_3$, is often cited as an oxo-wall violation. It is not: the complex has trigonal symmetry, producing a reordering of the metal *d*-orbitals with two nonbonding levels below the degenerate MO π* pair. In threefold symmetric complexes, then, multiple MO bonding is allowed for as many as 7 *d* electrons. As expected, reducing the number of ancillary ligands frees up orbitals for MO multiple bonding, which means that crossing the oxo wall is allowed.

In 2008, David Milstein and coworkers reported evidence for the formation of a terminal Pt(IV)-oxo complex [36]. The $d^6$ electronic configuration is incompatible with oxo formation in a tetragonal complex, but the loss of two equatorial ligands in this 4-coordinate (PCN)Pt(O) molecule (PCN=$C_6H_3[CH_2P(t\text{-Bu})_2](CH_2)_2N(CH_3)_2$) greatly stabilizes the $x^2-y^2$ orbital (it may even drop below the Pt-O π* *xz* and *yz* orbitals). With 6 *d*-electrons, then, two would occupy the PtO π* orbital, leaving one net PtO π bond. DFT calculations suggest a 1.8 Å PtO distance, consistent with a double bond.

The two-coordinate Ni(II)-imido complex reported by Greg Hillhouse and coworkers does not violate the oxo wall [37]. The ground state is a spin triplet, consistent with population of one electron in each of two nearly degenerate NiNR π* orbitals. With a $d^8$ electronic configuration, one more low-energy orbital (in addition to *xy* and $x^2-y^2$) is required. This orbital is often referred to as $z^2$, but we suggest that this assignment is incorrect. With its axial lobes pointing directly at the imido σ orbital, a pure $d_{z^2}$ orbital should lie above the Ni-NR π* orbitals.

But, mixing with the 4*s* level produces a $d_{z^2}$-*s* hybrid orbital with reduced density along the axis and increased density in the doughnut-shaped lobe in the equatorial plane that is devoid of ligands. The low coordination number in the molecule leads to $z^2$-*s* hybridization, allowing strong NiNR π bonding.

## 5 Concluding Remarks

Realistic targets that would constitute B&G approved oxo-wall violations include tetragonal *trans*-dioxo Ir(VII) complexes. Josh Palmer, a recent member of our research group, made several unsuccessful attempts to prepare such (and related nitrido) species.

As we approach the 50th anniversary of the publication of the B&G bonding model, inorganic chemists still have not found a stable tetragonal oxo complex of a group 9 or 10 metal!

**Acknowledgments** We dedicate this paper to the memory of Carl Ballhausen, a great scientist and a dear friend (Fig. 5). We note in closing that the B&G model is providing a firm foundation for structure/reactivity correlations in our current work on oxo-metal complexes [oxidative

**Fig. 5** Carl Ballhausen visited the Beckman Institute at Caltech on several occasions. This photograph from one visit in the early 1990s shows (*left to right*): Gary Mines, Jay Winkler, Bo Malmström, Harry Gray, Carl Ballhausen, Danilo Casimiro, I-Jy Chang, Jorge Colón, Zhong-Xian Huang, and Deborah Wuttke

enzymes P450 and nitric oxide synthase (NIH DK019038, GM068461); water oxidation catalysts (NSF CCI Solar Program, CHE-0947829); and *trans*-dioxo osmium(VI) electrochemistry and photochemistry (BP)]. We thank the Gordon and Betty Moore Foundation and the Arnold and Mabel Beckman Foundation for support of our research programs.

## References

1. Nugent WA, Mayer JM (1988) Metal-ligand multiple bonds. John Wiley & Sons, New York
2. Koppel J, Goldmann R (1903) Z Anorg Allg Chem 36:281
3. International Union of Pure and Applied Chemistry (1960) J Am Chem Soc 82:5523
4. International Union of Pure and Applied Chemistry (1970) Nomenclature of inorganic chemistry. Butterworths, London
5. Ballhausen CJ, Gray HB (1962) Inorg Chem 1:111
6. Furman SC, Garner CS (1950) J Am Chem Soc 72:1785
7. Furman SC, Garner CS (1951) J Am Chem Soc 73:4528
8. Jørgensen CK (1957) Acta Chem Scand 11:73
9. Hartmann H, Schlafer HL (1954) Angew Chem Int Ed Engl 66:768
10. Ballhausen CJ (1962) Introduction to ligand field theory. McGraw-Hill, New York
11. Bennett RM, Holmes OG (1960) Can J Chem 38:2319
12. Hartmann H, Schlafer HL (1951) Z Naturforsch A 6:754
13. Hartmann H, Schlafer HL (1951) Z Naturforsch A 6:760
14. Ilse FE, Hartmann H (1951) Z Naturforsch A 6:751
15. Rossotti FJC, Rossotti HS (1955) Acta Chem Scand 9:1177
16. Taube H (1982) In: Rorabacher DB, Endicott JF (eds) Mechanistic Aspects of Inorganic Reactions, vol 198, ACS Symposium Series. American Chemical Society, Washington DC, p 151
17. Selbin J, Ortolano TR, Smith FJ (1963) Inorg Chem 2:1315
18. Ballhausen CJ, Djurinski BF, Watson KJ (1968) J Am Chem Soc 90:3305
19. Gray HB, Hare CR (1962) Inorg Chem 1:363
20. Patterson HH, Nims JL (1972) Inorg Chem 11:520
21. Horner SM, Tyree SY (1963) Inorg Chem 2:568
22. Weber J, Garner CD (1980) Inorg Chem 19:2206
23. Winkler JR, Gray HB (1981) Comments Inorg Chem 1:257
24. Hopkins MD, Miskowski VM, Gray HB (1986) J Am Chem Soc 108:6908
25. Mohammed AK, Fronczek FR, Maverick AW (1994) Inorg Chim Acta 226:25
26. Mohammed AK, Maverick AW (1992) Inorg Chem 31:4441
27. Winkler JR (1984) Spectroscopy and Photochemistry of Metal-Oxo Complexes, Ph.D. California Institute of Technology, California
28. Strickler SJ, Berg RA (1962) J Chem Phys 37:814
29. Tanabe Y, Sugano S (1954) J Phys Soc Jpn 9:753
30. Tanabe Y, Sugano S (1954) J Phys Soc Jpn 9:766
31. Winkler JR, Rice SF, Gray HB (1981) Comments Inorg Chem 1:47
32. Sinnecker S, Svensen N, Barr EW, Ye S, Bollinger JM Jr, Neese F, Krebs C (2007) J Am Chem Soc 129:6168
33. Riggs-Gelasco PJ, Price JC, Guyer RB, Brehm JH, Barr EW, Bollinger JM, Krebs C (2004) J Am Chem Soc 126:8108
34. Price JC, Barr EW, Tirupati B, Bollinger JM, Krebs C (2003) Biochemistry 42:7497
35. Hay-Motherwell RS, Wilkinson G, Hussain-Bates B, Hursthouse MB (1993) Polyhedron 12:2009
36. Poverenov E, Efremenko I, Frenkel AI, Ben-David Y, Shimon LJW, Leitus G, Konstantinovski L, Martin JML, Milstein D (2008) Nature 455:1093
37. Laskowski CA, Miller AJM, Hillhouse GL, Cundari TR (2011) J Am Chem Soc 133:771

Struct Bond (2012) 142: 29–48
DOI: 10.1007/430_2011_54

Published online: 6 October 2011

# Early Days in Kemisk Laboratorium IV and Later Studies

**Colin D. Flint**

**Abstract** A brief review of my work with Carl Ballhausen in 1967–1968 and subsequent work. The assignments of the vibronic sidebands in the emission spectra of chromium ammine complexes are given with some comments on the Jahn–Teller effect in the emissive state. Energy transfer and cross relaxation phenomena are discussed and the shell model for this processes in lanthanide elpasolites is presented

**Keywords** Chromium ammines · Energy transfer · Permanganate ion

## Contents



C.D. Flint (✉)
Department of Chemistry, School of Chemical and Biological Sciences, Birkbeck College (University of London), Malet Street, London WC1, UK
e-mail: chigwell999@hotmail.com

## 1 Historical Background

My first knowledge of Carl Ballhausen occurred in 1963. I had just been awarded a small book prize as an undergraduate at Imperial College. I asked my future PhD supervisor, Dr. Margaret Goodgame to recommend a text that would be useful to me in my Ph.D. studies. She immediately recommended Carl's classic text "Introduction to Ligand Field Theory" [1]. I started to struggle through this text, which was not easy reading for an undergraduate brought up on little more than the Bohr–Sommerfeld atom. I eventually bought Cotton's "Chemical Applications of Group Theory" [2], Griffith's "The Theory of Transition Metal Ions" [3], Condon and Shortley's "Theory of Atomic Spectra" [4] and Paul Dirac's "Principles of Quantum Mechanics" [5]. As it turned out, my Ph.D. involved a lot of synthesis and vibrational spectroscopy so I did not have much opportunity to use the knowledge that I was painfully acquiring. In 1964, Ray Dingle who was working as a post-doc in Kemisk Laboratorium IV gave a seminar at Imperial College, and that convinced me that the kind of work that was going on in Laboratorium IV was what I wanted to do. Then in 1965, Carl gave a lecture at Imperial and that confirmed my views. So in 1967, I arrived in Copenhagen as a very inexperienced post-doc with a NATO Fellowship.

## 2 Early Work in Copenhagen

The project Carl gave me was to build a sensitive instrument to search for luminescence from the permanganate ion, which had been the subject of a series of experimental single crystal absorption spectral studies and theoretical studies in the laboratory [6]. The spectrometer was built, but after repeated attempts using a range of crystals, excitation conditions and temperatures, no luminescence was detected. All subsequent efforts by others have confirmed this failure [7], under laser irradiation in iodide lattices some emission has been detected, but this is derived from the manganese ion $MnO_4^{2-}$ produced by a photoredox process [8]. This left me without many results to show for my year's work. I made some measurements on the intensely luminescent alkali metal platinocyanides but this did not lead to any new insights.

Looking for a new system to study, I came across some crystals of $2Cr(en)_3Cl_3$. $KCl.6H_2O$ for which the absorption spectra had been studied previously by Martin Vala and Paul McCarthy [9]. This proved a complex system to study. The problem was that at low temperatures, a photochemical reaction occurs leading to new emission centres with a much broader emission band (termed B sites in the original paper [10]) some 400 $cm^{-1}$ to low energy of the sharper $^2E \rightarrow {}^4A_2$ emission from the $2Cr(en)_3{}^{3+}$.(Termed N sites in the paper) The number of these centres decreases as the temperature is raised so that by warming to 140 K or above they are removed from the system. The crystals may be repeatedly recycled through this behaviour. Energy transfer occurs from the N-sites to the B sites, which results in a complex temperature dependence of the emission intensities, lifetimes and rise times. In 1969, I had no idea what these B sites were, but it is likely that are due to the photoproduction of sites where one of the nitrogen ligators of the ethylenediamine has been replaced by a chloride ion. Above 140 K, these sites relax to the tris (ethylenediamine)chromium ion, expelling the chloride ion from the coordination sphere of the Chromium. The broadness of the B emission is due to a statistical distribution of a large number of slightly different sites At the end of 1969, I returned to London where I stayed for 35 years, building on my experience in Laboratorium IV. and set up a luminescence system based on my experience in Copenhagen. Two sayings of Carl greatly influenced my future work.

> An experimental result that is not corrected for experimental uncertainty is not a result of value.
>
> The only thing we cannot afford to waste is our time.

# 3 Work in London Under Carl's Influence

## 3.1 The $Cr(NH_3)_6{}^{3+}$ Ion

Just before I left Copenhagen, I started work on two other systems, the octahedral $Cr(NH_3)_6{}^{3+}$ ion in the cubic $Cr(NH_3)_6(ClO_4)_3$ crystal and the cubic $Cs_2MnF_6$. I had realized that the extra (i.e. those that were not present in the infrared spectrum) bands in the vibronic sidebands of the $^2E_g \rightarrow {}^4A_{2g}$ transition measure in emission were due to the infrared and Raman inactive $t_{2u}$ vibrations of the octahedral ions. I sent an early version of what became a short paper [10] to Carl for his comments. Whilst not disputing the assignment, Carl was very critical of my theoretical argument and was kind enough to send me a preprint of his superb paper with Aage Hansen [11]. That greatly clarified my thoughts on vibronic coupling and I rewrote the paper. That started a series of papers on the vibronic analysis of the $^2E_g \rightarrow {}^4A_{2g}$ transitions in octahedral Chromium(III) complex ions [12–15]. One of the problems with understanding the assignment of the $t_{1u}$ and $t_{2u}$ Cr–N–H rocking mode was to account for the high intensity of the modes in the vibronic spectrum. After all from a vibrational spectroscopic point of view, they were essentially hydrogen motions involving the second coordination sphere and this was confirmed by their large deuteration shifts. (Figs. 1 and 2)

They were, for example, very much more intense that the H–N–H bending modes and the N–H stretching modes For this reason, my assignments were disputed at a number of conferences. Sometime later, I argued that this intensity arose because the hydrogen motion moved the nitrogen lone pair out of the metal–nitrogen axis [16], this was later confirmed by a crude but instructive calculation [17].

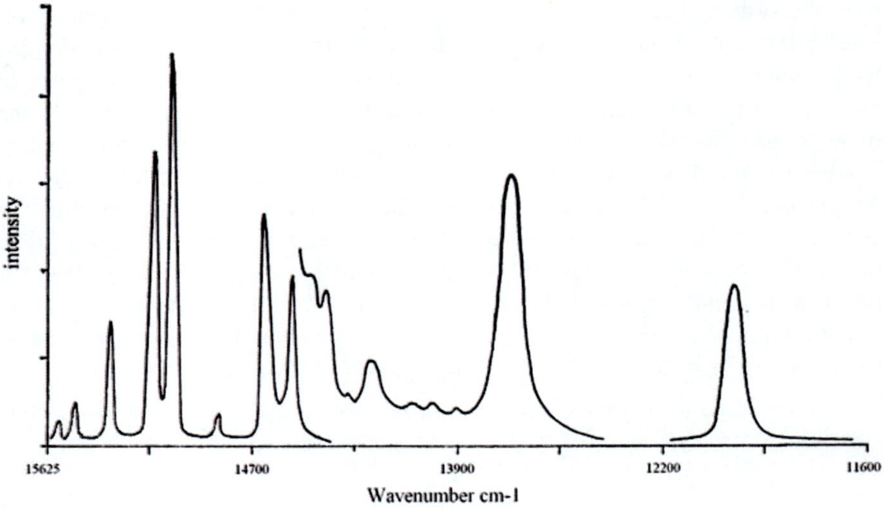


**Fig. 1** The 436 nm excited luminescence spectrum of $Cr(NH_3)_6(ClO_4)_3$ at 80 K

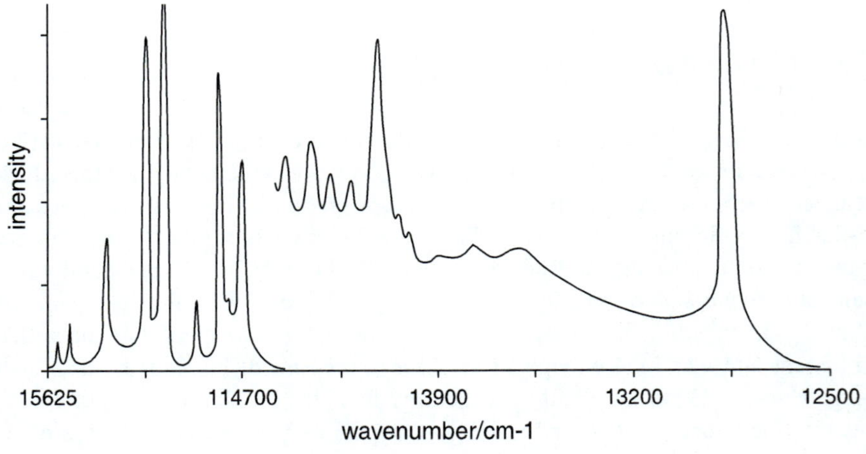


**Fig. 2** The 436 nm excited luminescence spectrum of $Cr(ND_3)_6(ClO_4)_3$ at 80 K

## 3.2 *$Cs_2MnF_6$*

The work on $Cs_2MnF_6$ lead to the only time when I won a theoretical argument with Carl. The luminescence of this compound is particularly intense and well resolved [18]. The vibronic origins due to the $t_{1u}$ bend, $t_{2u}$ bend and $t_{1u}$ stretching modes of the $MnF_6^{2-}$ dominate the luminescence, but based on these vibronic origins there are well-developed but much weaker progressions in the $a_{1g}$ and $e_g$ stretching modes. I speculated that the $e_g$ modes showed the presence of a weak Jahn–Teller in the $^2E$ initial state leading to a displacement and splitting of the state along the $e_g$ vibrational coordinate. A little later Ferguson published the luminescence spectrum of $CrF_6^{3-}$ [19], which is isoelectronic with $MnF_6^{2-}$ but the spectra are quite different. There was a strong progression in a mode of 481 $cm^{-1}$, which I assigned as due to the $e_g$ Cr –F stretching mode as well as the $a_{1g}$ symmetric stretching mode of 561 $cm^{-1}$. In the chromium compound, the $^2E_g$ state is much closer to the $^4T_{2g}$ state due to the weaker crystal field in the trivalent metal compound and the interaction between these states leads to a greater displacement and splitting of the $^2E_g$ state. This seemed to confirm the assignment. I sent off a short paper on this to Chemical Physics Letters [20]. Some weeks later, I received a rejection on the grounds of symmetry arguments, but I was able to point out that the fourfold degenerate $^2E_g$ state (which transforms as $\Gamma_8$ in the $O_h$ double group) [21]) does not obey the usual symmetry arguments. I pointed this out to the referee via the Editor and the paper was immediately accepted. Early in 1972, I attended a conference at Oxford as a very junior participant. As I walked into the St John's College, I saw Carl on the other side of the lawn. Carl immediately strode across the grass and apologized most profusely, saying "Colin! How could I have been so stupid" Indeed his apology was so profuse and public that this very young and inexperienced scientist found it quite embarrassing and I remember it clearly to this day. This is very much the measure of the man.

## 3.3 Energy Transfer

In the early 1970s there was a debate in the literature as to whether significant emission could be detected from the $^4T_{2g}$ state in those relatively strong ligand field complexes where this state lies well above the emissive $^2E_g$ state. In our work on the $Cr(NH_3)_6^{3+}$. I had noticed that the emission between the vibronic origins did not return exactly to the baseline. I therefore asked Peter Mathews, an undergraduate project student in my group to measure the decay lifetimes of this weaker broader. It immediately became clear that the broader emission was due to an impurity, Cr $(NH_3)_5Cl^{2+}$. This was confirmed by doping experiments. In fact it proved extremely difficult to produce $Cr(NH_3)_6(ClO_4)_3$ free from minute traces of $Cr(NH_3)_5Cl^{2+}$, several recrystalisations of the starting $Cr(NH_3)_6Cl_3$ before precipitation of the perchlorate salt being necessary under a red photographic lamp rather than daylight or room light. The $^2E_g$ state of the minute trace of $Cr(NH_3)_5Cl^{2+}$ was populated from the $^2E_g$ state of $Cr(NH_3)_6^{3+}$ by energy transfer and this is beautifully demonstrated by the slow rise of the broader, weaker emission which paralleled the decay of the majority $Cr(NH_3)_6^{3+}$ ions [22] (Fig. 3).

This work lead to an interest in the luminescence of lower symmetry chromium ammine complexes. In these complexes, the $^2E_g$ state is split into two components by the lower symmetry ligand field. The tetragonal ligand field parameters for many of these compounds were well known or easily available from optical spectroscopy, but the splitting of the $^2E_g$ as measured by the absorption and emission spectroscopy

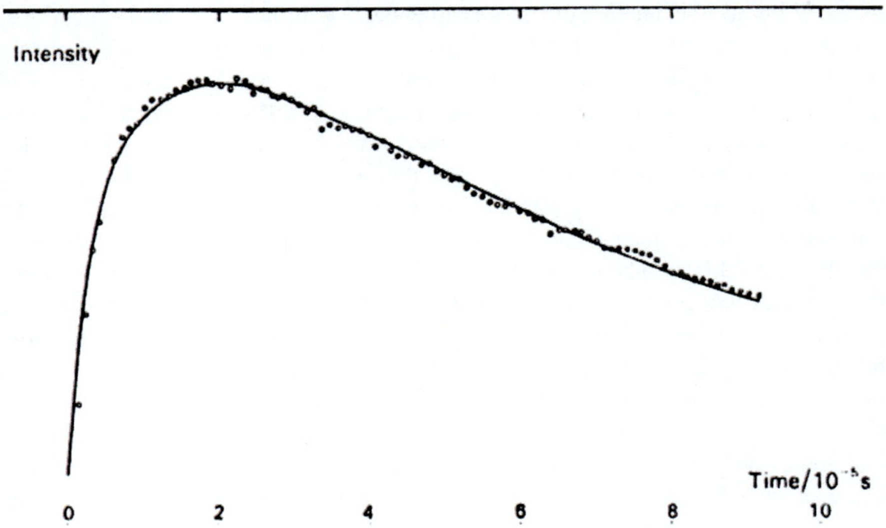


**Fig. 3** Time dependence of the $^2E \rightarrow {}^4A$ emission of a trace of $Cr(NH_3)_5Cl^{2+}$ in $Cr(NH_3)_6(ClO_4)_3$ at 80 K followed pulsed excitation. The solid line is the calculated curve and the agreement with the observed kinetics proves that the $^2E$ state of $Cr(NH_3)_5Cl^{2+}$ is populated by energy transfer from the $^2E_g$ state of $Cr(NH_3)_6{}^{3+}$

were invariably about 10 times larger. Following Griffith [3], I proposed that this was due to a failure of the spherical parameterization of the electron–electron interaction due to metal–ligand overlap [22]. This was described by Gűdel as very worrying [23], but seems to be a general phenomenon [24, 25]. The effect appears similar to the failure of the simple crystal field model in lanthanide complexes and has been attributed to electron correlation induced by the crystal field (CCF). In the lanthanide complexes $LnX_6{}^{3-}$ The percentage deviation of each multiplet from the one-electron model decreases in the order X = Br > Cl > F. This indicates the importance of covalency and/or ligand polarizability contributions to the CCF [26, 27]. Carl would have been very happy with this interpretation since, although he is best known for his work in ligand field theory, he was always very unhappy about the quantitative application of the non-symmetry determined aspects of crystal field theory.

## 4 A Shell Model for Energy Transfer and Cross Relaxation

By the late 1970s I had looked at the luminescence spectra of a large number of chromium(III) compounds and I was looking for new systems to study. One approach was to look at the well-known emission from uranyl compounds, but just after we started work on them Denning and colleagues solved the most pressing problem of the identity of the lowest (i.e. luminescent) state in a series of beautiful papers on the low temperature absorption spectra of $Cs_2UO_2Cl_4$ and my group was left to fill in the details of the nature of the emission in a variety of more complex systems. We also looked at the emission from halide complexes of the second and third row transition series, this was greatly enhanced by the construction of dedicated luminescence spectrometer for the 3,000–10,000 $cm^{-1}$ region. We tried to work on lanthanide complexes but could not prepare salts of the $LnX_6^{3-}$ ion by wet chemistry. At this time Bob Schwartz spent a year in London and introduced us to the Bridgeman technique for growing the crystals from the melt in sealed tubes, he provided some samples and we were soon able to grow our own.

## 4.1 Lanthanide Elpasolites

One of the first systems we studied was the perfectly octahedral $EuCl_6^{3-}$ ion in the cubic "elpasolite" $Cs_2NaEuCl_6$.

We were attracted to this system by a paper [28] which claimed to have been to achieve a good agreement between the calculated intensities of the vibronic origins due to the $\upsilon_6$, $\upsilon_4$ and $\upsilon_3$ vibrational modes of the $EuCl_6^{3-}$ group in spite of an incorrect assignment of the $\upsilon_6$ mode. Explicit calculations of the relative intensities of these modes [29] in the $MnF_6^{2-}$ ion using both the formalism due to Liehr and Ballhausen [30] and the closure method of Koide and Price [31] had shown the sensitivity of the vibronic intensity distribution to the assumed force field. The $\upsilon_4$ and $\upsilon_3$ vibronic origins of $EuCl_6^{3-}$ clearly show the transverse optic – longitudinal optic splitting [32] as expected in this pure compound. Explicit calculations of vibronic intensity distributions in the lanthanide ions proved to be laborious and we turned our attention to the study of decay curves. It seemed likely that the high symmetry of the lattices and the well understood energy levels of these ions would permit exact evaluations of the shape of the decay curves in the presence of energy transfer to neighbouring lanthanide ions and thus avoid the crude approximations of existing models [33]

We therefore developed a much simplified, but mathematically exact, version of a shell model which is particularly suitable for the analysis of cross relaxation in high symmetry systems such as the lanthanide elpasolites. Since the vibronic structure in the emission and absorption spectra of these compounds is both intense and broad (relative to the electronic origins), there are many cases where the vibronic structure of one electronic transition in emission overlaps with the vibronic structure of another electronic transition of a chemically identical ion in absorption. The emission of the excited ion may then be partially quenched by energy transfer. Implicit in this formulation is the assumption that the interionic coupling is weak compared with the vibronic coupling; this is certainly true for the lanthanide elpasolites where the lanthanide ions are separated by distances of more than 0.7 nm. We refer to this process as cross relaxation by the electric dipole vibronic–electric dipole vibronic (EDVEDV) mechanism.

The first application of our model was to the EDV–EDV cross relaxation involving the $^3P_0$ and $^1G_4$ states of $Cs_2\,NaY_{1-x}Pr_xCl_6$ ($x = 0.001$–$1$),

The room temperature structure of $Cs_2NaPrCl_6$ is shown in Fig. 4.

The energy level diagram of $PrCl_6^{3-}$ relevant to this study is shown in Fig. 5.

The intensity of the emission resulting from the decay of a set of identical excited isolated ions in a crystalline lattice may be written as

$$I(t) = I(0)\exp(-kt),$$

where $k$ is the sum of the radiative and non-radiative decay processes of those isolated ions. Consider a lattice where there are other ions chemically identical to the excited ion, which surround the excited ions and may act as acceptors for cross relaxation processes. These may be collected in individual shells labelled as $n = 1$ (nearest neighbour), $n = 2$ (next nearest neighbours), etc. The number of lattice sites that may be occupied by acceptor ions in the $n$th shell is $N_n$ these integers are determined by the symmetry of the crystal lattice and are given in Table 1.

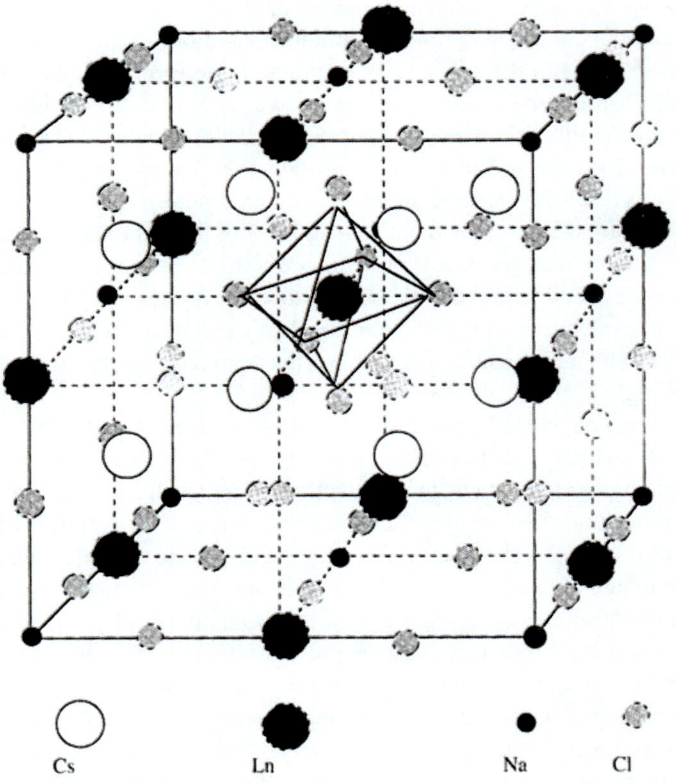


**Fig. 4** The structure of $Cs_2NaLnCl_6$ at room temperature

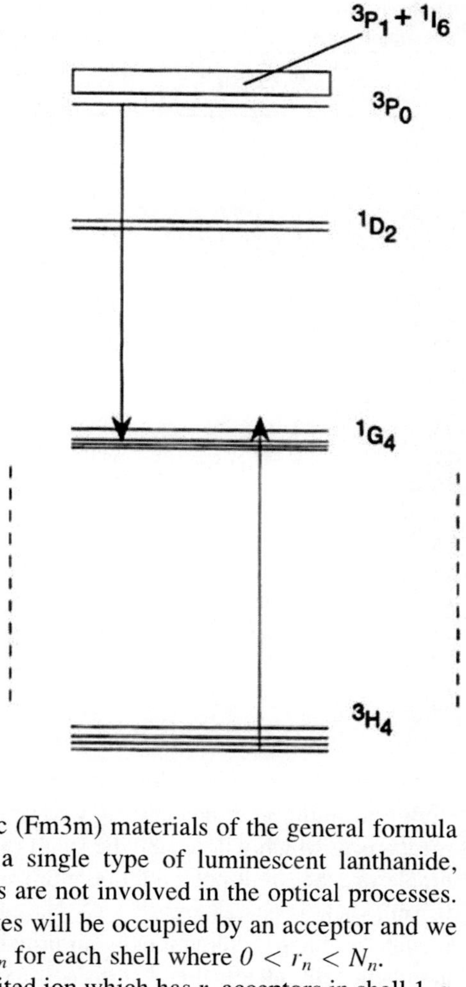


**Fig. 5** Simplified energy level diagram for $PrCl_6^{3-}$ showing the cross relaxation processes considered in this work. The $^3H_5$, $^3H_6$ and 3F$_1$ levels states in the region 1,000–9,000 cm$^{-1}$ have been omitted for clarity

We are concerned with a set of cubic (Fm3m) materials of the general formula $A_2A_{1-x}$'$MCl_{-6}$ $Ln_XC1_6$ where Ln is a single type of luminescent lanthanide, transition metal or other ion, the M ions are not involved in the optical processes. When $x \neq 1$, not all the available $N_n$ sites will be occupied by an acceptor and we introduce an integer occupancy factor $r_n$ for each shell where $0 < r_n < N_n$.

The decay of a particular type of excited ion which has $r$, acceptors in shell 1, $r$, acceptors in shell 2, etc., will be

$$I(t) = I(0)\exp - (k + \sum_{i}^{\text{acceptors}} k_i)t, \tag{1}$$

where $k_i$ is the transfer rate to the $i$th acceptor. Energy migration between the donors, energy migration between the acceptors and back transfer from the acceptors to the donors have been neglected. Neglecting processes involving more than two ions, this becomes

$$I(t) = I(0)\exp - (k + \sum_{n}^{\text{shells}} r_n k_n)t. \tag{2}$$

**Table 1** Donor–acceptor distance, maximum occupation number and relative dipole–dipole interaction term in the first 20 shells of the fcc lattice

| $n$ | $R_n/a_0$ | $N_n$ | $N_n/(R_n/R_1)^6$ | $\sum_n N_n/(R_n/R_1)^6$ |
|---|---|---|---|---|
| 1 | 0.707 | 12 | 12.0000 | 12.0000 |
| 2 | 1.000 | 6 | 0.7500 | 12.7500 |
| 3 | 1.225 | 24 | 0.8889 | 13.6389 |
| 4 | 1.414 | 12 | 0.1875 | 13.8264 |
| 5 | 1.581 | 24 | 0.1920 | 14.0184 |
| 6 | 1.732 | 8 | 0.0370 | 14.0554 |
| 7 | 1.871 | 48 | 0.1399 | 14.1953 |
| 8 | 2.000 | 6 | 0.0117 | 14.2070 |
| 9 | 2.121 | 36 | 0.0494 | 14.2564 |
| 10 | 2.236 | 24 | 0.0240 | 14.2804 |
| 11 | 2.345 | 24 | 0.0180 | 14.2984 |
| 12 | 2.449 | 24 | 0.0139 | 14.3123 |
| 13 | 2.550 | 72 | 0.0328 | 14.3451 |
| 14 | 2.646 | 0 | 0.0000 | 14.3451 |
| 15 | 2.739 | 48 | 0.0142 | 14.3593 |
| 16 | 2.828 | 12 | 0.0029 | 14.3622 |
| 17 | 2.915 | 48 | 0.0098 | 14.3720 |
| 18 | 3.000 | 30 | 0.0051 | 14.3771 |
| 19 | 3.082 | 72 | 0.0105 | 14.3876 |
| 20 | 3.162 | 24 | 0.0030 | 14.3906 |

For simplicity of notation, we now restrict the energy transfer processes to multipole–multipole interactions (i.e. we neglect energy transfer by the exchange mechanism which will usually be small compared with the electric dipole–electric dipole or other multipole–multipole process at interionic distances of greater than 0.7 nm such as characterize the $Cs_2NaLnC1_6$ lattices). Extension of our model to include exchange interactions is, in principle, straightforward. If all the acceptors are chemically identical and at distances $R_{,,}$ we may write

$$k_n = \sum_{\mathrm{p}} G_n^{\mathrm{p}} \frac{\alpha_{\mathrm{p}}}{(R_n)^{s_{\mathrm{p}}}} \tag{3}$$

where the summation is over all the individual energy transfer processes occurring at a specific donor ion, each with a coupling constant $\alpha_{\mathrm{p}}$. is the geometric (angular) factor describing the interaction between the donor transition multipole and the acceptor transition multipole, $s_{\mathrm{p}} = 6$, 8 or 10 for dipole–dipole, dipole–quadrupole and quadrupole–quadruple interactions, respectively. Again, for simplicity of notation, we assume that only a single dominant energy transfer process occurs and, in the lanthanide elpasolites, the dominant cross relaxation mechanism is usually the EDV–EDV process. We therefore consider only the term with $s_p = 6$ in (3). For a set of identical ions, within this model, the decay of the excited state is purely single exponential in time. It has become common to define a "critical distance" $\boldsymbol{R_o}$ for the energy transfer when the rate of the energy transfer equals the rate of the intrinsic decay [33]. This quantity is not well defined within shell models.

## 4.2 The Geometric Factor

The role of the geometric factor has frequently been ignored. However, in this special case it takes a simple form. The interaction Hamiltonian between two electric dipoles $\boldsymbol{\mu}_A$ and $\boldsymbol{\mu}_B$ separated by a distance $R$ may be written as

$$H_{AB} = \left(4\pi\varepsilon_0 R^3\right)^{-1} \mu_A \bullet \tilde{C} \bullet \mu_B, \tag{4}$$

where $\tilde{C}$ is the symmetric dipolar coupling tensor. The elements of $\tilde{C}$ are

$$C_{ij} = \left(\delta_{ij} - 3\frac{r_1 r_2}{R^2}\right). \tag{5}$$

The rate of energy transfer from an ion A to an ion B may be written as

$$W_{AB} = \tfrac{2\pi}{h}\langle a'b|H_{AB}|ab'\rangle \int g(a \to a')g(b \to b')\mathrm{d}W, \tag{6}$$

where the integral is the normalized line shape function. Substituting (4) and (5) in the matrix element in (6) we get

$$\langle a'b|H_{AB}|ab'\rangle^2 = \left(4\pi\varepsilon_0 R^3\right)^{-2} \times \left(\sum_{i,j=x,y,z} \left(\mu^i_{aa'}\right)^2 (C_{ij}) \left(\mu^j_{bb'}\right)^2\right) \tag{7}$$

here $\mu^i_{aa'}$, is the $i$th component of the transition dipole between states a and a' etc.

## 4.3 Specialization to Face-Centred Cubic Lattices

In a cubic lattice, the components of the transition dipoles are independent of orientation so we may write $(\mu^i{}_{aa'})^2 = (\mu_{aa'})^2$ The geometric factor in (3) may then be written as

$$\langle a'b|H_{AB}|ab'\rangle^2 = \frac{G_{aa''bb'}}{R^6(12\pi\varepsilon_0)} \left(\mu^i{}_{aa'}\right)^2 \left(\mu^j{}_{bb'}\right)^2 \tag{8}$$

so

$$G_{aa''bb'} = \tfrac{1}{9} \sum_{i,j=x,y,z} (C_{ij})^2. \tag{9}$$

In the face-centred cubic elpasolite lattice (Fig. 1), the 12 nearest neighbours to lattice point (0, 0, 0) are arranged at the points ($\pm a_o/\surd 2$, $\surd \pm a_O/\surd 2$, 0), ($\pm a_o/\surd 2$, 0. $\pm a_o/\surd 2$), (0, $\pm a_O/\surd 2$. $\pm a_O/\surd 2$). A typical tensor for an individual lattice site in the first nearest neighbour shell is then

$$\tilde{C}_1^1 = \begin{pmatrix} -\frac{1}{2} & -\frac{3}{2} & 0 \\ -\frac{3}{2} & -\frac{1}{2} & 0 \\ 0 & 0 & 1 \end{pmatrix}.$$

From the form of the Euler angles and the symmetry of the lattice, the *absolute* values of the matrix elements in the other matrices of the first shell $\tilde{C}_1^r$, $r = 2$–$12$ are all equal. From (9), the geometric factors are all 2/3. Similarly, the geometric factors may be calculated for all shells; the first few are:

$$\tilde{C}_2 = \begin{pmatrix} -2 & 0 & 0 \\ 0 & 1 & 0 \\ 0 & 0 & 1 \end{pmatrix}$$

$$\tilde{C}_3 = \begin{pmatrix} +\frac{1}{2} & -\frac{1}{2} & -1 \\ -\frac{1}{2} & +\frac{1}{2} & -1 \\ -1 & 0 & -1 \end{pmatrix}$$

$$\tilde{C}_4 = \begin{pmatrix} -\frac{1}{2} & 0 & -\frac{3}{2} \\ -\frac{3}{2} & 1 & 0 \\ -\frac{3}{2} & 0 & -\frac{1}{2} \end{pmatrix},$$

where $N_4 = 6, N_3 = 24, N_4 = 12$ at distances of $a_o$, ($\surd 3/\surd 2$) $a_o$ and 2 $a_o$ so that, *for dipole–dipole coupling*, the geometric factors are independent of *n*, a result that may be deduced from symmetry. These results assume that the microscopic effective dielectric constant in the region of the excited ion is isotropic. Inspection of Fig. 1 suggests that this is unlikely.

For a large number of $Ln^{3+}$ ions, randomly distributed in a crystal lattice of composition $A_2A{\cdot}M_{1-X}Ln_X\ Cl_6$ each ion will have a different set of integer values for "*r*", $r, \ldots r_\infty$. Each of these ions will have an exponential decay rate, but the sum of all these exponential decay processes with different decay rates will be non-exponential. To determine the form of this macroscopic decay curve, it is necessary to average over all possible distributions of the set $r_1, r_2 \ldots r_\infty$. For a particular $Ln^{3+}$ ion, the statistical probability of a specific shell *n* surrounding $r_n$ sites occupied by a $Ln^{3+}$ ion and $N_r - r_n$ sites occupied by $M^{3+}$ ions may be written as

**Table 2** Percentage occupancy factors $O^{12}_{r_n}(x)$

| x | 0 | 1 | 2 | 3 | 4 | 5 | 6 | 7 | 8 | 9 | 10 | 11 | 12 |
|---|---|---|---|---|---|---|---|---|---|---|---|---|---|
| 0.01 | 88.6 | 10.7 | 0.6 | 0.0 | – | – | – | – | – | – | – | – | – |
| 0.02 | 78.5 | 19.2 | 2.2 | 0.1 | 0.0 | – | – | – | – | – | – | – | – |
| 0.05 | 54.0 | 34.1 | 9.9 | 1.7 | 0.2 | 0.0 | – | – | – | – | – | – | – |
| 0.10 | 28.2 | 37.7 | 23.0 | 8.5 | 2.1 | 0.4 | 0.0 | – | – | – | – | – | – |
| 0.15 | 14.2 | 30.1 | 29.2 | 17.2 | 6.8 | 1.9 | 0.4 | 0.1 | 0.0 | – | – | – | – |
| 0.25 | 3.2 | 12.7 | 23.2 | 25.8 | 19.4 | 10.3 | 4.0 | 1.1 | 0.2 | 0.0 | – | – | – |
| 0.50 | 0.0 | 0.3 | 1.6 | 5.4 | 12.1 | 19.3 | 22.0 | 19.3 | 12.1 | 5.4 | 1.6 | 0.3 | 0.0 |

$$O^{N_n}_{r_n}(x) = \frac{N_n!}{(N_n - r_n)!r_n!}(1-x)^{N_n - r_n}x^{r_n}. \tag{10}$$

The factors for the elpasolite lattice are given in Table 2:

The probability of a particular distribution with $r_1$ $Ln^{3+}$ ions in shell 1, $r_2$ $Ln^{3+}$ ions in shell 2 etc is

$$P^{N_1N_2\ldots.N_\infty}_{r_1r_2\ldots\ldots r_\infty}(x) = \prod_n O^{N_n}_{r_n}(x).$$

If we consider a cross relaxation process in which the $Ln^{3+}$ ions surrounding an excited ion may act as acceptors,

$$I(t) = I(0)\exp(-kt)\prod_n^{\text{all shells}} \exp\left[-r_n\left(\frac{R_1}{R_n}\right)^6 k^{\text{ET}}t\right],$$

where $k^{\text{ET}}$ is the energy transfer rate to a single ion at the position of the nearest neighbour acceptor.

This includes the geometric factor. Summing over a large number of donor ions, we obtain

$$I(t) = I(0)\exp(-kt)\sum_{\text{all } r_n}^{\text{all } N_n}\left\{P^{N_1N_2\ldots.N_\infty}_{r_1r_2\ldots\ldots r_\infty}(x) \times \prod_n^{\text{all shells}} \exp\left[-r_n\left(\frac{R_1}{R_n}\right)^6 k^{\text{ET}}t\right]\right\}. \tag{11}$$

For computational purposes, this is more conveniently written as

$$I(t) = I(0)\exp(-kt)\prod_n^{\text{all shells}}\sum_{r_n}^{N_n} O^{N_n}_{r_n}(x)\exp\left[-r_n\left(\frac{R_1}{R_n}\right)^6 k^{\text{ET}}t\right]. \tag{12}$$

These equations assume that the probability of two nearby ions being excited is negligibly small; this, in turn, requires that very low excitation powers are used.

For the elpasolite lattice, the successive terms in the product of (15) rapidly approach unity because of the factor $R^6$ in the exponential term, so that even at

$x = 1$ some 98% of the energy transfer to the donors is included up to $n = 3$.The factors involved are included in Table 1. However, the detailed shape of the curve requires more shells to be included if it is to be accurately reproduced. When $x = 0.01$ and the $n = 1$ shell $O_r^{12}(0.01) = 0.886$, 0.107, and 0.006 for $r = 0$, 1, 2, respectively the contributions from the other shells are smaller. Deviations from exponential decay become discernible only when $k^{\mathrm{ET}}$ is several times greater than $k$. Moreover, at large values of $k^{\mathrm{ET}}$ the $n = 1$ term dominates and the decay is, to a fair approximation, double exponential, the faster process giving $k^{\mathrm{ET}}$ directly and the ratio of the fast process to the slow process slightly greater than (0.107/0.886) = 12%.

For $x = 1$, only the $r = N_i$ terms contribute and the decay is exponential with a decay constant of $k + (12 + 6/8 + 24/27 + 12/64\ldots)k^{\mathrm{ET}} = k + 14.4k^{\mathrm{ET}}$. At intermediate values of $x$, the decay curves will be non-exponential, especially when $k^{\mathrm{ET}} \cong k$ but the apparent deviations from exponential decay may be small due to the large number of terms that contribute to the decay. For example, at $x = 0.5$ there are 13, 7 and 18 terms greater than 1% of the largest value of $O_r^{12}(0.5)$) for $r = 1$. 2 and 3, respectively. The semi-log plots are then smooth curves rather than clearly bi- or multi-exponential.

This behaviour is shown in Fig. 6

For measurements over the whole concentration range $x = 0$–1 at constant temperature, the experimental exponential decay constant at very low dilutions gives the value of $k$ directly whilst that of the $x = 1$ material will be $k$ + *14.4*$k^{E'I}$. At intermediate concentrations, there are *no free parameters to describe the non-exponential curves obtained at other concentrations;* these curves may be calculated directly and there are no approximations in the mathematical solution of the model once defined. Any difference between calculated and experimental curves

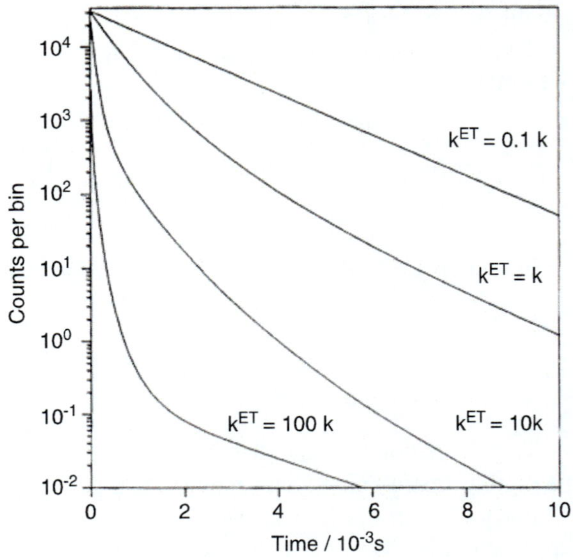


**Fig. 6** Computed decay curves for different ratios of $k^{\mathrm{ET}}$ and $k$ ($x = 0.25$, $k = 500\ \mathrm{s}^{-1}$)

must be interpreted as due to effects that have not been included in the model. We emphasize that this is a much more exacting' test of our understanding of energy transfer processes than the traditional approach of introducing one or more parameters which are determined by curve fitting procedures. It must be expected that real systems will normally show deviations from these idealized models. We illustrate this procedure by reference to experimental data on the emission from the $^3P_0$ state of $Cs_2NaY_{1-x}Pr_xCl_6$.

## 4.4 *Energy Levels, Structural and Vibrational Properties of $Cs_2NaPrCl_6$*

$Cs_2NaYCl_6$ and $Cs_2NaPrCl_6$, (above 158 K) belong to the face-centred cubic Bravais lattice, the unit cell dimensions are $a_0 = 1.09118(13)$ nm for $Cs_2NaPrCl_6$ and 1.07315(15) for $Cs_2NaYCl_6$. This difference arises because the $Pr^{3+}$ ion is appreciably larger than the $Y^{3+}$ ion.

The electronic energy levels of $Cs_2NaPrCl_6$ have been studied by Richardson [34] and by Flint and Tanner [35]. It is now generally agreed that the emissive state in the blue region at low temperatures is the $^3P_0$ at 20,602 $cm^{-1}$. see Fig. 5. The next lower states are the $^1D_2$, at 16,670 $cm^{-1}$ and the $^1G_4$ with crystal field components at $\Gamma_1$ 9,853: $\Gamma_3$ (not observed, expected position about 9,870): $\Gamma_4$ 9,900, $\Gamma_5$ 10,325 $cm^{-1}$. The ground state $^3H_4$ with crystal field components at $\Gamma_1$ 0; $\Gamma_4$ 236; $\Gamma_3$ 422: $\Gamma_5$ 701$cm^{-1}$. To high energy of the $^3P_0$ state the $^3P_1$ state is near 21,200 $cm^{-1}$ and the components of the $^1I_0$ state occur between 21,164 and 22,000 $cm^{-1}$. The remaining levels are not involved in the processes discussed here but are given in [34] and [35].

$Cs_2NaPrCl_6$ crystallizes in the Fm3M-$0_h^5$ space group at room temperature where individual well separated, perfectly octahedral $PrCl_6^{3-}$ ions occupy the anionic sites in the antifluorite structure. At 158 K, the repulsion between the chloride ions is sufficient to cause a phase transition to the 14/m-$C_{4h}$ structure, the anions rotating slightly from their positions in the cubic lattice. Whilst this phase transition causes splitting of a few wavenumbers in the degenerate electronic and vibrational levels we have been unable to detect any abrupt changes in the intensity or decay constants of the emission at about this temperature. We note that the energy transfer processes discussed in this work are all mediated by the EDV–EDV mechanism involving dispersionally broadened vibronic processes, the splittings of the electronic and vibrational levels due to the phase transition are much smaller than this broadening and the effect of the phase transition will be negligible.

$Cs_2NaY_{1-x}Pr_xCl_6$ ($x < 0.5$) does not undergo a phase transition over the temperature range studied.

The vibrational properties of this lattice are also well established, the important internal modes with odd parity at the Pr site are $v_3$ 240–276, $v_4$ 98–120 and $v_5$

70–85 $cm^{-1}$. The range given encompasses the TO-LO splittings at the zone centre for $\nu_3$ and $\nu_4$ together with the vibrational dispersion away from the zone centre for all three vibrations. There are lattice vibrations between 22 and 66 $cm^{-1}$ plus a mode which is essentially the motion of the sodium ion at 175 $cm^{-1}$. All of these modes, selection rules permitting, contribute appreciable vibronic intensity to many of the observed electronic transitions.

## 4.5 Cross Relaxation from the $^3P_O$ State of $PrCl_6^{3-}$

The emission from the $^3P_0$ state of $Cs_2NaY_{0.99}Pr_{0.01}$ is relatively intense at all temperatures between 10 and 293 K. The emission from the same state of the pure compound is much weaker even at low temperatures and is virtually undetectable at room temperature. The luminescence lifetimes decrease by more than two orders of magnitude over this concentration range. It therefore seems probable that a cross relaxation is involved. Noting that at low temperatures, all possible electronic transitions that are allowed by the electric dipole mechanism must involve the creation of an odd parity phonon at both the donor and acceptor sites, the only possible cross relaxations are those involving the $^1G_4$ (state. Taking the $^3P_o$ level at 20,602 $cm^{-1}$, the donor pathway is likely to be $^3P_o \rightarrow {}^1G_4(\Gamma_5, \Gamma_4) + \Delta E_{vib}$ of wavenumber 10277 – $\nu_{vib}$ and 10702 – $\nu_{vib}$ $cm^{-1}$ respectively. The acceptor process will be $^3H_4 \rightarrow 1G_4\ (\Gamma_1, \Gamma_3, \Gamma_4, \Gamma_5) + \Delta E_{vib}$ of wavenumber 9,853, 9,870, 9,900 and 10,325 $cm^{-1}$ each plus a vibrational wavenumber $\nu_{vib}$. Clearly at low temperatures there are several possibilities involving the vibrational modes of wavenumber about 255, 175, 110 and 80 $cm^{-1}$. At higher temperatures, many further possibilities occur due to the presence of hot bands in the absorption and emission spectra and the general spectral broadening.

The 80 K luminescence decay curves [36] of the $^3P_0$ state of $Cs_2NaY_{1-x}Pr_xCl_6$) $x = 0.001$–1 are shown in Fig. 5. The curves for $x = 0.00003$, 0.0001 and 0.01 are essentially identical to the $x = 0.001$ showing that this curve corresponds to isolated $Pr^{3+}$ ions. Both the $x = 0.001$ and $x = 1$ curves are exactly exponential over eight half-lives with decay constants of 3,610 and 23,000 $s^{-1}$, respectively. At intermediate $x$ values, the decay curves are just perceptibly non-exponential, the deviation being greatest at $x = 0.25$ and $x = 0.5$, the decay rates increasing steadily with concentration. Rather similar data are obtained at 20 K with decay constants of 2,110 $s^{-1}$ ($x = 0.001$) and 9,220 $s^{-1}$ ($x = 1$). This behaviour is quite inconsistent with the traditional Inokuti–Hiryama kinetics that predict that the decay rate approaches that of the isolated ion at long times. It might be explicable assuming migration in the $^3P_0$ state (i.e. using the Yokoto–Tanimoto model) but since the ground state and the emissive state are both $\Gamma_1$, energy migration at low temperatures is most unlikely.

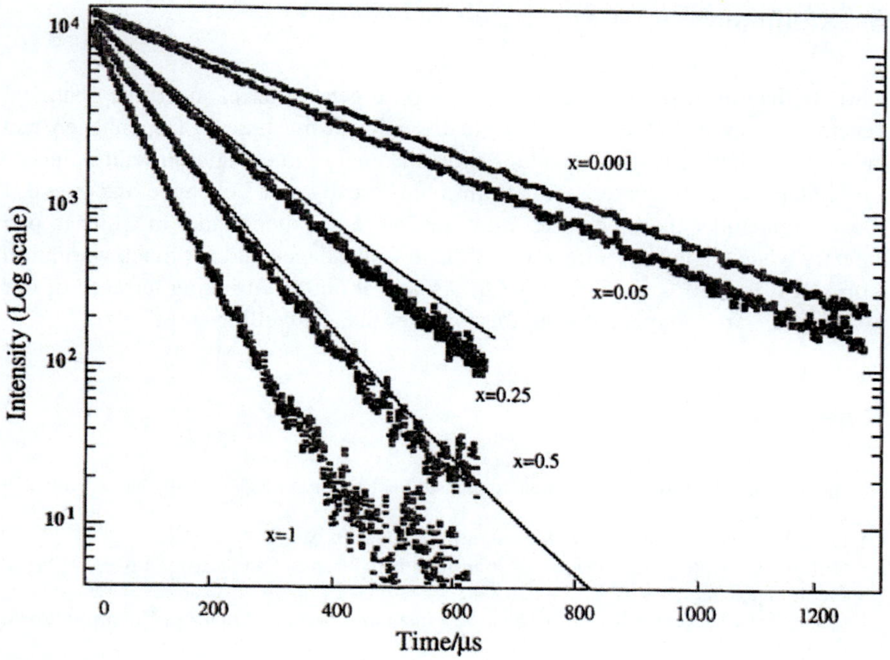


**Fig. 7** 468 nm excited decay curves of $Cs_2NaY_{1-x}Pr_xCl_6$ at 80 K for different x for $x = 0.25$ and $x = 0.5$ the computed curves are based on $k = 3{,}610\ s^{-1}$ and $k^{ET} = 23{,}000\ s^{-1}$

The data for $x = 1$ enable the value of $k = 1{,}350\ s^{-1}$ to be derived. The solid lines for $x = 0.25$ and $x = 0.5$ in Fig. 6 correspond to the calculated curves from (12) using all contributions from the first seven shells (*no curve fitting is involved*). The agreement is encouraging and certainly better than could be achieved by use of any other widely used model *with no free parameters,* but not perfect. Essentially perfect agreement can be obtained by treating $k^{ET}$ as a free parameter to be determined by curve fitting and values in the region 1,500–1,600 $s^{-1}$ are obtained. This does not provide any information on the source of the discrepancy. Somewhat unexpectedly, better agreement is obtained between the calculated and experimental curves is obtained if (17) is truncated after the $n = 1$ shell. This implies that the energy transfer is of shorter range than is predicted by (12). It is unlikely that higher order transition multipoles are involved. Similarly, the exchange pathway through at least two anions and a cation is implausible. A possible explanation is that the assumption of an isotropic dielectric around the donor atom is a contributing factor. The interaction pathway between the donor and the $n = 1$ acceptor involves interstices between the other anions (Fig. 1), whilst the direct line pathways from the donor to the $n = 2$ and $n = 3$ acceptors is through other ions. The shielding effect of these ions may be significant, especially in the $x = 1$case. These ideas have been discussed more extensively elsewhere [37] (Fig. 7).

## 5 Summary

Carl Ballhausen had a profound influence on a generation of inorganic chemical spectroscopists including myself. I greatly enjoyed my time in Copenhagen and subsequent visits. He always had the time and energy to explain difficult points to his students and co-workers. I shall miss him greatly as a colleague and friend. I always remember the time when we organized a summer school in Chile in one January when the temperature was −25°C in Copenhagen and not much warmer in London. It was 30°C in Vina del Mar. As we sat on the "drinking terrace" of our apartment, Carl commented that "this is better than shovelling snow".

## References

1. Ballhausen CJ (1962) Introduction to ligand field theory. McGraw-Hill, New York and London, 35
2. Cotton FA (1963) Chemical applications of group theory. Wiley, New York
3. Griffith JS (1962) The theory of transition-metal ions. Cambridge University Press, Cambridge, UK
4. Condon EU, Shortley GH (1959) The theory of atomic spectra. Cambridge University Press, Cambridge, UK
5. Dirac PAM (1958) The principles of quantum mechanics, 4th edn. Oxford University Press, Oxford, UK
6. Holt SL, Ballhausen CJ (1967) Theor Chim Acta 7:313
7. Maksimova T, Hermanowicz K, Hanuza J, Happek U (2002) J Alloys Compd 341:239–243
8. Maksimova T, Hermanowicz K, Macalik L, Hanuza J (2001) J Mol Struct 563–564:353–357
9. McCarthy PJ, Vala MT (1973) Mol Phys 25:17
10. Flint CD (1969) J Mol Spectrosc 32:514
11. Ballhausen CJ, Hansen AE (1972) Annu Rev Phys Chem 23:15
12. Flint CD, Greenough P (1972) J Chem Soc Faraday Trans II 68:897
13. Flint CD, Greenough P (1974) J Chem Soc Faraday Trans II 70:815
14. Flint CD, Matthews AP (1974) J Chem Soc Faraday Trans II 70:1301
15. Flint CD (1974) Coord Chem Rev 14:47
16. Flint CD (1976) J Chem Soc Faraday Trans II 72:7
17. Flint CD, Acevedo R (1985) Mol Phys 54:619
18. Flint CD (1971) J Mol Spectrosc 37:414
19. Ferguson J, Guggenheim HJ, Wood DL (1971) J Chem Phys 54:504
20. Flint CD (1971) Chem Phys Lett 11:27
21. Sturge MD (1967) In: Seitz F, Turnbull D, Ehrenreich H (eds) Solid state physics, vol 20. Academic, New York, p 91
22. Flint CD, Matthews AP (1973) J Chem Soc Faraday Trans II 69:419
23. Decurtins S, Guedel HU, Neuenschwander K (1977) Inorg Chem 16:796
24. Flint CD, Matthews AP, O'Grady PJ (1977) J Chem Soc Faraday Trans II 73:655
25. Flint CD, Matthews AP (1980) J Chem Soc Faraday Trans II 76:1381
26. Denning RG, Berry AJ, McCaw CS (1998) Phys Rev B 57:2021
27. Thorne JRG, McCaw CS, Denning RG (2000) Chem Phys Lett 319:185
28. Faukner TR, Richardson FS (1978) Mol Phys 35:1141
29. Acevedo R, Flint CD (1983) Mol Phys 49:1065
30. Ballhausen CJ, Liehr AD (1959) Mol Phys 2:123
31. Koide S, Pryce MHL (1958) Phil Mag 3:607

32. Flint CD, Stewart-Darling FL (1981) Mol Phys 44:61
33. Inokuti M, Hirayama F (1965) J Chem Phys 43:1978
34. Richardson FS, Reid M, Dellara J, Smith RD (1985) J Chem Phys 83:3813
35. Tanner PA (1986) Mol Phys 57:697
36. Vasquez SO, Flint CD (1995) Chem Phys Lett 238:378
37. Luxbacher T, Fritzer HP, Sabry-Grant R, Flint CD (1995) Chem Phys Lett 241:113

Struct Bond (2012) 142: 49–90
DOI: 10.1007/430_2011_52

Published online: 14 September 2011

# Transition Metal Corrole Coordination Chemistry

## A Review Focusing on Electronic Structural Studies

**Joshua H. Palmer**

**Abstract** The transition metal complexes of the non-innocent, electron-rich corrole macrocycle are discussed. A detailed summary of the investigations to determine the physical oxidation states of formally iron(IV) and cobalt(IV) corroles as well as formally copper(III) corroles is presented. Electronic structures and reactivity of other metallocorroles are also discussed, and comparisons between corrole and porphyrin complexes are made where data are available. The growing assortment of second-row corrole complexes is discussed and compared to first-row analogs, and work describing the synthesis and characterization of third-row corroles is summarized. Emphasis is placed on the role of spectroscopic and computational studies in elucidating oxidation states and electronic configurations.

**Keywords** Corroles · Electronic structure · Porphyrins

## Contents



J.H. Palmer (✉)
California Institute of Technology, 1200 E California Blvd., Pasadena, CA 91125, USA
e-mail: joshuap@caltech.edu

# 1 Introduction

## *1.1 Dedication*

Carl Johan Ballhausen (Fig. 1) was a longtime rival of Christian Klixbüll Jørgensen, the progenitor of the *Structure and Bonding* series. However, the two men were also scientific collaborators, working together on numerous studies examining the spectroscopy of coordination complexes [1–3]. Although Ballhausen and Jørgensen butted heads at times, they both strove to increase the chemical community's understanding of metal–ligand bonding and to develop the theoretical underpinnings of structural inorganic chemistry. It is therefore wholly appropriate that Ballhausen's pioneering contributions are remembered in this special volume of *Structure and Bonding*, a series that Jørgensen helped found and edited for many years.

C. J. Ballhausen contributed mightily to coordination chemistry with his groundbreaking 300 page book on ligand field theory (LFT) [4], which was written in

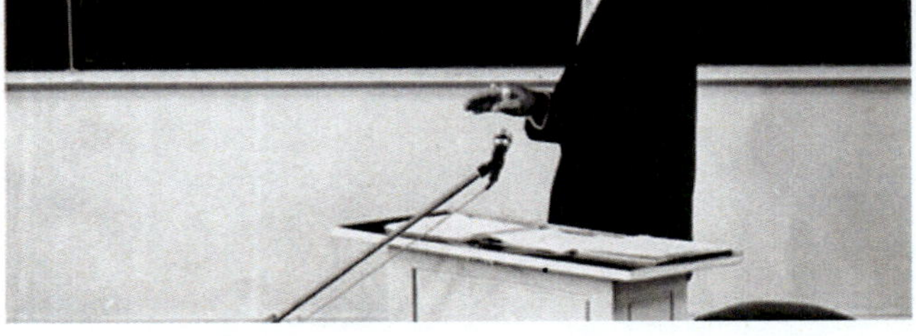

**Fig. 1** C. J. Ballhausen teaches in Copenhagen ca. 1972

English and has been translated into both Russian and Japanese [5]. In this book, still an enlightening read today, Ballhausen attempted to give "an introduction to that field of chemistry which deals with the spectral and magnetic features of inorganic complexes." The preface testifies to Ballhausen's bold personality. His insistence upon acting as "a chemist writing for chemists" and his willingness to use "inelegant" notation to explain difficult concepts are major reasons for the continued popularity of this seminal book.

Ballhausen retains a hold on my generation through a power greater than his written words and derived equations. His laboratory in Copenhagen was a popular place for young American scientists to spend time as postdoctoral fellows, and Ballhausen's early work with his fellows Harry Gray [6–10], and Tom Spiro [11], as well as his collaborations with other American scientists, such as Ed Solomon [12–14], Fred Basolo [15], and Al Cotton [16], allowed his mode of thought to spread to the USA decades ago. Ballhausen worked particularly intensively with Harry Gray, my graduate advisor; not only did they publish a number of papers together, they also co-wrote two books [17, 18] and remained close friends until Ballhausen's passing.

Harry Gray's connection to the Danish chemistry community, formed through his close relationship with C. J. Ballhausen, has remained intact for decades. Jesper Bendix, currently of the University of Copenhagen (and a student of Ballhausen's contemporary Claus Schäffer), was a postdoctoral fellow in Gray's laboratory at Caltech from 1999 to 2000, and collaboration between the two continues into the present. For his part, Bendix carries on the chemical tradition of C. J. Ballhausen, studying the chemistry of transition metals in unique coordination environments [19–21].

C. J. Ballhausen was one of the founders of theoretical inorganic chemistry, developing a description of the LFT that was easily applied to coordination compounds and was highly predictive of their molecular and electronic properties. His books and papers are still relevant today, a mammoth accomplishment in a field that moves as rapidly as ours. Ballhausen's influence was not limited to his written work; a number of his students are now high-profile researchers, and his ideas have been passed down through multiple generations. This chapter is dedicated to Dr. Carl Johan Ballhausen, a chemist for chemists.

## 1.2 *From Ballhausen to Bendix*

The early collaboration between Gray and Bendix, like that between Ballhausen and Gray, involved the development of electronic and molecular descriptions of novel transition metal complexes. During his fellowship at Caltech, Bendix examined the manganese [22, 23] and gallium [24] complexes of a class of porphyrinoid macrocycles known as corroles [25–28]. Corroles were once difficult to prepare in the laboratory, but they have become more accessible due to the development of new syntheses [29–37]. According to a Scifinder [38] search performed on June 5, 2011, there were seven journal articles, reviews, or book chapters published on corroles in 1996 and 51 in 2006.

While porphyrins are famed for their substitutional flexibility [39], available corroles were initially restricted to a few compounds with small alkyl substituents at the β-pyrrole positions. However, the library of free-base corrole complexes has expanded greatly since the advent of new syntheses, [40–42] especially those which allowed for the synthesis of corroles with aryl or alkyl substituents at the *meso* positions [29, 43] with no β-substituents. These can be derivatized by a variety of electrophilic substitution reactions [44–47] to place nitro, formyl, sulfonate, and halide groups on the ring.

## *1.3 The Corrole Macrocycle*

Corroles are similar to the famed porphyrins, but they form trianionic ligands when fully deprotonated, have smaller binding cavities, and are stronger σ-donors. These properties allow formation of stable complexes with metals in unusually high formal oxidation states. The corrole macrocycle was first synthesized by Johnson and Price [25] as a corrin analog (formally, corroles are pentadehydrocorrins), with the hope that it would be a good model complex for Vitamin $B_{12}$ chemistry. However, the corrole (Fig. 2) is an aromatic macrocycle with three acidic pyrrole N–H protons, while corrins are only partially conjugated and have only one acidic pyrrole moiety.

As noted above, the corrole coordination environment tends to favor higher formal metal oxidation states. This owes both to the smaller binding pocket of the corrole and the greater density of negative charge in the trianionic corrolate core than in the corresponding dianionic porphyrinates. Corroles also stabilize metals in formally high oxidation states through mixing of filled corrole π-orbitals with empty metal d orbitals. This has led to a reassessment of the correlation between formal and physical oxidation states in many metallocorroles, and it has been demonstrated that numerous formally high-valent corrole complexes possess

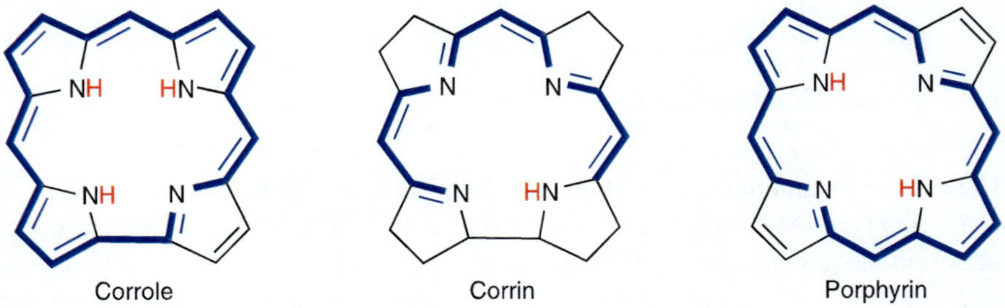


**Fig. 2** The corrole, corrin, and porphyrin macrocycles. The aromatic pathway is highlighted in *dark blue*. Note that both corroles and porphyrins contain 18 π-electron conjugation pathways, while corrin is a 12 π-electron "half-aromatic" macrocycle

significant hole density on the macrocycle and are therefore better described as lower-valent metals bound to macrocyclic radicals. Such oxidized corroles are referred to as "π-cation radicals" throughout this review (in keeping with the literature), although it should be noted that they are actually dianionic ligands.

The photophysical properties of corrole complexes are also quite similar to those of porphyrins. The "four-orbital model" developed by Gouterman to describe porphyrin electronic absorption spectra [48, 49] has been shown to apply to corroles as well [50] (corrins have been given a similar treatment by Day [51, 52]). They possess a distinctive set of high-energy $S_0 \rightarrow S_2$ transitions (β, or "Soret" bands) and another set of lower-energy $S_0 \rightarrow S_1$ transitions (α, or "Q" bands). For reasons that have not been thoroughly explored, corroles tend to have lower molar extinction coefficients. The well-studied Al(III) and Ga(III) corroles are more fluorescent than the analogous Mg(II) and Zn(II) porphyrins [53, 54]; this has occasionally been attributed to the rigid planar structure preferred by the corrole complexes, but this explanation fails to explain the unusually intense fluorescence of free-base corroles [55]. The absorption spectra of main group corroles and analogous porphyrins are compared below (Fig. 3). This chapter is primarily concerned with transition metal corrole complexes; the reader interested in main group corrole chemistry is referred to a 2011 review by Aviv-Harel and Gross [56].

The numbering schemes of porphyrins and corroles are essentially the same (Fig. 4). Throughout this chapter, the protonated macrocycles are denoted by the addition of "$H_n$" to the appropriate abbreviation, where $n$ = the number of protons in the $N_4$ core. The abbreviations "cor" and "por" are used to denote the bare corrolate and porphyrinate macrocycles, respectively. The 5, 10, 15, and 20 positions on the ring are referred to as *meso* positions, while the 2, 3, 7, 8, 12, 13, 17, and 18 ring positions are referred to by the shorthand "β" (they are β-pyrrole carbon atoms).

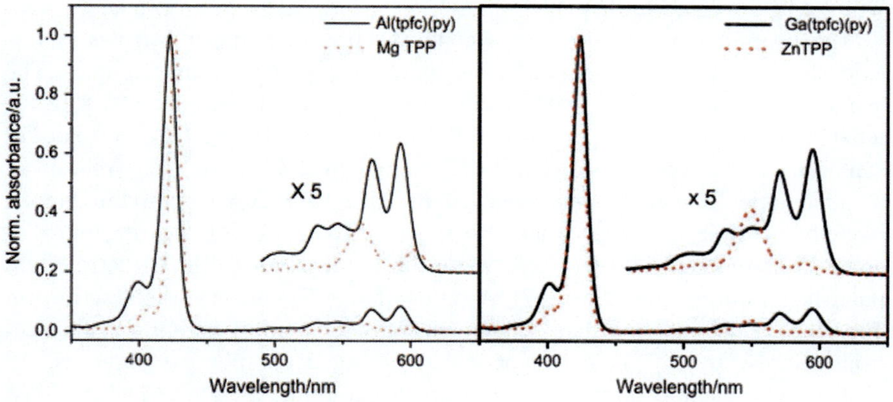


**Fig. 3** UV-vis absorption spectra of Al and Ga *meso*-triarylcorroles compared with the absorption spectra of Mg and Zn *meso*-triarylporphyrins. Reprinted with permission from [53]. Copyright 2008 Elsevier

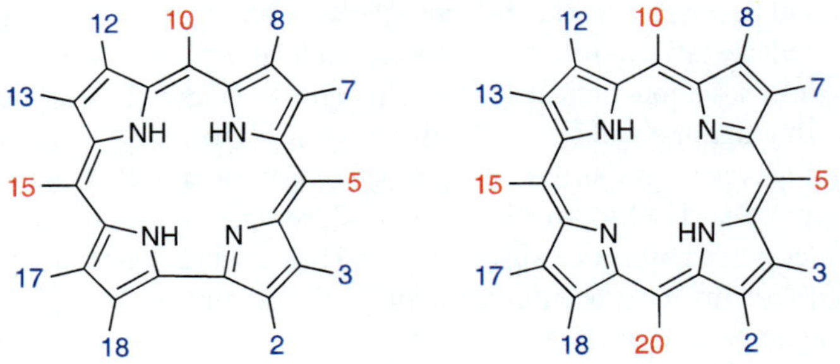


**Fig. 4** The numbering schemes for corroles and porphyrins

## *1.4 The Periodic Table of Metallocorroles*

Although their coordination chemistry is not as well developed as that of porphyrins [57], an appreciable variety of corrole transition metal complexes have been reported. Some early examples are briefly discussed below to convey the early history of the field.

Reports of first-row transition metal corroles are numerous: The synthesis of Ti and V (oxo)M(cor) complexes, the only early transition metallocorroles [58], was disclosed in 1995. The 1980 report of an (oxo)Cr corrole was the first publication describing a well-characterized Cr(V) complex in an $N_4$-macrocyclic framework, testifying to the σ-donating ability of the macrocycle [59]. Mn(omc) (omc = β-octamethylcorrolate) was synthesized in 1990 by heating manganese chloride at reflux in dmf (dimethylformamide) along with $H_3$omc [60]. Co(omc) was synthesized in 1973 along with Fe(dehmc) (dehmc = 8,12-diethyl-2,3,7, 13,17,18-hexamethylcorrolate) [61]. Corrole complexes of Cu, Ni, and Zn were synthesized in 1965 [62]; the Zn complexes of corroles are generally unstable and will not be discussed further.

The coordination chemistry of second-row transition metal corroles is less extensive than for first-row metals, but there are now numerous examples. The synthesis of brick red (oxo)Mo(mec) (mec = 2,3,17,18-tetramethyl-7,8,12,13-tetraethylcorrolate) was reported in a 1977 letter [63]. The first Ru corrole was synthesized as a cofacial Ru(III)-Ru(III) dimer in 2000 [64]; the synthesis of a monomeric Ru corrole was achieved in 2003 [65]. A dicarbonylrhodium(I) *N*-methyl corrole complex, where Rh(I) is bound to two of the ring nitrogen atoms, was reported in 1976 [66], while the synthesis of a Rh(III) corrole was published a dozen years later [67]. A series of Ag(III) corroles was described in 2003 [68]. This oxidation state assignment was supported by an X-ray photoelectron spectroscopy (XPS) comparison to cationic Ag(III) porphyrin complexes.

Third-row metallocorrole complexes are scarce in the literature. The synthesis of (oxo)Re(tfmc) (tfmc = 5,10,15-(tris)trifluoromethylcorrolate), procured via a porphyrin ring contraction in the presence of $Re_2(CO)_{10}$, was reported in 1988 [69], and the synthesis, electrochemistry, and spectral properties of a number of Au

(III) β-octabromocorroles were described in 2011 [70]. A report detailing the synthesis, electrochemistry, and molecular structures of Ir(III) triarylcorroles was published in 2008 [71].

# 2 The First-Row Metallocorroles

As noted above, the complexes of corroles with metals of the first transition series comprise the majority of reported metallocorroles. Many investigations have focused on the usual bioinorganic suspects, Cu and Fe. A number of computationally heavy reports on the electronic and molecular structures and spectroscopic properties of Cu corroles have been produced over the last few years, while the field has seen numerous attempts to formalize consensus electronic structural descriptions of Co and Fe corroles. There is also a rich literature describing the catalytic utility of high-valent Cr and Mn complexes with oxo, imido, or nitrido ligands, and a small amount of work has been performed on Ti and V corroles. Ni corroles have been reported in the literature as well, but investigations of their properties have often been folded into larger studies.

## 2.1 *Titanium and Vanadium: Early Transition Metal Oxo Complexes*

The sole report detailing the characteristics of early transition metal corrole complexes is a 1995 article by Paolesse and coworkers [58]. The classic technique used to obtain titanyl and vanadyl porphyrinates involves boiling the starting M (acetylacetonate)$_3$ complex with the macrocycle at high temperature [72]. Titanyl and vanadyl corrole syntheses were attempted in ethylene glycol, but decomposition led to yields below 5%. Synthesis in phenol, on the other hand, proceeded rapidly, allowing for the isolation of both titanyl and vanadyl corroles in 60–70% yield. The metals are tetravalent, bound to three of the corrole nitrogen atoms and an oxo ligand. The fourth corrole nitrogen remains protonated in these complexes (Fig. 5).

The spectral properties of titanyl and vanadyl oep (oep = β-octaethylporphyrinate) are known [73] and allow for comparison with the corroles. In the case of Ti, the IR absorbance peaks corresponding to the M–O bond stretching mode are located at 960 and 965 cm$^{-1}$ for the porphyrin and corrole complexes, respectively. For V, the difference is larger: the bond stretches appear at 990 and 970 cm$^{-1}$, respectively. V(Homc)(O) displays an isotropic electron paramagnetic resonance (EPR) signal at room temperature when treated with base, which eliminates tautomerism by deprotonating the fourth pyrrole nitrogen. Its $g_{iso}$ value of 1.99 compares favorably with the value of 1.98 in V(oep)(O). Despite the apparent similarity between the oep and omcH complexes, their electronic absorption spectra

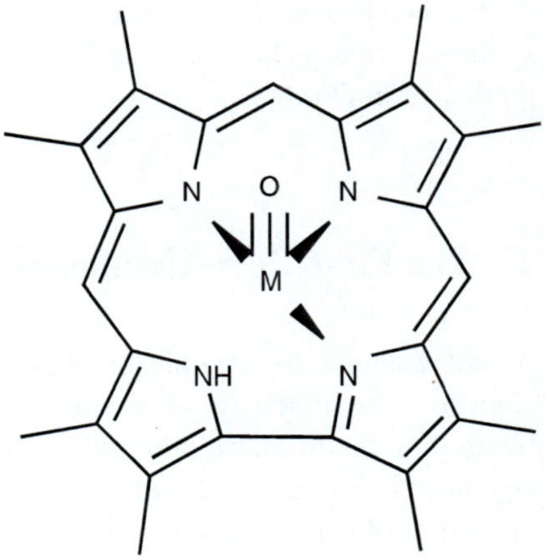


**Fig. 5** M(Homc)(O), where M = Ti or V

are markedly different. The Soret transitions of both sets of complexes appear within a few dozen wavenumbers of each other; however, the porphyrins have molar extinction coefficients ten times as large as those of the corroles.

## *2.2 Chromium and Manganese: Oxo, Imido, and Nitrido Group Transfer Reagents*

Cr and Mn corroles have been studied intensively as atom transfer catalysts. They are isolable as M(III) complexes, but can easily be converted to high-valent species stabilized by multiply bonded oxygen or nitrogen donors (Fig. 6). Cr(III) corroles spontaneously form (oxo)Cr(V) species in aerated solution [59, 74], whereas Mn(III) corroles can be converted to isolable (oxo)Mn(V) species by treatment with oxidants such as ozone [75]. Pentavalent (oxo)Mn(tpfc) (tpfc = 5,10,15-tris (pentafluorophenyl)corrolate) is the presumed active species in Mn-corrole-catalyzed epoxidation reactions. The enhanced stability of high-valent Cr corroles corresponds to relatively sluggish activity as group-transfer catalysts. A thorough commentary on the early discoveries related to high-valent chromium, manganese, and iron corroles was penned in 2001 [76], and a focused mini-review on the catalytic properties of Cr and Mn corroles was written in 2004 [77].

### 2.2.1 Chromium Corroles

Chromium corroles have been isolated as 6-coordinate Cr(III) and 5-coordinate Cr(IV/V/VI) oxo, nitrido, and imido complexes. The synthesis of Cr(tpfc)$(py)_2$ (py = pyridine), which is oxidized to (oxo)Cr(tpfc) on standing in aerated solution, was reported in 2001 [78]. Cr(tfpc)(O) could be reduced by cobaltocene to give

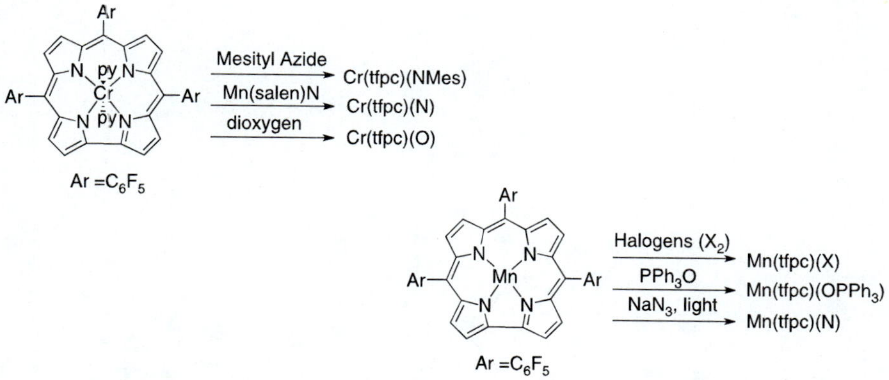


**Fig. 6** The varied reactivity of chromium and manganese corrole complexes

the Cr(IV) complex [$Cp_2Co$][Cr(tpfc)(O)]; treatment with dioxinium hexachloroantimonate, on the other hand, led to oxidation of the corrole ring to give [Cr(O)(tpfc)][$SbCl_6$]. Additionally, Cr(tpfc)(N) complexes in both the Cr(V) and Cr(VI) oxidation states have been reported [79]. The Cr(V) complex was synthesized by nitrogen atom transfer to Cr(tpfc)$(py)_2$ from Mn(salophene)(N) [80]. Meanwhile, the Cr(VI) corrole was obtained via treatment of Cr(tpfc)$(py)_2$ with $NH_4OH$ and NaOCl in acetonitrile. This same reaction typically leads to Cr(V) complexes when performed on porphyrins, exemplifying the ability of corroles to stabilize high-valent metals.

Abu-Omar and coworkers published a 2005 report describing (imido)M(V) corrolates, where M = Cr or Mn [81]. The synthesis of the Cr(tpfc)(imido) compounds was achieved by heating either 2,4,6-trichlorophenyl or mesityl azide with (tpfc)Cr$(py)_2$ in dry toluene at 85°C. The mesitylimide complex possesses a Cr–N (imido) bond length of 1.635 (2) Å and a Cr–N(imido)-C(mesityl) angle of 169.59 (18)°, consistent with a Cr–N triple bond. Its magnetic moment is 1.90 $\mu_B$, consistent with the reported values for other (imido)Cr(V) complexes [82, 83]. Its room temperature EPR spectrum is isotropic, with $g_{iso} = 1.985$, $A^{53}\text{Cr} = 21.8$, and $A^{14}\text{N} = 3.1$ G. This $^{53}$Cr hyperfine coupling constant is between those of the Cr (V) nitrido (26.7) and oxo (16.4 G) complexes of tpfc, a property that was attributed to the intermediate π-donicity of the imido ligand.

The synthesis of a chiral *meso*-ABC (oxo)Cr(V) corrole complex (Fig. 7) was recently described [84]; the free-base starting material was obtained using the method of Gryko [85]. A *meso*-ABC corrole possesses different substituents at each of the *meso*-positions on the ring and is prochiral. This complex is not as efficacious an oxygen atom transfer catalyst as Cr(tpfc)(O) [86]. However, it could be a starting point for a new generation of chiral corrole catalysts.

A report was recently communicated elucidating a counterintuitive property of (oxo)Cr(tpfc) complexes: namely, that the chromyl bond is stronger than the perchromyl bond [87]. This can be demonstrated by the examination of resonance Raman (RR) data: the Cr–O stretching frequencies are 986 and 1,002 $cm^{-1}$

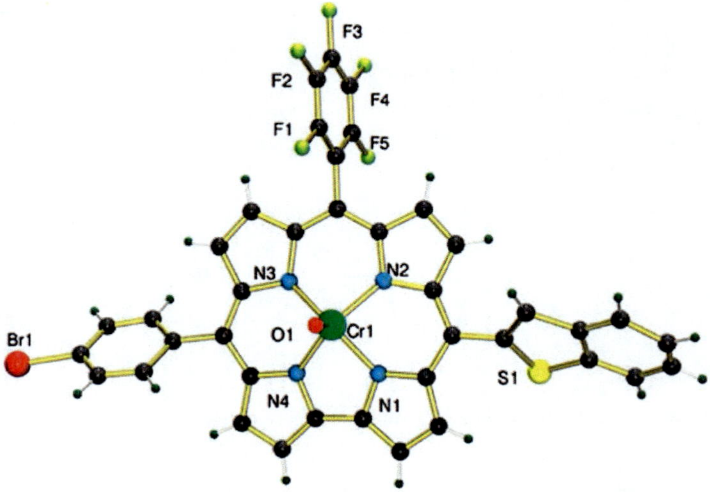


**Fig. 7** Structure of the Cr(O) ABC corrole. Reprinted with permission from [82]. Copyright 2009 American Chemical Society

for Cr(tpfc)(O) and $[Cr(tpfc)(O)]^-$, respectively, in $CS_2$ solution. By way of an explanation, the authors point to the observation that the vanadyl bond in porphyrins is weaker than the analogous chromyl bond [88]. In the porphyrin study, this was explained by invoking a simple LFT argument. The low-energy $d_{xy}$ orbital is nonbonding, and thus M–O bonding is not perturbed by the presence of electrons in this orbital. The higher effective nuclear charge of the chromyl complex therefore results in a stronger bond. An alternative explanation for this phenomenon is that the $d_{xy}$ orbital participates in bonding by undergoing an antibonding interaction with the pyrrole nitrogen atoms. This introduces a "push–pull" effect on the system, strengthening the chromium–oxo bond when the $d_{xy}$ orbital is filled. DFT calculations are consistent with this explanation: the Cr–N(pyrrole) bond lengths increase by about 0.03 Å upon reduction to Cr(IV), while the out-of-plane displacement of the chromium ion increases by 0.02 Å.

### 2.2.2 Manganese Corroles

Mn corroles have been isolated as Mn(III) [22, 60, 89, 90], Mn(IV) halogenato [91], and Mn(V) oxo [75, 92–94], imido [81, 95], and nitrido complexes [96, 97] (oxidation of (nitrido)Mn(V) corrolates to Mn(VI) complexes has been reported as well). Mn(III) corroles represent rare examples of four-coordinate Mn(III) complexes. Mn(IV) corroles are unusually stable; in fact, Mn(dpoec)(I) (dpoec = 5,15-diphenyl-2,3,7,8,12,13,17,18-octaethylcorrolate) was the first well-characterized Mn(IV)-I complex reported in the literature [98]. Meanwhile, the high-valent oxo, imido, and nitrido complexes have proven to be competent catalysts for a variety of atom- and group-transfer reactions, such as the formation of epoxides and aziridines from olefins. A review of Mn and Fe corrole atom transfer chemistry was published in early 2011 [99].

A number of Mn(III) corroles have been crystallographically characterized and subjected to EPR spectroscopy and SQUID. Licoccia and coworkers found that Mn(omc) has a magnetic moment of 4.76 $\mu_B$, consistent with a high-spin electronic configuration [100]. Meanwhile, they crystallographically characterized Mn(hedmc) (hedmc = 2,3,7,13,17,18-hexaethyl-8,12-dimethylcorrolate), which has an average Mn–N bond length of 1.879 Å and possesses a nearly square-planar geometry around the metal and a planar corrole macrocycle. Mn(III) porphyrins display longer Mn–N bond distances, out-of-plane displacement of the Mn ion, and a ruffled macrocyclic ring. In addition, many Mn(III) porphyrins contain a stabilizing axial ligand [101]. When pyridine was added to a solution of Mn (hedmc), the UV-vis and low-temperature EPR spectra of the corrole changed dramatically. The solution-state magnetic moment of this complex was found to be highly temperature dependent, with $\mu_{eff} = 3.60\ \mu_B$ at 293 K and 3.02 $\mu_B$ at 235 K, suggesting an antiferromagnetically coupled intermediate-spin Mn(II) π-cation radical electronic configuration. On the other hand, the *meso*-triarylcorrole complex Mn(tpfc) [22] is not as inert as the octaalkylcorrole complexes, interacting strongly with solvent molecules to form various 5- and 6-coordinate solvento complexes. It could be isolated as the five-coordinate complex Mn(tpfc)($OPPh_3$) upon standing in solution with triphenylphosphine oxide. Mn(tpfc)($OPPh_3$) has a magnetic moment of 4.88 $\mu_B$, close to that of Mn(hedmc). The Mn ion sits 0.29 Å out of the $N_4$ plane, and the average Mn–N bond distance is 1.916 Å. High-field EPR measurements suggested that the deviation from an ideal axial electronic structure is very small for this complex.

The synthesis and properties of Mn(tpfc)(Cl) and Mn(tpfc)(Br) were reported in a 2001 communication that also disclosed the synthesis of Mn(tpfc)(N) [23]. The out-of-plane displacement of the metal is 0.21 and 0.065 Å for the Br and Cl complexes, respectively, while the average Mn–N bond distances are, respectively, 1.932 and 1.925 Å. Addition of excess elemental bromine to Mn(tpfc) results in the formation of Mn($Br_8$tpfc) ($Br_8$tpfc = β-octabromo-5,10,15-tris(pentafluorophenyl) corrolate), whose metal center is too electron-poor to be easily converted to the Mn (IV) state. The first anodic half-wave potentials in $CH_2Cl_2$ for Mn(tpfc) and Mn ($Br_8$tpfc) are located at 0.71 and 1.08 V vs. SCE, respectively. Presumably owing to the greater oxidizing power of the catalytically active species in the perhalogenated system, Mn($Br_8$tpfc) is a significantly superior olefin epoxidation catalyst. The Mn(dpoec)(I) complex reported in 2007 displayed an out-of-plane metal displacement of 0.3798 Å, much larger than that of the Mn(tpfc)(halide) complexes. The Mn–I bond is long, 2.6626 Å versus 2.428 for Mn–Br and 2.312 Å for Mn–Cl. An extensive study examining the redox chemistry and optical spectroscopy of Mn corroles in the presence of various coordinating species was published in 2008 [102]. Coordination of axial ligands was found to stabilize higher oxidation states of the central metal in all cases while destabilizing the products of reduction.

## 2.3 *Iron and Cobalt Corroles: High-Valent or Non-innocent?*

The history of Fe and Co corroles, particularly the former, has been rife with discussion regarding the assignment of oxidation states. An initial wave of reports disclosed the syntheses, structures, and properties of numerous five-coordinate Fe corrolates with X-type axial ligands, often described as low-spin Fe(IV) corrolates. Nuclear magnetic resonance (NMR) spectroscopic and theoretical investigations suggested that the physical oxidation state of the Fe in 5-coordinate chloroiron corrolates was intermediate-spin Fe(III); this metal center would then be coupled antiferromagnetically to a π-cation radical, accounting for the $S = 1$ configuration of the complexes. A series of papers by numerous investigators brought together NMR and Mössbauer spectroscopy, temperature-dependent magnetic moment measurements taken using a superconducting quantum interference device (SQUID), and computational techniques to show that the more accurate electronic structural description is the one invoking an Fe(III) oxidation state. Nearly all neutral Co corroles have been unambiguously assigned low-spin Co(III) configurations. However, their one-electron-oxidized cations have been described alternatively as either Co(IV) corroles or Co(III) π-cation radicals; these competing descriptions have been applied to Co corroles with X-type axial ligands as well.

### 2.3.1 Iron Corroles: Rich Electronic Structure

The earliest article describing a formal Fe(IV) corrole complex was published in 1994 [103]. Vogel and coworkers synthesized the μ-oxo dimer $Fe_2(oec)_2(O)$ from $H_3oec$ and $Fe_2(CO)_9$ under aerobic conditions. $Fe_2(oec)_2(O)$ could be treated with HCl in $CH_2Cl_2$ to give Fe(oec)(Cl), which could be further converted to Fe(oec)(Ph) (Ph = phenyl) by treatment with PhMgBr. Both of these monomeric complexes were assigned Fe(IV) oxidation states on the basis of SQUID and Mössbauer measurements. They have $S = 1$ electronic configurations, with room temperature magnetic susceptibilities of 2.97 and 2.89 $\mu_B$, respectively. Their Mössbauer spectra showed small isomer shifts and moderate quadrupole splittings, in line with known Fe(IV) heme compounds [104].

A follow-up study examined the electrochemically oxidized and reduced forms of these formally Fe(IV) corroles (Fig. 8) and the Fe(III) complexes Fe(oec)(py) and $Fe(oec)(py)_2$. The one-electron reduced forms of (oec)Fe(Cl) and (oec)Fe(phenyl) were assigned respective intermediate- and low-spin Fe(III) configurations on the basis of their EPR spectra, while the mono- and bis-pyridine complexes of Fe(oec) were assigned intermediate- and low-spin configurations, respectively, on the same basis. Fe(oec)(Ph) remains stable for long periods subsequent to one-electron oxidation, and was therefore crystallized and subjected to X-ray diffraction (XRD) after treatment with $Fe(ClO_4)_3$. Compared to the neutral complex, the cation displays a 0.02 Å shorter Fe–C bond and a metal center 0.03 Å closer to the $N_4$ plane. On the basis of Mössbauer, UV-vis, and EPR spectra, both $[Fe(oec)(Ph)]^+$

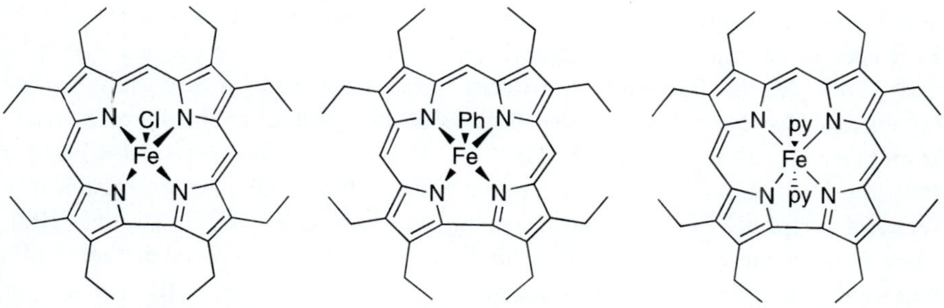


**Fig. 8** Iron octaethylcorrolates in formal Fe(IV) (*left*, *middle*) and Fe(III) oxidation states

and $[Fe(oec)(Cl)]^+$ were formulated as having Fe(IV) centers antiferromagnetically coupled to π-cation radicals. This behavior is distinct from that of Fe(IV) porphyrin π-cation radicals, which generally display ferromagnetic coupling.

The assignment of an Fe(IV) electronic configuration to Fe(oec)(Cl) was first challenged in a 2000 paper reporting NMR data that suggested an antiferromagnetically coupled intermediate-spin Fe(III) π-cation radical was a more accurate description of the electronic configuration of chloroiron corrolates [105]. The presence of a π-cation radical was inferred from the large downfield shifting of the protons at the unsubstituted *meso*-positions, while strong antiferromagnetic coupling was confirmed by SQUID measurements. A combined electrochemical and computational study by a different group provided further support for this assignment [106]. The first oxidation processes of the chloroiron complexes occur at much higher potentials than in the μ-oxo dimers, implying that the locus of oxidation could be different. Additionally, spin-unrestricted DFT calculations showed that the chloroiron complexes possess significant spin density across the macrocycle, whereas the spin is localized on the metal center in the case of the μ-oxo dimer. The $^{57}Fe$ Mössbauer isomer shift of the dimer is also much smaller than that observed for the chloroiron corrolate. In a subsequent study, Walker and coworkers used a combined experimental and theoretical approach to further examine the electronic configuration of formally Fe(IV) corroles [107]. The temperature dependence of the magnetic moment, the presence of extremely downfield-shifted *meso*-protons, and the calculated spin densities of $-0.65$ to $-0.79$ on the corrole ligand were all suggestive of an antiferromagnetically coupled intermediate-spin Fe(III) π-cation radical configuration in the chloroiron complex; meanwhile, the phenyliron corrole, which has quite different Mössbauer parameters as well as less dramatically shifted *meso*-protons, was described as containing a low-spin Fe(IV) center. In both cases there is a large degree of covalency, making a definitive oxidation state assignment less meaningful than in complexes with more ionic bonding character.

Gross and coworkers published a paper in 2002 describing the spectroscopic properties of some Fe corroles with electron-withdrawing *meso*-aryl groups [108]. $Fe(tpfc)(py)_2$ and $Fe(tpfc)(OEt_2)_2$ were unambiguously characterized as low-spin and intermediate-spin iron(III) complexes, respectively, on the basis of their EPR spectra and magnetic susceptibilities. Fe(tpfc)(Cl) and Fe(tdcc)(Cl) (tdcc = 5,10,15-tris(2,6-dichlorophenyl)corrolate) were described as low-spin Fe(IV) complexes based on comparisons of their NMR spectra to Fe(IV) porphyrins; from there, it was argued that the Fe(IV) assignment was also correct for Fe(oec)(Cl). Simkhovich and Gross also reported the synthesis and characterization of a series of Fe(tpfc)(X) and Fe(tdcc)(X) complexes, where X = F, Cl, Br, or I [109]. They assigned these complexes low-spin Fe(IV) configurations based on their NMR spectra. Meanwhile, Walker and coworkers continued to present NMR evidence for the presence of an open-shell macrocycle in the chloroiron corrolates [110, 111], culminating in the publication of a lengthy review article wherein the chloroiron complexes were assigned antiferromagnetically coupled intermediate-spin Fe(III) π-cation radical configurations and the phenyliron complexes were assigned low-spin Fe(IV) configurations [112]. In 2008, Gross, Thiel, Neese, and coworkers published a combined Mössbauer, SQUID, and computational study supporting this analysis and making some comparisons to iron porphyrin compounds [113] (Fig. 9). The same year, a report was published detailing complete active space self-consistent field (CASSCF) calculations that also supported the assignment of an Fe (III) configuration for the chloroiron corrolates [114].

A 2009 article reported the X-ray absorption spectroscopy (XAS) of the low-spin Fe(III) corrole-Fe(tdcc)$(py)_2$, and compared its electronic structure with that of the cationic Fe(III) porphyrin [Fe(tpp)$(ImH)_2$]Cl (tpp = 5,10,15,20-tetraphenylporphyrinate) [115]. The data presented, which were accompanied by DFT calculations, suggested that the covalency between the Fe and macrocyclic orbitals is slightly greater in corroles than in porphyrins, one of the contributing factors to the difficulty of determining metallocorrole physical oxidation states. Additionally, the increased σ-donating ability of the corroles and the calculated concentration of

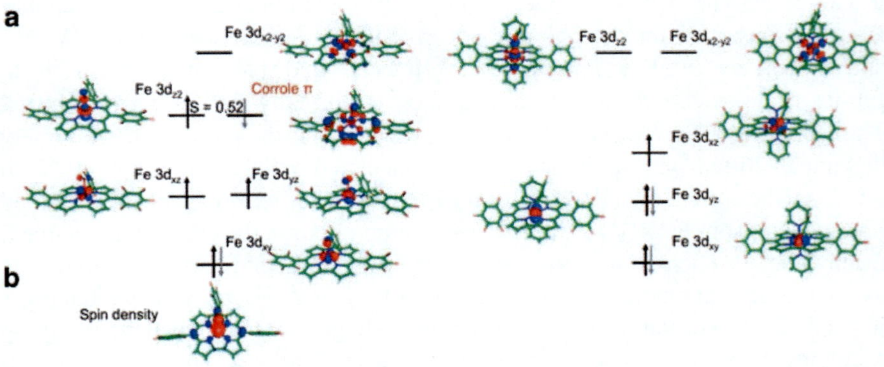


**Fig. 9** Electronic structures of Fe(tpfc)Cl (*left*) and Fe(tpfc)$(py)_2$ (*right*). Part (**a**) on the *left* and the figure on the *right* show the location of spin up and spin down electrons in the orbitals, while part (**b**) on the *left side* shows the spin density across Fe(tpfc)Cl. Reprinted with permission from [113]. Copyright 2008 American Chemical Society

LUMO density on one side of the macrocycle were suggested to be the primary factors leading to the experimentally observed axial lability of Fe(III) corrolates.

The preponderance of the evidence now points to an intermediate-spin Fe(III) description for the chloroiron corrolates and a low-spin Fe(IV) description for the phenyliron corrolates and μ-oxo corrole dimers. Meanwhile, the five- and six-coordinate Fe(III) corroles with pyridine ligands can be unambiguously assigned intermediate- and low-spin electronic configurations, respectively. In all cases, a high degree of covalency confounds assignment of meaningful oxidation states.

### 2.3.2 Monomeric Cobalt Corroles: High-Valent and Covalent

Early studies of cobalt corroles focused on their redox chemistry [116, 117], optical and EPR spectra [118–120], and XRD structures [121]. Co corroles usually have a small "electrochemical HOMO-LUMO gap" (the separation between the first anodic and first cathodic waves observed by voltammetric methods), reflecting the likelihood that the first anodic and cathodic processes do not both occur on the macrocycle. The normal electrochemical HOMO-LUMO gap in tetrapyrrole complexes where both the processes occur on the macrocycle is greater than 2 V. EPR spectroscopic traces of reduced cobalt(III) corroles show large axial hyperfine coupling to a 7/2-spin nucleus, suggesting that reduction gives Co(II) species. This is consistent with the relatively high reduction potential for the first cathodic process of, e.g., Co(mec)$(py)_2$ ($E_{1/2} = -0.53$ V vs. SCE in $CH_2Cl_2$). The XRD structure of Co(cor)$(PPh_3)$, published in 1976, was one of the first reported crystallographic structures of a five-coordinate Co(III) complex (the crystal structure of the methylcobalt(III) complex of *N,N′*-ethylene-bis(acetylacetoneimine) [122] represents one earlier example). Co(cor)$(PPh_3)$ possesses an axial Co–P bond of 2.210(5) Å, with an average Co–N distance of 1.87 Å. Hitchcock and McLaughlin note that this is shorter than the average Co–N bond of 1.98 Å observed in structures of Co(tpp)$(L)_2$, where tpp = tetraphenylporphyrin and L = imidazole or piperidine [123], probably reflecting the smaller $N_4$ core of the corrole macrocycle.

The synthesis, characterization, and electrochemistry of various cobalt(III) corrole complexes were reported throughout the 1990s [124–126]. This surge of interest can be traced to an advance in synthetic methodology: much of the work performed on cobalt corroles during this period was facilitated by the development of a means to synthesize corroles from simple precursors using a cobalt template reaction [31, 32, 127]. This phenomenon perfectly exemplifies the concept that new synthetic techniques can be a powerful driving force in a relatively young field of inquiry.

In 1996, the neutral and cationic forms of Co(oec)(Ph) were synthesized and structurally characterized [128]. The neutral complex has a total spin of ½, with a $^1$H NMR spectrum showing sizable paramagnetic shifting, while the cationic complex, obtained via oxidation by $Fe(ClO_4)_3$, is diamagnetic and displays a well-resolved $^1$H NMR spectrum. Both the neutral and the cationic complex have essentially planar macrocyclic skeletons, with minor out-of-plane displacement of the cobalt ion [0.185 and 0.165 Å for the neutral and cationic forms, respectively, versus 0.380 Å in Co(cor)($PPh_3$)] and correspondingly short Co–N bonds (ca. 1.85 Å in both complexes). The Co–Ph bond is shorter in the formally Co (IV) complex (1.937 Å) than in the formally Co(V) cation (1.970 Å).

Co(oec)(Ph) displays a rhombic EPR spectrum at 77 K ($g_z = 2.12$, $g_y = 2.01$, $g_x = 1.94$), with $A^{53}$Co = 14–16 G. Based on the rhombicity of the spectrum, the neutral complex was assigned a low-spin Co(IV) configuration; however, the modest magnitude of the hyperfine coupling, the significant paramagnetic shifting observed in the $^1$H NMR spectrum, and the presence in the solid state of strong antiferromagnetic coupling between π-stacked dimeric complexes led to the conclusion that the true electronic configuration was a resonance hybrid between Co (IV) and a Co(III) π-cation radical. A subsequent study by a different group used pulsed EPR, electron-nuclear double resonance, and DFT calculations to argue for a delocalized electronic structural model wherein the unpaired electron resides in an orbital with 65% π-cation and 35% Co $d_{yz}$ character [129].

Vogel and coworkers assigned the formally Co(V) cation a Co(III) π-dication electronic configuration based on the NMR spectrum of the complex. [Co(oec) (Ph)]$^+$ possesses a paramagnetic ring current, implying a 16 π-electron conjugation pathway and therefore a doubly oxidized corrole ring. This electronic structural assignment is consistent with the crystallographically determined Co–Ph bond distances.

The synthesis of Co(tpfc)($PPh_3$) was reported in 2000 [130]. This complex, and the analogous Co(tpfc)$(py)_2$, have a strong tendency to form $C_3$–$C_3$-linked dimers in aerated solution [131]. The twist angle between the two corrole rings in [Co(tpfc) $(py)_2]_2$ is 45° (Fig. 10), with a 1.479(13) Å C–C bond between the monomeric units.

Gross and coworkers have used Co(tpfc)($PPh_3$) as a test bed for ring-substitution reactions; it was on this compound that they first demonstrated the facility of electrophilic chlorosulfonation as a corrole functionalization method [132]. They also found that doubly reduced [(tpfc)Co]$^{2-}$ can catalyze the reduction of $CO_2$ over sodium amalgam in acetonitrile, although the catalytic reaction

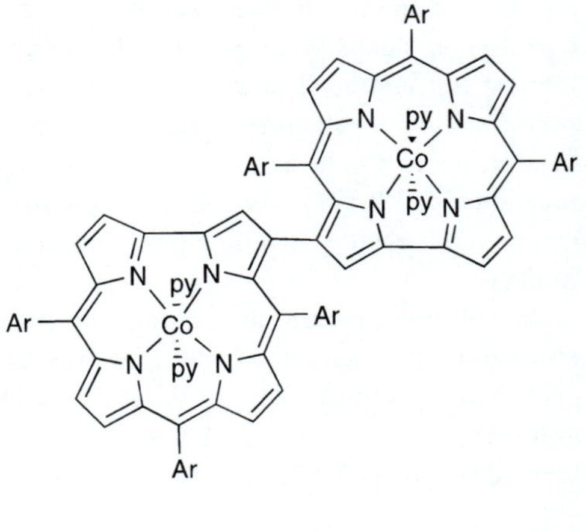


**Fig. 10** The C–C dimer $[(tpfc)Co(py)_2]_2$. Ar = $C_6F_5$

is quite slow [133]. In a related vein, numerous studies have suggested that monomeric Co(III) corroles could have applications in CO sensing [134, 135]. A recent paper detailed the rich reactivity of Co(oec) with ethyl diazoacetate (EDA), uncovered during an investigation of the mechanism of catalytic olefin cyclopropanation by Co corroles [136]. Gross and coworkers have studied similar reactions with Rh corrolates (*vide infra*, Section 3.3).

### 2.3.3 Cobalt Corrole–Corrole and Porphyrin–Corrole Dimers

A number of reports have been published detailing the synthesis and properties of "Pacman" face-to-face Co corrole–corrole and corrole–porphyrin dimers [137–142]. These complexes have been posited as oxygen reduction catalysts and were reviewed in 2003 [143]. An interesting 2005 study detailed the synthesis and properties of heterobimetallic face-to-face bismacrocyclic complexes where a formally Co(IV) corrole is attached via a rigid aromatic linker to either a Mn(III) or Fe(III) porphyrin (Fig. 11); in all cases, the metal complexes possess axial chlorides [144]. These corroles were assigned a low-spin Co(IV) configuration based on their lack of a room temperature EPR signal and their moderate hyperfine coupling constants, but definitive characterization requires further study. The crystal structure of the iron–cobalt biphenylene-linked dimer shows a Fe–Co distance of 4.13 Å, suggesting that M–M interactions are weak or nonexistent. Indeed, cyclic voltammetric measurements show half-wave potentials very similar to those observed for the separated monomers.

It was observed in a 2005 article that Co(II) porphyrin-Co(III) corrole dimers are more effective dioxygen reduction electrocatalysts than analogous Co(III)–Co(III) corrole dimers or monomeric Co(III) corroles [145]. The heterodimers operated effectively at lower overpotentials and promote complete reduction to water (the average number of electrons transferred per $O_2$ molecule approaches 4 in the best porphyrin–corrole catalyst). It was suggested that the inferior catalytic performance of the corrole homodimers could be due to a reduction in the basicity of the activated intermediate when two Co(III) moieties are involved, leading to a less favorable 4-electron reduction. Heterobimetallic catalysts containing formally Co (IV) corroles were also examined as potential dioxygen reduction catalysts [146].

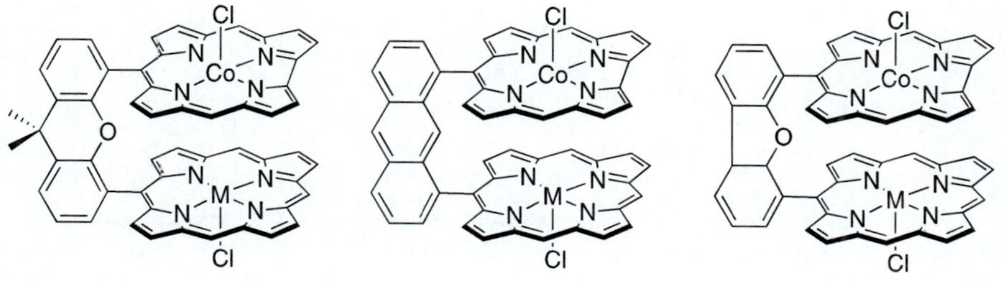


**Fig. 11** Co corrole-Fe/Mn porphyrin heterodimers. Ring substituents removed for clarity

While the Co corrole-Fe/Mn porphyrin dyads were active electrocatalysts, the homobimetallic Co porphyrin–corrole dyads operate at lower overpotentials and favor 4-electron reduction to a greater extent. This was attributed to the locus of reduction in each complex: apparently, the porphyrin is reduced first in the homometallic dimer, whereas the corrole is the site of the first reduction in the heterometallic dimers. A later article presented EPR and spectroelectrochemical evidence supporting the assignment of a Co(III) π-cation electronic configuration for the oxidized derivatives of monomeric Co triarylcorrolates, suggesting that the dimeric five-coordinate Co corrole catalysts are also Co(III) π-cation radicals [147]. Further study is needed to elucidate this point.

### 2.3.4 Recent Studies on Hangman Corroles

Nocera and coworkers have used the hangman motif to allow facile proton shuttling by placing a moiety capable of rapid proton transfer (generally a carboxylic acid) in close proximity to a catalytic site [148, 149]. This has proven to be of particular utility in O–O bond-breaking reactions, where coupling of multi-electron bond-breaking events to proton transfer allows them to proceed with far greater facility. To date, the chemistry of Fe and Co hangman corroles (Fig. 12) has been studied and compared with that of porphyrin analogs [150]. The Fe hangman corroles were synthesized with a built-in xanthene-based hangman moiety, and could be isolated as either the μ-oxo dimer or the chloroiron monomer [151]. In line with previous studies, the chloroiron complexes were assigned a Fe(III) π-cation radical electron configuration. These FeCl hangman corrole complexes, like other Fe(III) macrocycles, catalyze the dismutation of $H_2O_2$ to $H_2O$ and $O_2$, though not as efficiently as the best hangman Fe porphyrin catalysts [152]. Based on the analysis of the reaction mixture after catalysis, the likely reason for this is decomposition of the corrole. Additionally, the hangman-containing corrole catalyst did not perform significantly better than a control analog; the authors attributed this to decomposition limiting corrole catalyst performance as well.

An article published in 2011 explored the oxygen reduction reactivity of Co hangman corroles [153]. Along with novel catalytic activity, this paper described some potentially useful synthetic procedures. For example, $H_3$tpfc was obtained in

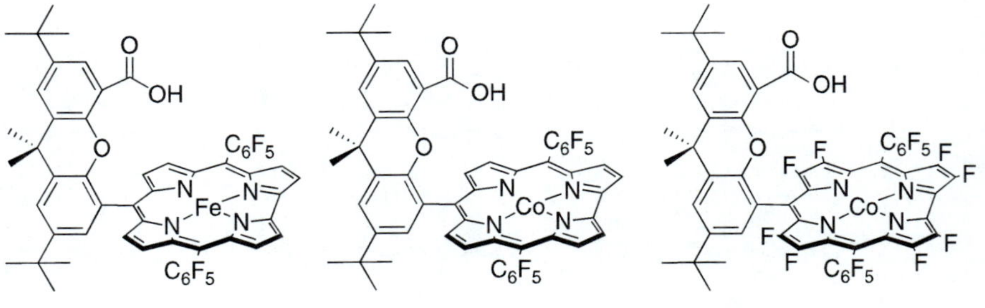


**Fig. 12** Iron and cobalt hangman corrole complexes

7% isolated yield using a modified Lindsey porphyrin synthesis [154]. Additionally, cobalt was cleanly and rapidly inserted into the corrole framework in a microwave reactor. The hangman corrole with pentafluorophenyl *meso*-substituents was found to be a slightly more efficient catalyst for the electrochemical 4-electron reduction of dioxygen than either Co(tpfc)(Cl) or a more closely analogous complex where a methyl ester replaced the carboxylate hangman moiety. However, these Hangman catalysts do not promote the 4-electron pathway as selectively as the most effective Co porphyrin–corrole Pacman dimers [145]. The syntheses described by the authors could likely be used to create Co porphyrin–corrole heterodimers with built-in hangman functionalities. Such binuclear complexes might promote the 4-electron reduction pathway with heretofore unobserved specificity.

In addition to the article on oxygen reduction, Nocera and coworkers published a 2011 communication detailing the ability of fully fluorinated hangman corroles to catalyze electrochemical water oxidation [155]. These complexes were designed to be extremely electron withdrawing so as to raise their oxidation potentials to a point where they would be useful for water oxidation; an additional effect of the fluorination is protection of the ligand from oxidative degradation. The overpotential of the best catalyst is approximately 0.6 V, while its TOF for $O_2$ evolution is 0.81 $s^{-1}$. This catalyst produces $O_2$ with high selectivity during the water oxidation reaction and exhibits minimal apparent decomposition under catalytic conditions. Based on the position of the catalytic wave at a higher reduction potential than that of the formal $Co^{IV/III}$ process, the authors posited an active species with an electron configuration of Co(IV) coupled to a π-cation radical. Considering the finding by Vogel and coworkers that oxidized Co(tpfc)(Ph) is a Co(III) π-dication [128], further evidence is needed before the electronic configuration of the active species can be assigned with confidence. The impressive catalytic water oxidation ability of the β-octafluorinated Co corroles hints at the type of novel reactivity still undiscovered in the relatively youthful metallocorrole field.

## 2.4 *Copper Corroles: Non-innocent Distortions*

Cu was among the first of the transition metals to be placed into the corrole core [62], and numerous early reports of Cu corrole syntheses and structures exist [156–158]. Cu corroles exhibit interesting electronic structures, with low-lying, energetically close Cu(III) and Cu(II) π-cation states. They tend to be strongly saddled due to interactions between filled corrole π and empty metal $3d_{x2-y2}$ orbitals, exhibiting exceptionally substituent-dependent optical spectra because of this interaction.

### 2.4.1 Copper Corrole Oxidation States: Experiment and Theory

In the early literature, Cu corroles are usually described as Cu(II) complexes [159], with the implicit assumption that they are bound in a similar fashion to the Cu(II) ions found in porphyrinic complexes. It was not until 1997 that investigators first suggested that Cu corroles were better described as containing Cu(III) [160] (the same report supported the assignment of Ni(II) as the oxidation state of Ni octaalkylcorrolates). Later DFT and configuration interaction calculations were consistent with this assignment [50]. The calculated Cu(III) configuration is separated by roughly 1,300 $cm^{-1}$ from the lowest-lying ferromagnetically coupled (triplet) Cu(II) π-cation radical configuration, consistent with the paramagnetic broadening observed in high-temperature NMR spectra of copper corrole complexes [161]. An investigation into the spectroelectrochemical and EPR parameters of nine diverse Cu corrole complexes resulted in the finding that only the complex Cu($Br_8$tpfc) appeared to possess a triplet Cu(II) π-cation electronic configuration [162]. The authors attributed this anomalous behavior to electronic destabilization of Cu(III) by the electron-withdrawing $Br_8$tpfc ligand, but it should be noted that the same electronic factors that cause $Br_8$tpfc to destabilize higher oxidation state metal centers also make it significantly more difficult to oxidize. Saddling distortions of the corrole macrocycle increase overlap between the Cu $3d_{x2-y2}$ and corrole π orbitals [163]. Therefore, Cu($Br_8$tpfc) may favor the triplet configuration due to the favorability of saddling when the corrole is heavily substituted at the β-position. Gross [164] and Sarkar [165] described, in separate reports, the synthesis of Cu corroles with electron-withdrawing *meso*-substituents that reside in a Cu(III) electronic configuration. In both of these papers, triplet states were found to be very close in energy to the $S = 0$ ground state electronic configuration.

In 2007, a brief paper presented evidence that the electronic configurations of Cu(III) corrole complexes are better described as antiferromagnetically coupled (singlet) Cu(II) π-cation radicals than as Cu(III) closed-shell corroles [166]. The authors pointed out two salient facts to support this conclusion: for one, they found that the geometries of (dpoec)copper and its 10-oxacorrole counterpart (an authentic Cu(II) complex) are virtually identical in the Cu–N core, whereas one might expect shorter bond distances for a Cu(III) complex; additionally, quantum chemical calculations showed that the antiferromagnetically coupled Cu(II) π-cation singlet state of a Cu(5,15-diphenylcorrolate) model is more stable than the Cu (III) state by 1,670 $cm^{-1}$. However, the calculations also predicted saddling angles for the corrole much larger than those observed experimentally, and the calculated singlet-triplet gap of 670 $cm^{-1}$ diverges substantially from the experimentally observed value of 2,010 $cm^{-1}$. Additionally, only one corrole was examined, and it bears little chemical resemblance to the *meso*-triarylcorroles that have been examined by other groups. In a 2010 theoretical study using advanced CASSCF techniques [167], it was found that the favored electronic structure is heavily dependent upon the chosen computational methodology. The overall theoretical picture appears to be of an electronic configuration with hybrid Cu(III) and Cu(II) π-cation character, with majority Cu(III) character. As the saddling angle increases, theory suggests that the degree of overlap between corrole π and Cu $3d_{x2-y2}$ orbitals should grow, leading to increased Cu(II) π-cation character.

### 2.4.2 Spectroscopy and Saddling

Ghosh reported the spectroscopic and electrochemical properties of a series of Cu(III) corroles in a combined experimental and theoretical study [163]. This investigation found that copper triarylcorroles possess strong "marker" bands in their RR spectra which decrease in energy upon bromination; unfortunately, these bands are not consistent across a broad range of substituents, and their relative intensities change in seemingly random fashion in the brominated derivatives. This article reported an electrochemical HOMO-LUMO gap for Cu corrolates that was quite low, lending further support to the notion that oxidation occurs on the corrole ring while the reduction occurs at the copper center. It was also found that the Soret absorption systems of Cu corroles are highly *meso*-substituent dependent, unlike those of free-base corroles. It was suggested that this owes to mixing of LMCT character into the $\pi\pi^*$ transition manifold that comprises the Soret system. As to the specifics of the substituent effects upon the spectra, a perusal of the data presented suggests a general trend toward red-shifted absorption maxima when the substituents are more electron-donating, but no overwhelming relationship is in evidence. The brominated corroles have Soret bands that present a multitude of bandshapes and intensities; their behavior possesses no obvious relationship to the electronics of the *meso*-substituents. Perhaps this owes to the saddling preferences showed by highly substituted copper corroles (Fig. 13). Although numerous theoretical studies suggest that copper corroles in either diamagnetic configuration will retain a moderate degree of saddling [163, 166, 167], this distortion was predicted to be significantly enhanced in octabrominated copper corrole complexes [168].

A number of short papers published over the last decade describe highly saddled copper corrole complexes with esoteric β-substituents. A 2003 article describing the synthesis of a variety of β-octafluorocorroles also reported the properties of their Cu complexes, which appear to have higher-lying paramagnetic states than the octabrominated analogs [169]. This article suggested that some broadening

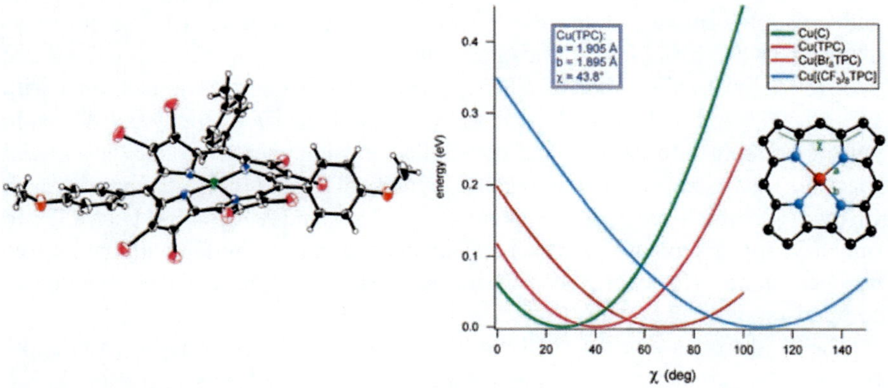


**Fig. 13** The crystal structure of a highly saddled copper β-octabrominated corrole complex (*left*) and the predicted saddling angles for a number of *meso*-triaryl-substituted copper corroles. Reprinted with permission from [171]. Copyright 2010 American Chemical Society

observed in the $^{19}F$ NMR above room temperature hinted at low-lying excited states. However, all other evidence points to a heavily favored diamagnetic electronic configuration, so this broadening might be better ascribed to dynamic molecular processes than to thermal population of an excited state. In the latter half of the decade, a report described the synthesis of 5,10,15-triaryl-β-octakis (trifluoromethyl)corroles, unprecedentedly bulky macrocycles whose copper complexes exhibit severely red-shifted absorption and were predicted by DFT to undergo extreme saddling distortions [170]. These large predicted saddling angles were confirmed by XRD measurements 2 years later [171].

## 3 The Second-Row Metallocorroles

Significant work has been performed on the corrole complexes of four second-row transition metals: Mo, Ru, Rh, and Ag. The spectroscopic properties of a number of (oxo)Mo(V) corroles have been examined in detail. Meanwhile, a number of dimeric Ru corroles have been synthesized and their electrochemical properties have been reported; monomeric Ru–NO corroles have also been disclosed. The syntheses of both Rh(I) and Rh(III) corroles have been reported by a number of groups, and the excited-state EPR spectroscopy of the Rh(III) derivatives has been presented. High-valent Ag(III) corroles have been studied extensively with regards to their facile demetalation.

### *3.1 Molybdenum Corroles: Second-Row Oxometallocorrole Complexes*

The first reported (oxo)metallocorrole was (oxo)Mo(mec) [63]. As with the titanyl and vanadyl corrolates, its synthesis requires harsh conditions. Mo(mec)(O) was obtained in ca. 40% yield by heating either $MoCl_5$ or $Mo(CO)_6$ with $H_3$mec at 170–180°C in decalin. The RT EPR spectrum of this complex is isotropic, with $g_{iso} = 1.967$, $A^{95/97}$Mo = 41.8, and $A^{14}$N = 2.3 G. At 77 K, the $g$ and $A^{95/97}$Mo tensors are split into parallel and perpendicular components, producing an axial spectrum consistent with a tetragonal pyramidal ligand field. Compared with Cr(tpfc)(O), which has $g_{iso} = 1.986$, $A^{53}$Cr = 16.4, and $A^{14}$N = 3.0 G at RT, Mo (mec)(O) appears to have greater metal character and less overlap with the pyrrole nitrogen atoms. However, the differing electronics of tpfc and mec complicate comparison.

The syntheses of a variety of (oxo)Mo triarylcorrolates were reported in 2004, but the EPR parameters of the complexes were mentioned only in passing [172]. The RT EPR parameters of Mo(tpfc)(O), reported by Gross and coworkers [173], are as follows: $g_{iso} = 1.975$, $A^{95/97}$Mo = 46.6, and $A^{14}$N = 2.6 G. These are quite similar to those reported for Mo(mec)(O), and they show the same trends compared with the Cr(V) analog. The Mo–O bond length of 1.684(2) Å is significantly longer than the Cr–O bond length of 1.5700(17) reported for Cr(tpfc)(O) [74], as are the Mo–N bond lengths. These changes are likely owing to the different atomic radii of the 3d and 4d M(V) ions.

The RR spectral properties of Mo(tbtc)(O) (tbtc = 5,10,15-tris[3,5-bis(trifluoromethyl)phenyl]corrolate) were reported in a 2007 paper [174]. It was found that the Mo–O bond-stretching frequency correlated with Swain's "A + B" solvent polarity parameters (essentially, a combined measure of a given solvent's ability to solvate both anionic and cationic species) [175], and that the Mo–O bond strength is diminished in highly polar solvents owing to obviation of $Mo^{5+}$ acceptivity and $O^{2-}$ donicity. The correlation between Mo–O bond strength and Gutmann's "acceptor number," a measure of the electron-accepting capability of the solvent [176], is not as strong, implying that solvation of Mo is an important factor in the determination of the Mo–O bond strength (not so in the case of, e.g., vanadyl porphyrins) [177]. Perhaps this owes to the relatively large solvent-exposed surface area of the Mo(V) ion in these complexes. In a separate report, Czernuszewicz and coworkers described the solvent-dependent UV-vis and RR spectral properties of Mo(tpc)(O) (tpc = 5,10,15-triphenylcorrolate) [178]. This complex also displays a strongly solvent-dependent Mo–O bond-stretching frequency. Additionally, its UV-vis spectral bands undergo significant bathochromic shifting when the compound is dissolved in solvents with high indices of refraction, implying that there is significant CT character present in these transitions. The resonance enhancement of the Mo–O Raman signal was found to be more pronounced upon excitation into the Soret band than was the enhancement of Raman-active macrocycle vibrational modes (Fig. 14), implying that the Soret manifold includes CT transitions from

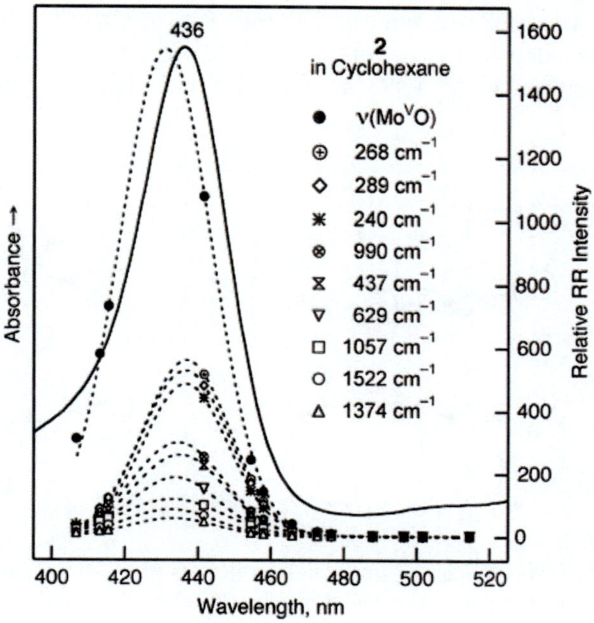


**Fig. 14** RR intensities of nine skeleton vibrations and the metal-oxo stretch of Mo (tpc)(O), all overlaid on the UV-vis absorption spectrum. The Mo–O stretch is the most enhanced, and the blue shifting of the peak implies CT character in the high-energy portion of the Soret manifold. Reprinted with permission from [176]. Copyright 2009 American Chemical Society

filled corrole π orbitals into empty Mo $4d_{xz,yz}$ orbitals. This finding suggests that LMCT transitions could contribute intensity to the Soret absorption manifold in other metallocorroles as well; this LMCT character could be exploited to develop novel reactivity via photo-induced weakening of the bonds between the metal center and axial ligands. Time-resolved RR experiments would provide further insight into the degree of CT into the metal orbitals, perhaps setting the stage for, e.g., photocatalytic oxygen atom transfer reactions using Mo corroles.

## 3.2 *Ruthenium Corroles: Dimers and Monomers*

The earliest report of a ruthenium corrole complex appeared in the literature in 2000 [64], and the most recent in 2003 [179]. The initial report described the synthesis of the face-to-face dimer $[Ru(hedmc)]_2$ (Fig. 15), synthesized by reaction of $H_3$hedmc with $[Ru(cod)Cl_2]_2$ (where cod = 1,5-cycloöctadiene) in 2-methoxyethanol in the presence of trimethylamine. This complex displays an extremely blue-shifted Soret electronic absorption system, with minimal Q-band intensity. The Ru–Ru bond distance is 2.166(1) Å; the Ru atoms are displaced toward one another and out of the corrole plane by 0.514(4) Å. Based on the MO description of Collman and Arnold [180], a Ru–Ru bond order of 3 was posited. This assignment is consistent with available XRD data for Ru–Ru dimers in tetrapyrrole coordination environments: In $[Ru(oep)]_2$, which has a bond order of 2, the Ru–Ru bond length is 2.408(1) Å [181]. $[Ru(tpp)]_2^+$, bond order 2.5, possesses a Ru–Ru distance of 2.296(2) Å [182].

A subsequent report on Ru–Ru corrole dimers explored the electrochemistry of $[(hedmc)Ru]_2$ and $[(dehmc)Ru]_2$ [183]. Each dimer exists in six electrochemically accessible oxidation states (Fig. 16). In $CH_2Cl_2$ solution, the first one-electron oxidation ($E_{1/2}$ = 0.14 V vs. SCE) results in a species with an isotropic EPR signal ($g_{iso}$ = 1.998) at RT and a rhombic EPR signal ($g_z$ = 2.049, $g_y$ = 1.991,

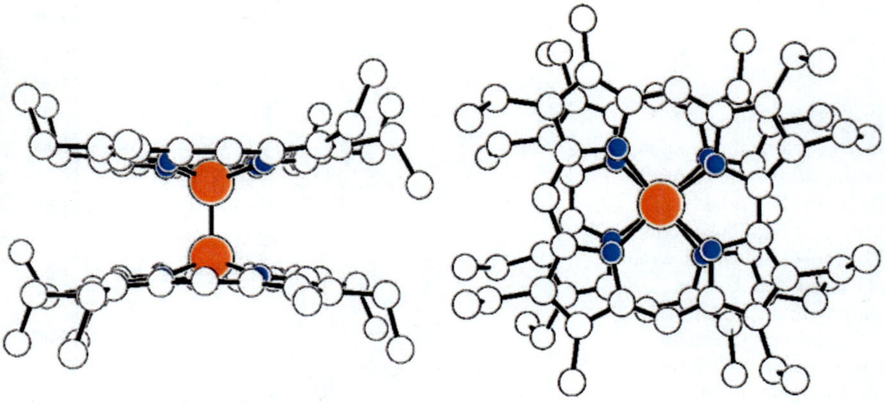

**Fig. 15** $[Ru(hedmc)]_2$ viewed from the side (*left*) and down the Ru–Ru axis (*right*). Reprinted with permission from [64]. Copyright 2000 John Wiley and Sons

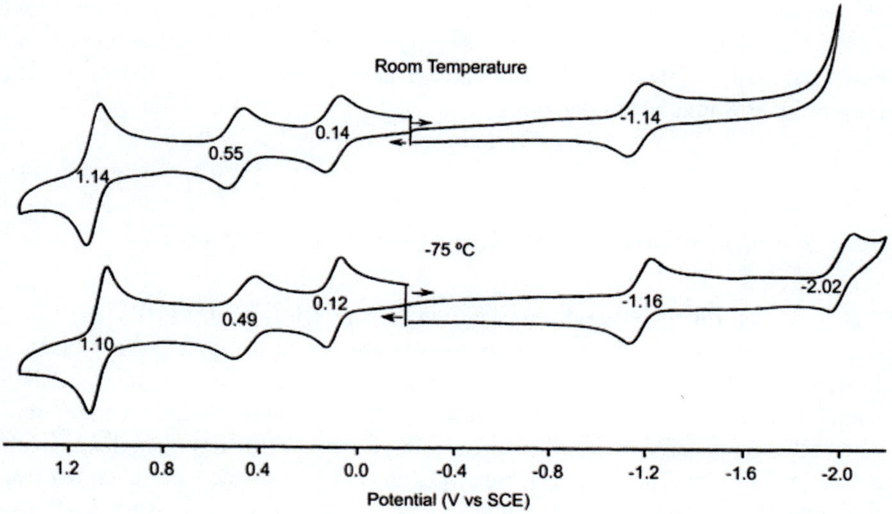


**Fig. 16** Cyclic voltammograms of $[Ru(hedmc)]_2$ in $CH_2Cl_2$ at two different temperatures. Reprinted with permission from [183]. Copyright 2002 Elsevier

$g_x = 1.928$) at 77 K, implying an Ru(III,IV) oxidation state. The second one-electron oxidation ($E_{1/2} = 0.55$ V vs. SCE) results in very small changes to the UV-vis spectra, implying a second Ru-centered oxidation to an Ru(IV,IV) state. The final accessible oxidation state is the result of a two-electron oxidation at 1.14 V vs. SCE that is reversible for the $[(hedmc)Ru]_2$ but irreversible for $[(dehmc)Ru]_2$. On the basis of the substantial changes in the UV-vis spectra upon oxidation, this state was assigned as Ru(IV,IV) with two π-cation radical corrole macrocycles. Two reversible one-electron reductions could be observed for the dimers, at −1.15 and −2.00 V vs. SCE, but the products of these reductive processes could not be definitively assigned. It was hypothesized that the successive reductions result in the formation of Ru(II,III) and Ru(II,II) states based on the behavior of the porphyrinate analog $[Ru(oep)]_2$ [184].

In 2003, the first examples of monomeric ruthenium corrole species, which could be stabilized using NO as an axial ligand [179], were reported. Using $H_3$tpfc and $H_3$tdcc as the corrole sources, the authors used the conditions of Guilard [64], but added NO gas at the end of the reaction and could isolate NO-bound monomers along with dimeric species. Both Ru(tpfc)(NO) and Ru(tdcc)(NO) are diamagnetic at room temperature. XRD structural characterization showed essentially linear Ru–N–O vectors with Ru–N axial bonds of 1.72 Å. The out of plane displacement of the Ru ion is 0.54 Å. Both complexes are able to take on a second axial ligand, so their IR spectra were measured in pyridine to ensure the presence of only one species. The N–O stretching frequencies of the complexes are 1,835 and 1827 $cm^{-1}$ for Ru(tpfc)(NO)(py) and Ru(tdcc)(NO)(py), respectively, reflecting stronger Ru–N bonding than in the isoelectronic porphyrin $[Ru(tpp)(NO)(py)]^+$, which has an N–O stretch located at 1879 $cm^{-1}$. However, $H_2$tpp has more electron-donating *meso*-substituents than either $H_3$tpfc or $H_3$tdcc, which could easily lead to less opportunity for back-bonding from the ligand; as Gross and coworkers note in their article, comparison to a more electron-withdrawing Ru porphyrinate would be enlightening.

## 3.3 Rhodium Corroles: Multiple Coordination Modes

Stable Rh corrolates have been reported in the Rh(I) and Rh(III) oxidation states (Fig. 17), where the former complexes are bound in square-planar fashion to two internal corrole nitrogen atoms [66] and the latter possess the usual $N_4$-bound geometry [67]. Early work focused on the synthesis, characterization, and electrochemistry of five-coordinate (octaalkylcorrolato)Rh(III) complexes with arsine [185] or phosphine [124] ligands. Later work extended the chemistry of rhodium(III) corroles with the synthesis and characterization of Rh(tpfc)($PPh_3$) [130]. Gross and coworkers reported that these complexes are excellent catalysts for the cyclopropanation of electron-rich olefins by carbenoids such as ethyl diazoacetate; in fact, they are significantly better catalysts than the corresponding porphyrins [186]. Collman and coworkers reported the apparent ability of Rh(III) corrolates to activate certain C–N and C–C bonds [187]. Trialkylamines lose one alkyl group when reacted with $[Rh(cod)Cl]_2$ and $H_3$tpfc under aerobic conditions, forming Rh(bis-dialkylamine)(tpfc) complexes. Bulky aromatic bases such as 2,6-lutidine lose one methyl group under the same conditions. Under anaerobic conditions, no rearrangements were observed, emphasizing the necessity of oxidation to Rh(III) for this reaction to proceed. Additionally, β-hydride elimination was postulated to play a key role in the reaction: no bond activation was observed for substrates without accessible β-protons.

A 2002 paper reported the synthesis of a Rh(I) complex with the $N^{22}$-picolyl-substituted derivative of $H_3$tpfc [188]. This structurally unusual metallocorrole, Rh($N^{22}$-picolyl-Htpfc)(CO)($PPh_3$), features an out-of-plane Rh displacement of 1.51 Å, compared with a displacement from the $N_4$-plane of 0.2763(5) Å in the Rh(III) complex Rh(tpfc)($PPh_3$). The rhodium ion is bound to the $N^{23}$ and

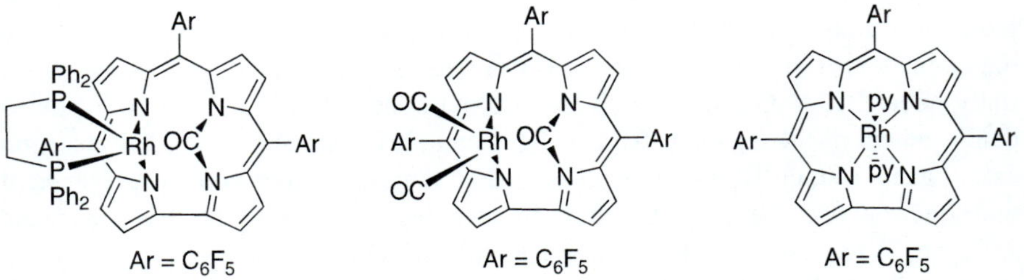


**Fig. 17** Rh(I) carbonylated corroles (*left, middle*) and a Rh(III) corrole (*right*)

$N^{24}$ atoms, with Rh–N bond lengths of 2.095 (compared to 1.968 Å for the five-coordinate Rh(III) corrole), an Rh–C bond length of 1.832, and an Rh–P bond distance of 2.257 Å. The syntheses of the dicarbonylrhodium(I) complex of an $N^{23}$,$N^{24}$-carbonylated derivative of tpfc, as well as a monocarbonyl mono-triphenylphosphine Rh(I) complex of the same derivative [189], have also been reported. Compared with the geometrically similar $N^{22}$-picolyl-Htpfc complex, the $N^{23}$,$N^{24}$-carbonylated complex has slightly shorter Rh–N bonds and a slightly elongated Rh–P bond distance. These compounds are poor catalysts for carbene-transfer reactions compared to the Rh(III) corrolates. Gross, Goldberg, and coworkers have also reported the synthesis of chiral Rh(I) $N^{23}$,$N^{24}$-carbonylated corroles with bulky diphosphine ligands [190]. They could spontaneously resolve the enantiomers by crystallization, resulting in a strong circular dichroism signal in the Soret band region.

The time-resolved EPR spectroscopy of photo-excited Rh(III) corroles has been examined at length in two articles [191, 192]. Compared with the excited states of main group corroles such as Sn(tpfc)(Cl) [193] and Ga(tpfc)(py) [194], the excited states of Rh(III) corroles display broadened signals and greater $g$ tensor anisotropy. This was attributed to spin–orbit coupling effects that effectively admix $^3$dd and $^3\pi\pi^*$ excited states.

## 3.4 Silver: High-Valent Late-Metal Corrole Complexes

Ag(III) corroles, synthesized by the treatment of free-base corroles with three equivalents of Ag(I) acetate [68], exemplify the ability of these macrocycles to promote the formation of unusual, high-valent metal complexes. As Brückner pointed out in a 2004 educational commentary, Ag(II) porphyrins [195] are themselves representative of relatively high oxidation state Ag [196], Ag(I) salts being far more common. The evidence for the Ag(III) oxidation state is quite strong: the Ag corroles reported by Brückner are diamagnetic and XPS comparison of Ag(tpc) to the Ag(II) and Ag(III) complexes of tpp shows that the Ag $3d_{3/2}$ and $3d_{5/2}$ binding energies of the Ag corrolate closely resemble those of the Ag(III) porphyrinate.

Ag(III) corrole complexes can be cleanly demetalated through a variety of routes, and the sundry means of performing the demetalation were explored in a 2009 report [197]. The authors compared the efficiency and scope of four demetalation reaction conditions: $H_2SO_4$ in $CHCl_3$; $NaBH_4$ in $CH_2Cl_2$/MeOH; NaOH in EtOH/toluene; and 1,8-Diazabicyclo[5.4.0]undec-7-ene (DBU) in thf. The most efficient methodology was the DBU technique, resulting in recovery of demetalated corrole in 62% yield (after treatment of the reaction mixture with hydrazine). The $NaBH_4$ reaction was also shown to be quite effective, allowing isolation of free-base corrole in 54% yield. Facile demetalation reactions of corroles are much needed to synthesize new free-base corroles; certain substitutions proceed with great facility on metallocorroles but decompose the corresponding free-base compounds, so the substituted free-base corroles can only be obtained via demetalation reactions.

## 4 Third-Row Metallocorroles

Only a handful of articles describing third-row metallocorrole complexes have been published. An early report of the synthesis and crystal structure of an (oxo) Re(V) corrolate was published in 1995, and an article describing the synthesis and spectroscopic properties of a variety of *meso*-triaryl-substituted Au corroles was published in 2011. Most of the work on third-row metallocorroles has described the synthesis, characterization, electrochemistry, and spectroscopic properties of the phosphorescent Ir corrolates.

### *4.1 Rhenium and Gold: Chemical Rarities*

The first well-characterized example of a third-row metallocorrole was Re(tfmc)(O) [69], synthesized fortuitously via ring contraction of the corresponding porphyrin in the presence of $Re_2(CO)_{10}$. Re(tfmc)(O) was characterized by XRD; the Re(V) oxidation state was assigned based on the diamagnetism of the complex. The Re–O bond length in this complex is 1.662(2) Å and the bond stretching frequency is 994 $cm^{-1}$, consistent with an Re–O triple bond. The Re ion sits 0.701 Å above the plane of the nitrogen atoms, slightly closer than the Mo ion in Mo(tpfc)(O) (0.729 Å). Corroles are well known for stabilizing high-valent metals, so it would be unsurprising if stable Re(VI) or Re(VII) corrolates were accessible. The development of a more practical synthesis of Re corrolates would open doors to new high-valent Re chemistry.

An article published in early 2011 reported the synthesis and characterization of Au(III) corrolates [70]. Commercially available gold starting materials tend to either oxidize or halogenate free-base corroles, producing intractable mixtures. However, Ghosh and coworkers found that fully brominated free-base corroles (which were obtained via demetalation of the corresponding copper [198] or manganese [199] complexes in this case) gave Au(III) corrolates upon treatment with chloroauric acid and trimethylamine in $CH_2Cl_2$. Unlike the Cu octabromocorrolates, which possess electrochemical HOMO-LUMO gaps of just 0.99–1.06 eV, the gold complexes display gaps of 2.27–2.34 eV. This suggests that while Cu corroles undergo metal-based reduction before the corrole ring is reduced, the corresponding Au complexes are first reduced to the π-radical anion. The first oxidation in both cases is assigned to the formation of the π-cation, and falls in the same range for both Cu and Au β-octabromocorrolates.

## 4.2 *Iridium: Well-Characterized Third-Row Corroles*

The synthesis, XRD, and electrochemical properties of two iridium corroles were communicated in 2008 [71]. Prepared by reaction of $H_3$tpfc with [Ir(cod)Cl]$_2$ in tetrahydrofuran, followed by treatment with trimethylamine *N*-oxide, Ir(tpfc)(NMe$_3$)$_2$ and Ir(Br$_8$tpfc)(NMe$_3$)$_2$ are planar and have essentially identical Ir-N (axial) bond lengths of 2.185(9) and 2.189(3) Å, respectively. Ir(tpfc)(NMe$_3$)$_2$ undergoes two reversible oxidations at +0.66 and +1.28 V vs. SCE in $CH_2Cl_2$, while Ir(Br$_8$tpfc)(NMe$_3$)$_2$ displays one reversible oxidation at +1.19 and one reversible reduction at −1.21 V vs. SCE. On the basis of literature comparisons, the first oxidation was tentatively assigned to formation of an Ir(IV) corrolate and the second to formation of an Ir(IV) π-cation radical. Meanwhile, the reduction product of Ir(Br$_8$tpfc)(NMe$_3$)$_2$ was identified as the π-anion radical on the basis of its very anodic reduction potential and quasi-reversibility. The tentative assignment of the Ir(IV) oxidation state for [Ir(tpfc)(NMe$_3$)$_2$]$^+$ was not supported by primary experimental evidence until the publication of a subsequent article examining the electrochemistry, EPR, and UV-vis spectra of five- and six-coordinate Co(III), Rh(III), and Ir(III) tpfc complexes and their one-electron oxidized cations [200]. In this report, [Co(tpfc)(py)$_2$]$^+$ and [Rh(tpfc)(py)$_2$]$^+$ were definitively assigned π-cation radical electronic configurations based on their isotropic *g* tensors and the similarity of their UV-vis spectra to known π-cation radical metallocorroles. Meanwhile, [Ir(tpfc)(NMe$_3$)$_2$]$^+$ and [Ir(tpfc)(py)$_2$]$^+$ were assigned Ir(IV) configurations based on the anisotropy of their EPR signals (Fig. 18) and the hypsochromic shifts of their Soret bands compared to the neutral complexes.

Recently published DFT calculations on the electronic structures of the Co, Rh, and Ir corrolate cations suggested that all three oxidized complexes share a common π-cation radical electronic structure [201]. However, very low-lying excited states with significant metal character were found in the case of the Ir corrolate cations, and the investigators who performed the calculations suggested that spin–orbit coupling effects, which were not examined computationally, could lead to stabilization of states with metal character in accord with the experimentally observed EPR anisotropy. DFT calculations including spin–orbit coupling effects have been used by Palmer and Lancaster to model the EPR parameters of a series of one-electron oxidized Ir(tpfc)(L)$_2$ complexes; a manuscript describing these results is in preparation as of this writing.

The Ir(III) corroles also possess a quality unique among metallocorrole complexes: at room temperature, they phosphoresce in the near-infrared region (Fig. 19) with lifetimes up to 5 μs [202]. According to a 2010 report on the phosphorescence of Ir(tpfc)(NMe$_3$)$_2$, Ir(Br$_8$tpfc)(NMe$_3$)$_2$, and Ir(tpfc)(py)$_2$, their quantum yields of phosphorescence range from 0.0001 to 0.01; interestingly, this low quantum yield appears to stem not from particularly rapid non-radiative decay but rather from slow radiative decay rates. A report by Palmer and coworkers describing cationic iridium(III) monoazaporphyrins, which have similar emission properties, will be published in late 2011 [203]. Borisov and coworkers detailed the photophysical properties of some

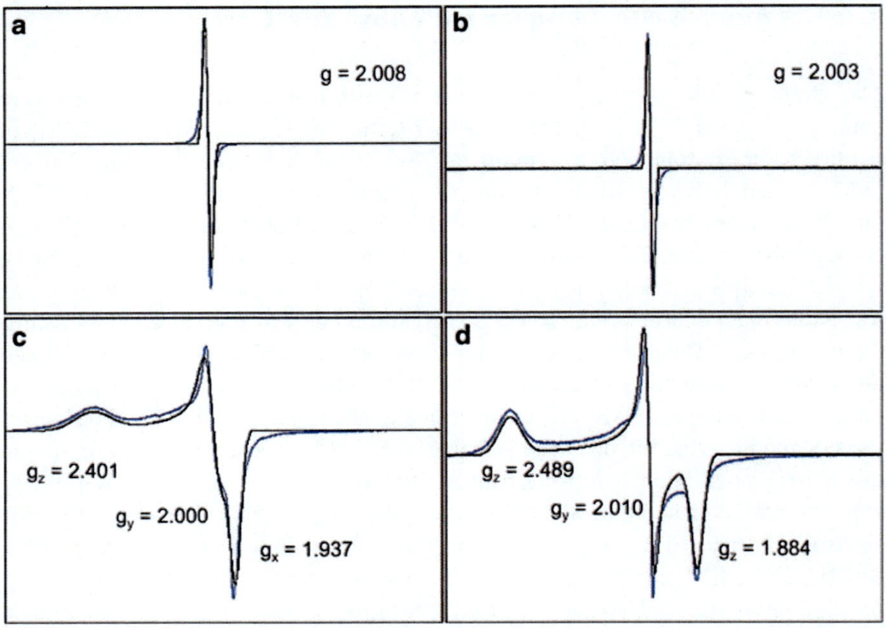


**Fig. 18** The EPR spectra of chemically oxidized Co(tpfc)$(py)_2$ (**a**), Rh(tpfc)$(py)_2$ (**b**), Ir(tpfc)$(py)_2$ (**c**), and Ir(tpfc)$(NMe_3)_2$ (**d**) in $CH_2Cl_2$, with spectra in *blue* and simulations in *black*. Reprinted with permission from [200]. Copyright 2009 American Chemical Society

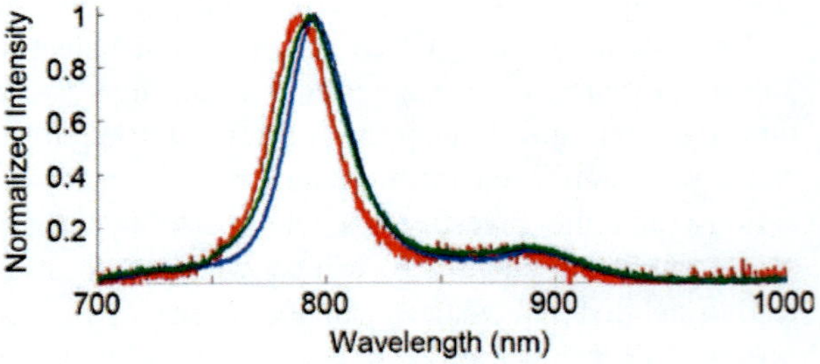


**Fig. 19** Luminescence spectra of Ir(tpfc)$(NMe_3)_2$ (*red*), Ir($Br_8$tpfc)$(NMe_3)_2$ (*blue*), and Ir(tpfc)$(py)_2$ (*green*) in toluene, excited at 496.5 nm. Reprinted with permission from [202]. Copyright 2010 American Chemical Society

Ir(III) porphyrins and benzoporphyrins in a recent article, reporting that these compounds have lifetimes on the order of tens of microseconds and luminescence quantum yields of 14–30% [204]. The large discrepancy between Ir(III) porphyrin and corrole photophysics warrants further investigation.

## 5 Present Successes and Future Challenges

The recent surge in corrole research, spurred on by synthetic advances that allow for preparation of large quantities of material, has resulted in the isolation, characterization, and study of many novel transition metal complexes with interesting and even unique reactivity and electronic structure. The ability of the corrole macrocycle to support high oxidation states has proven difficult to exploit due to the presence of low-lying filled corrole π orbitals, but contemporary studies suggest that protection of the corrole ring with electron-withdrawing substituents could allow corrole-based catalysts to become more robust. The development of interesting architectures such as the Pacman and Hangman cobalt corroles will also assist in achieving this goal. There are coordination chemistry challenges ahead as well: most of the third-row and many of the second-row metals have not been chelated by corroles. Additional investigation into the second-row and third-row metallocorroles seems inevitable. Over just a dozen years, corrole coordination chemistry has gone from a small specialty field populated by a few investigators to a large area involving some of the most recognizable inorganic chemists. In a few more years, as synthetic techniques continue to improve and new metallocorrole reactivity is uncovered, corroles may become as ubiquitous as porphyrins in the chemical literature.

**Acknowledgments** I thank Harry Gray and Zeev Gross for introducing me to corroles, and Kyle Lancaster, Alec Durrell, Theis Brock-Nannestad, and all my other collaborators for working with me throughout my investigations of these macrocycles. Kyle Lancaster and Jeff Warren provided helpful advice on the preparation of this chapter, which is very much appreciated. I am also grateful to the Editors of Structure and Bonding for allowing me the opportunity to participate in this Special Volume dedicated to one of the great modern chemists. Additionally, I would like to express my gratitude to Robert Grubbs for allowing me to undertake this time-consuming endeavor while on a postdoctoral fellowship in his laboratory group.

## References

1. Bjerrum J, Ballhausen CJ, Jørgensen CK (1954) Results of calculations on the spectra and configuration of copper(II) ions. Acta Chem Scand 8:1275
2. Ballhausen CJ, Jørgensen CK (1955) The spectra of cobalt(II) complexes. Acta Chem Scand 9:397
3. Ballhausen CJ, Jørgensen CK (1955) d-Electrons in crystal fields of different symmetries. Dan Mat Fys Medd 29: No.14
4. Ballhausen CJ (1962) Introduction to ligand field theory. McGraw-Hill, New York and London
5. Anders U (2003) Early ideas in the history of quantum chemistry. http://www.quantum-chemistry-history.com/Ball_Dat/BallPub1.htm
6. Gray HB, Ballhausen CJ (1961) The mixing of $3d^n$ and $3d^{n-1}4$ s states in crystalline field theory. Acta Chem Scand 15:1327
7. Gray HB, Ballhausen CJ (1961) The electronic structure of the vanadyl ion. Inorg Chem 1:111–122
8. Gray HB, Ballhausen CJ (1962) On the optical spectra of some transition metal pentacyanonitrosyl complexes. J Chem Phys 36:1151
9. Gray HB, Ballhausen CJ (1963) A molecular orbital theory for square planar metal complexes. J Am Chem Soc 85:260
10. Ballhausen CJ, Gray HB (1963) π-Bonding in transition metal complexes. Inorg Chem 2:426

11. Spiro TG, Ballhausen CJ (1961) Complexes of 1,3-diamino-2-aminomethylpropane with nickel(II). Equilibrium constants and visible spectra. Acta Chem Scand 15:1707
12. Solomon EI, Ballhausen CJ, Høg JH (1975) Pseudo-Stark splitting of charge transfer transitions of $KCrO_3Cl$. Chem Phys Lett 34
13. Solomon EI, Ballhausen CJ (1975) Identification of the structures of the $^3T_{1g}$ (I) $\leftarrow$ $^3A_{2g}$ band in the $Ni(H_2O)_6^{++}$ complex. Mol Phys 29:279–299
14. Trogler WC, Solomon EI, Trabjerg I, Ballhausen CJ (1977) Studies of the polarization behavior, temperature dependence and vibronic structure of the 23000 cm $^{-1}$ absorption system in the electronic spectra of $Mo_2(O_2CCH_3)_4$ and related compounds: emission spectrum of $Mo_2(O_2CCF_3)_4$ at 1.3 K. Inorg Chem 16
15. Basolo F, Ballhausen CJ, Bjerrum J (1955) Absorption spectra of geometrical isomers of hexacoordinated complexes. Acta Chem Scand 9:810
16. Cotton FA, Ballhausen CJ (1956) Soft X-ray absorption edges of metal ions in complexes I. J Chem Phys 25:617
17. Ballhausen CJ, Gray HB (1964) Molecular orbital theory. W A Benjamin Inc, New York, NY
18. Ballhausen CJ, Gray HB (1980) Molecular electronic structures: an introduction. Benjamin/ Cummings Publishing Co Inc, Reading, MA
19. Hedegaard ED, Schau-Magnussen M, Bendix J (2011) [Cr(N)(acac)2]: a simple chromium nitride complex and its reactivity towards late transition metals. Inorg Chem Commun 14:719–721
20. Schau-Magnussen M, Malcho P, Herbst K, Brorson M, Bendix J (2011) Synthesis and X-ray crystal structure of a novel organometallic (μ3-oxido)(μ3-imido) trinuclear iridium complex. Dalton Trans 40:4212–4216
21. Brock-Nannestad T, Hammershoi A, Schau-Magnussen M, Vibenholt J, Bendix J (2011) Electronic and molecular structure of [Cr(N)(NCO)4]2-, the first example of a mixed nitride-cyanate complex. Inorg Chem Commun 14:251–253
22. Bendix J, Gray HB, Golubkov G, Gross Z (2000) High-field (high-frequency) EPR spectroscopy and structural characterization of a novel manganese(III) corrole. Chem Commun 19:1957–1958
23. Golubkov G, Bendix J, Gray HB, Mahammed A, Goldberg I, DiBilio AJ, Gross Z (2001) High-valent manganese corroles and the first perhalogenated metallocorrole catalyst. Angew Chem Int Ed 40:2132–2134
24. Bendix J, Dmochowski IJ, Gray HB, Mahammed A, Simkhovich L, Gross Z (2000) Structural, electrochemical, and photophysical properties of gallium(III) 5,10,15-tris (pentafluorophenyl)corrole. Angew Chem Int Ed 39:4048–4051
25. Johnson AW, Price R (1960) Synthesis of derivatives of corrole (pentadehydrocorrin). J Chem Soc 1649–1653
26. Johnson AW, Kay IT (1961) The synthesis of derivatives of corrole. An amendment. Proc Chem Soc London 168–169
27. Johnson AW, Kay IT (1964) The pentadehydrocorrin (corrole) ring system. Proc Chem Soc London 168–169
28. Johnson AW, Kay IT (1965) Synthesis of corroles and related ring systems. Proc Royal Soc London A 288:334–341
29. Gross Z, Galili N, Saltsman I (1999) The first direct synthesis of corroles from pyrrole. Angew Chem Int Ed 38:1427–1429
30. Gross Z, Galili N, Simkhovich L, Saltsman I, Botoshansky M, Blaeser D, Boese R, Goldberg I (1999) Solvent-free condensation of pyrrole and pentafluorobenzaldehyde: a novel synthetic pathway to corrole and oligopyrromethenes. Org Lett 1:599–602
31. Paolesse R, Licoccia S, Bandoli G, Dolmella A, Boschi T (1994) First direct synthesis of a corrole ring from a monopyrrolic precursor crystal and molecular structure of (triphenylphosphine)(5,10,15-triphenyl-2,3,7,8,12,13,17,18-octamethylcorrolato)cobalt (III)-dichloromethane. Inorg Chem 33:1171–1176

32. Paolesse R, Tassoni E, Licoccia S, Paci M, Boschi T (1996) One-pot synthesis of corrolates by cobalt catalyzed cyclization of formylpyrroles. Inorg Chim Acta 241:55–60
33. Licoccia S, Di Vona ML, Paolesse R (1998) Acid-catalyzed cyclization of 1,19-unsubstituted a, c-biladienes. J Org Chem 63:3190–3195
34. Paolesse R, Mini S, Sagone F, Boschi T, Jaquinod L, Nurco DJ, Smith KM (1999) 5,10,15-triphenylcorrole: a product from a modified rothemund reaction. Chem Commun 14:1307–1308
35. Gryko DT (2000) A simple, rational synthesis of *Meso*-substituted A2B-corroles. Chem Commun 22:2243–2244
36. Guilard R, Gryko DT, Canard G, Barbe J-M, Koszarna B, Brandes S, Tasior M (2002) Synthesis of corroles bearing up to three different *meso* substituents. Org Lett 4:4491–4494
37. Gryko DT (2002) Recent advances in the synthesis of corroles and core-modified corroles. Eur J Org Chem 11:1735–1743
38. Scifinder® by Chemical Abstracts Service. Copyright 2010 American Chemical Society
39. Kadish KM, Smith KM, Guilard R (eds) (2000) The Porphyrin Handbook, Volume 1: Synthesis and Organic Chemistry, Academic Press – San Diego, CA and London, UK
40. Gryko DT (2008) Adventures in the synthesis of *meso*-substituted corroles. J Porphyr Phthalocyanines 12:906–917
41. Paolesse R (2008) Corrole: the little big porphyrinoid. Syn Lett 15:2215–2230
42. Aviv I, Gross Z (2007) Corrole-based applications. Chem Commun 20:1987–1999
43. Paolesse R, Nardis S, Sagone F, Khoury RG (2001) Synthesis and functionalization of *meso*-aryl-substituted corroles. J Org Chem 66:550–556
44. Saltsman I, Mahammed A, Goldberg I, Tkachenko E, Botoshansky M, Gross Z (2002) Selective substitution of corroles: nitration, hydroformylation, and chlorosulfonation. J Am Chem Soc 124:7411–7420
45. Aviv-Harel I, Gross Z (2009) Aura of corroles. Chem Eur J 15:8382–8394
46. Stefanelli M, Nardis S, Tortora L, Fronczek FR, Smith KM, Licoccia S, Paolesse R (2011) Nitration of iron corrolates: further evidence for non-innocence of the corrole ligand. Chem Commun 47:4255–4257
47. Paolesse R, Nardis S, Venanzi M, Mastroianni M, Russo M, Fronczek FR, Vicente MGH (2003) Vilsmeier formylation of 5,10,15-triphenylcorrole: expected and unusual products. Chem Eur J 9:1192–1197
48. Gouterman M (1961) Spectra of porphyrins. J Mol Spectrosc 6:138–163
49. Gouterman M, Wagniére GH, Snyder LC (1963) Spectra of porphyrins Part II: four orbital model. J Mol Spectrosc 11:108–127
50. Ghosh A, Wondimagegn T, Parusel ABJ (2000) Electronic structure of gallium, copper, and nickel complexes of corrole: high-valent transition metal centers versus non-innocent ligands. J Am Chem Soc 122:5100–5104
51. Day P (1967) A theory of the optical properties of vitamin $B_{12}$ and its derivatives. Theor Chem Acc 7:328–341
52. Day P (1967) The electronic structure and spectrum of vitamin $B_{12}$. Coord Chem Rev 2:109–116
53. Liu X, Mahammed A, Tripathy U, Gross Z, Steer RP (2008) Photophysics of soret-excited tetrapyrroles in solution III: porphyrin analogues: aluminum and gallium corroles. Chem Phys Lett 459:113–118
54. Kowalska D, Liu X, Tripathy U, Mahammed A, Gross Z, Hirayama S, Steer RP (2009) Ground- and excited-state dynamics of aluminum and gallium corroles. Inorg Chem 48:2670–2678
55. Ventura B, Esposti AD, Koszarna B, Gryko DT, Flamigni L (2005) Photophysical characterization of free-base corroles, promising chromophores for light energy conversion and singlet oxygen generation. New J Chem 29:1559–1566
56. Aviv-Harel I, Gross Z (2011) Coordination chemistry of corroles with focus on main group elements. Coord Chem Rev 255:717–736

57. Kadish KM, Smith KM, Guilard R (eds) (2000) The Porphyrin Handbook, Volume 3: Inorganic, Organometallic, and Coordination Chemistry, Academic Press – San Diego, CA and London, UK
58. Licoccia S, Paolesse R, Tassoni E, Polizio F, Boschi T (1995) First-row transition-metal complexes of corroles: synthesis and characterization of oxotitanium(IV) and oxovanadium (IV) complexes of β-alkylcorroles. J Chem Soc Dalton Trans 3617–3621
59. Matsuda Y, Yamada S, Murakami Y (1980) Preparation and characterization of an oxochromium(V) complex with a macrocyclic $N_4$-ligand: oxo-(2,3,17,18-tetramethyl-7,8,12,13-tetraethylcorrolato)chromium(V). Inorg Chim Acta 44:309–311
60. Boschi T, Licoccia S, Paolesse R, Tagliatesta P, Tehran MA (1990) Synthesis and characterization of novel metal(III) complexes of corrole: crystal and molecular structure of (2,3,7,8,12,13,17,18-octamethylcorrolato)(triphenylarsine)rhodium(III). J Chem Soc Dalton Trans 463–467
61. Conlon M, Johnson AW, Overend WR, Rajapaksa D, Elson CM (1973) Structure and reactions of cobalt corroles. J Chem Soc Perkin Trans 2281–2288
62. Johnson AW, Kay IT (1965) Corroles: Part 1: synthesis. J Chem Soc Res 1620–1629
63. Murakami Y, Matsuda Y, Yamada S (1977) Preparation and characterization of oxo (2,3,17,18-tetramethyl-7,8,12,13-tetraethylcorrolato)molybdenum(V). Chem Lett 6:689–692
64. Jérôme F, Billier B, Barbe JM, Espinosa E, Dahaoui S, Lecomte C, Guilard R (2000) Evidence for the formation of a RuIII-RuIII bond in a ruthenium corrole homodimer. Angew Chem Int Ed 39:4051–4053
65. Simkhovich L, Luobeznova I, Goldberg I, Gross Z (2003) Mono- and binuclear ruthenium corroles: synthesis, spectroscopy, electrochemistry, and structural characterization. Chem Eur J 9:201–208
66. Abeysekera AM, Grigg R, Trocha-Grimshaw J, Viswanatha V (1976) Structure and reactivity of Rh(I) complexes of polypyrrole macrocycles and their *N*-alkylated derivatives. Tet Lett 36:3189–3192
67. Boschi T, Liccocia S, Paolesse R, Tagliatesta P (1988) Synthetic routes to rhodium(III) corrolates. Inorg Chim Acta 141:169–171
68. Brückner C, Barta CA, Briñas RP, Krause-Bauer JA (2003) Synthesis and structure of [*meso*-triarylcorrolato]silver(III). Inorg Chem 42:1673–1680
69. Tse MK, Zhang Z, Mak TCW, Chan KS (1998) Synthesis of an oxorhenium(V) corrolate from porphyrin with detrifluoromethylation and ring contraction. Chem Commun 11:1199–1200
70. Alemayehu AB, Ghosh A (2011) Gold corroles. J Porphyr Phthalocyanines 15:106–110
71. Palmer JH, Day MW, Wilson AD, Henling LM, Gross Z, Gray HB (2008) Iridium corroles. J Am Chem Soc 130:7786–7787
72. Buchler JW (1975) In: Smith KM (ed) Porphyrins and metalloporphyrins. Elsevier, Amsterdam, Chapter 5
73. Buchler JW (1978) In: Dolphin D (ed) The porphyrins. Academic Press, New York, Vol. IA, Chapter 10
74. Meier-Callahan AE, Gray HB, Gross Z (2000) Stabilization of high-valent metals by corroles: oxo[tris(pentafluorophenyl)corrolato]chromium(V). Inorg Chem 39:3605–3607
75. Gross Z, Golubkov G, Simkhovich L (2000) Epoxidation catalysis by a manganese corrole and isolation of an oxomanganese(V) corrole. Angew Chem Int Ed 39:4045–4047
76. Gross Z (2001) High-valent metallocorroles. J Biol Inorg Chem 6:733–738
77. Gross Z, Gray HB (2004) Oxidations catalyzed by metallocorroles. Adv Synth Catal 346:165–170
78. Meier-Callahan AE, di Bilio AJ, Simkhovich L, Mahammed A, Goldberg I, Gross Z, Gray HB (2001) Chromium corroles in four oxidation states. Inorg Chem 40:6788–6793
79. Golubkov G, Gross Z (2003) Chromium(V) and chromium (VI) nitrido complexes of tris (pentafluorophenyl)corrole. Angew Chem Int Ed 42:4507–4510

80. Du Bois J, Tomooka CS, Hong J, Carreira EM (1997) Nitridomanganese(V) complexes: design, preparation, and use as nitrogen atom-transfer reagents. Acc Chem Res 30:364–372
81. Edwards NY, Eikey RA, Loring MI, Abu-Omar MM (2005) High-valent imido complexes of manganese and chromium corroles. Inorg Chem 44:3700–3708
82. Meyer K, Bendix J, Bill E, Weyhermüller T, Wieghardt K (1998) Molecular and electronic structure of nitridochromium(V) complexes with macrocyclic amine ligands. Inorg Chem 37:5180–5188
83. Che CM, Ma JX, Wong WT, Lai TF, Poon CK (1988) Synthesis and molecular structure of a nitridochromium(V) complex stabilized with dianionic organic amide ligand. Inorg Chem 27:2547–2548
84. Egorova OA, Tsay OG, Khatua S, Huh JO, Churchill DG (2009) A chiral *meso*-ABC-corrolatochromium(V) complex. Inorg Chem 48:4634–4636
85. Guilard R, Gryko DT, Canard G, Barbe JM, Koszarna B, Brandès S, Tasior M (2002) Synthesis of corroles bearing up to three different *meso* substituents. Org Lett 4:4491–4494
86. Mahammed A, Gray HB, Meier-Callahan AE, Gross Z (2003) Aerobic oxidations catalyzed by chromium corroles. J Am Chem Soc 125:1162–1163
87. Czernuszewicz R, Mody V, Czader A, Gałęzowski M, Gryko DT (2009) Why the chromyl bond is stronger than the perchromyl bond in high-valent oxochromium(IV, V) complexes of tris(pentafluorophenyl)corrole. J Am Chem Soc 131:14214–14215
88. Czernuszewicz RS, Su YO, Stern MK, Macor KA, Kim D, Groves JT, Spiro TG (1988) J Am Chem Soc 110:4158–4165
89. Kryztek J, Tesler J, Hoffman BM, Brunel LC, Liccocia S (2001) High frequency and field EPR investigation of (8,12-diethyl-2,3,7,13,17,18-hexamethylcorrolato)manganese(III). J Am Chem Soc 123:7890–7897
90. Fryxelius J, Eilers G, Feyziyev Y, Magnuson A, Sun L, Lomoth R (2005) Synthesis and redox properties of a [(*meso*-tris(4-nitrophenyl)corrolato]Mn(III) complex. J Porphyr Phthalocyanines 9:379–386
91. Ou Z, Erben C, Autret M, Will S, Rosen D, Lex J, Vogel E, Kadish KM (2005) Manganese (III) and manganese(IV) corroles: synthesis, spectroscopic, electrochemical, and X-ray structural characterization. J Porphyr Phthalocyanines 9:398–412
92. Liu HY, Lai TS, Yeung LL, Chang CK (2003) First synthesis of perfluorinated corrole and its Mn-O complex. Org Lett 5:617–620
93. Liu HY, Yam F, Xie YT, Li XY, Chang CK (2009) A bulky bis-pocket manganese(V)-oxo corrole complex: observation of oxygen atom transfer between triply bonded MnVtO and alkene. J Am Chem Soc 131:12890–12891
94. Kumar A, Goldberg I, Botoshansky M, Buchman Y, Gross Z (2010) Oxygen atom transfer reactions from isolated (oxo)manganese(V) corroles to sulfides. J Am Chem Soc 132:15233–15245
95. Eikey RA, Khan SI, Abu-Omar MM (2002) The elusive terminal imido of manganese(V). Angew Chem Int Ed 41:3592–3595
96. Golubkov G, Gross Z (2005) Nitrogen atom transfer between manganese complexes of salen, porphyrin, and corrole and characterization of a (nitrido)manganese(VI) corrole. J Am Chem Soc 127:3258–3259
97. Saltsman I, Goldberg I, Gross Z (2010) Water-soluble manganese(III) corroles and corresponding (nitrido)manganese(V) complexes. J Porphyr Phthalocyanines 14:615–620
98. Bröring M, Hell C, Brandt CD (2007) Iodomanganesecorrole: a stable $Mn^{IV}$-I species. Chem Commun 1861–1862
99. Abu-Omar MM (2011) High-valent iron and manganese complexes of corrole and porphyrin in atom transfer and dioxygen evolving catalysis. Dalton Trans 40:3435–3444
100. Licoccia S, Morgante E, Paolesse R, Polizio F, Senge MO, Tondello E, Boschi T (1997) Tetracoordinated manganese(III) alkylcorrolates. spectroscopic studies and the crystal and molecular structure of (7,13-Dimethyl-2,3,8,12,17,18-hexaethylcorrolato)manganese(III). Inorg Chem 36:1564–1570

101. Scheidt WR (1978) Porphyrin stereochemistry. In: Dolphin D (ed) The porphyrins, Vol 3 (Part A), pages 463–511 – published by Academic Press (New York, NY)
102. Shen J, El Ojaimi M, Chkounda M, Gros CP, Barbe JM, Shao J, Guilard R, Kadish KM (2008) Solvent, anion, and structural effects on the redox potentials and UV-visible spectral properties of mononuclear manganese corroles. Inorg Chem 47:7717–7727
103. Vogel E, Will S, Schulze A, Neumann L, Lex J, Bill E, Trautwein AX, Wieghardt K (1994) Metallocorroles with formally tetravalent iron. Angew Chem Int Ed Engl 33:731–735
104. Bill E, Ding XQ, Bominaar EL, Trautwein AX, Winkler H, Mandon D, Weiss R, Gold A, Jayaraj K, Hatfield WE, Kirk ML (1990) Evidence for variable metal-radical spin coupling in oxoferrylporphyrin cation radical complexes. Eur J Biochem 188:665–672
105. Cai S, Walker FA, Licoccia S (2000) NMR and EPR investigations of iron corrolates: iron (III) corrolate $\pi$ cation radicals or iron(IV) corrolates? Inorg Chem 39:3466–3478
106. Steene E, Wondimagegn T, Ghosh A (2001) Electrochemical and electronic absorption spectroscopic studies of substituent effects in iron(IV) and manganese(IV) corroles: do the compounds feature high-valent metal centers or noninnocent corrole ligands? Implications for peroxidase compound I and II intermediates. J Phys Chem B 105:11406–11413
107. Zakharieva O, Schünemann V, Gerdan M, Licoccia S, Cai S, Walker FA, Trautwein AX (2002) Is the corrolate macrocycle innocent or noninnocent? Magnetic susceptibility, mössbauer, $^{1}$H NMR, and DFT investigations of chloro- and phenyliron corrolates. J Am Chem Soc 124:6636–6648
108. Simkhovich L, Goldberg I, Gross Z (2002) Iron(III) and iron(IV) corroles: synthesis, spectroscopy, structures, and no indications for corrole radicals. Inorg Chem 41:5433–5439
109. Simkhovich L, Gross Z (2004) Halogeno-coordinated iron corroles. Inorg Chem 43: 6136–6138
110. Yatsunyk L, Walker FA (2002) 19 F isotropic shifts in paramagnetic iron(III) octaethyltetraphenylporphyrinate and tetraphenylporphyrinato complexes of a variety of electronic ground states: implications for the electron configuration of chloroiron tri-(pentafluorophenyl)corrolate. Inorg Chim Acta 337:266–274
111. Cai S, Licoccia S, D'Ottavi C, Paolesse R, Nardis S, Bulach V, Bertrand Z, Shokhivera TK, Walker FA (2002) Chloroiron *meso*-triphenylcorrolates: electronic ground state and spin delocalization. Inorg Chim Acta 339:171–178
112. Walker FA, Licoccia S, Paolesse R (2006) Iron corrolates: unambiguous chloroiron(III) (corrolate)$^{2-}$ $\pi$-cation radicals. J Inorg Biochem 100:810–837
113. Ye S, Tuttle T, Bill E, Simkhovich L, Gross Z, Thiel W, Neese F (2008) The electronic structure of iron corroles: a combined experimental and quantum chemical study. Chem Eur J 14:10839–10851
114. Roos BO, Verzazov V, Conradie J, Taylor PR, Ghosh A (2008) Not innocent: verdict from Ab initio multiconfigurational second-order perturbation theory on the electronic structure of chloroiron corrole. J Phys Chem B 112:14099–14102
115. Hocking RK, George SD, Gross Z, Walker FA, Hodgson KO, Hedman B, Solomon EI (2009) Fe L- and K-edge XAS of low-spin ferric corrole: bonding and reactivity relative to low-spin ferric porphyrin. Inorg Chem 48:1678–1688
116. Conlon M, Johnson AW, Overend WR, Rajapaksa D, Elson CM (1973) Structures and reactions of cobalt corroles. J Chem Soc Perkin Trans 2281–2289
117. Murakami Y, Aoyama Y, Hayashida M (1980) Hydroxide-promoted reduction of the corrole complexes of cobalt(III) and iron(III) in the presence of olefin. J Chem Soc Chem Comm 501–502
118. Hush NS, Woolsey IS (1974) Optical and electron spin resonance spectra of cobalt complexes related to vitamin $B_{12}$. J Chem Soc Dalton Trans 24–34
119. Murakami Y, Yamada S, Matsuda Y, Sakata K (1978) Transition-metal complexes of pyrrole pigments XV: coordination of pyridine bases to the axial sites of cobalt corroles. Bull Chem Soc Jpn 51:123–129

120. Murakami Y, Yamada S, Matsuda Y, Sakata K, Yamada S, Tanaka Y, Aoyama Y (1981) Transition-metal complexes of pyrrole pigments XVII: preparation and spectroscopic properties of corrole complexes. Bull Chem Soc Jpn 54:163–169
121. Hitchcock PB, McLaughlin GM (1976) Five-co-ordinate [sic] cobalt(III): crystal and molecular structure of corrole-(triphenylphosphine)cobalt(III). J Chem Soc Dalton Trans 1927–1930
122. Brückner S, Calligaris M, Nardin G, Randaccio L (1969) The crystal and molecular structure of N, N'-ethylene-bis(acetylacetoneiminato)methylcobalt(III). Inorg Chim Acta 3:308–312
123. Ibers JA, Lauher JW, Little RG (1974) Stereochemistry of cooperativity effects in the Prosthetic Group of Coboglobin. Acta Cryst B30:268–272
124. Kadish KM, Koh W, Tagliatesta P, Sazou D, Paolesse R, Licoccia S, Boschi T (1992) Electrochemistry of rhodium and cobalt corroles: characterization of (OMC)Rh($PPh_3$) and (OMC)Co($PPh_3$) where OMC is the trianion of 2,3,7,8,12,13,17,18-octamethylcorrole. Inorg Chem 31:2305–2313
125. Paolesse R, Liccocia S, Fanciullo M, Morgante E, Boschi T (1993) Synthesis and characterization of cobalt(III) complexes of *meso*-phenyl-substituted corroles. Inorg Chim Acta 203:107–114
126. Adamian VA, D'Souza F, Licoccia S, Di Vona ML, Tassoni E, Paolesse R, Boschi T, Kadish KM (1995) Synthesis, characterization, and electrochemical behavior of (5,10,15-tri-X-phenyl-2,3,7,8,12,13,17,18-octamethylcorrolato)cobalt(III) triphenylphosphine complexes, where X = *p*-OCH3, *p*-CH3, *p*-Cl, *m*-Cl, *o*-Cl, *m*-F, or *o*-F. Inorg Chem 34:532–540
127. Liccocia S, Tassoni E, Paolesse R, Boschi T (1995) The effect of steric hindrance in the synthesis of corrolates via the cobalt catalyzed cyclization of 2-(α-hydroxyalkyl)pyrroles. Inorg Chim Acta 235:15–20
128. Will S, Lex J, Vogel E, Adamian VA, Caemelbecke EV, Kadish KM (1996) Synthesis, characterization, and electrochemistry of σ-bonded cobalt corroles in high oxidation state. Inorg Chem 35:5577–5583
129. Harmer J, Doorslaer SV, Gromov I, Bröring M, Jeschke G, Schweiger A (2002) A pulse EPR and ENDOR investigation of the electronic structure of a σ-carbon-bonded cobalt(IV) corrole. J Phys Chem B 106:2801–2811
130. Simkhovich L, Galili N, Saltsman I, Goldberg I, Gross Z (2000) Coordination chemistry of the novel 5,10,15-tris(pentafluorophenyl)corrole: synthesis, spectroscopy, and structural characterization of its cobalt(III), rhodium(III), and iron(IV) complexes. Inorg Chem 39:2704–2705
131. Mahammed A, Giladi I, Goldberg I, Gross Z (2001) Synthesis and structural characterization of a novel covalently-bound corrole dimer. Chem Eur J 7:4259–4265
132. Mahammed A, Goldberg I, Gross Z (2001) Highly selective chlorosulfonation of tris (pentafluorophenyl)corrole as a synthetic tool for the preparation of amphiphilic corroles and metal complexes of planar chirality. Org Lett 3:3443–3446
133. Grodkowski J, Neta P, Fujita E, Mahammed A, Simkhovich L, Gross Z (2002) Reduction of cobalt and iron corroles and catalyzed reduction of $CO_2$. J Phys Chem A 106:4772–4778
134. Barbe JM, Canard G, Brandes S, Francois J, Dubois G, Guilard R (2004) Metallocorroles as sensing components for gas sensors: remarkable affinity and selectivity of cobalt(III) corroles for CO vs. $O_2$ and $N_2$. Dalton Trans 8:1208–1214
135. Barbe JM, Canard G, Brandes S, Guilard R (2007) Selective chemisorption of carbon monoxide by organic-inorganic hybrid materials incorporating cobalt(III) corroles as sensing components. Chem Eur J 13:2118–2129
136. Chattopadhyay P, Matsuo T, Tsuji T, Ohbayashi J, Hayashi T (2011) Thermal isomerization of *N*-bridged cobalt corrole complexes through a transiently formed axial carbenoid. Organometallics 30:1869–1873
137. Jérôme F, Gros CP, Tardieux C, Barbe JM, Guilard R (1998) Synthesis of a 'face-to-face' porphyrin-corrole: a potential precursor of a catalyst for the four-electron reduction of dioxygen. New J Chem 22:1327–1329

138. Jérôme F, Gros CP, Tardieux C, Barbe JM, Guilard R (1998) First synthesis of sterically hindered cofacial bis(corroles) and their bis(cobalt) complexes. Chem Commun 18:2007–2008
139. Guilard R, Jérôme F, Barbe JM, Gros CP, Ou Z, Shao J, Fischer J, Weiss R, Kadish KM (2001) Alkyl and aryl substituted corroles 2: synthesis and characterization of linked "face-to-face" biscorroles: X-ray structure of (BCA)Co2(py)3, where BCA represents a biscorrole with an anthracenyl bridge. Inorg Chem 40:4856–4865
140. Guilard R, Francois J, Gros CP, Barbe JM, Ou Z, Shao J, Kadish KM (2001) Synthesis of an anthracenyl bridged porphyrin–corrole bismacrocycle: physicochemical and electrochemical characterisation [sic] of the biscobalt μ-superoxo derivative. C R Acad Sci Paris Chimie 4:245–254
141. Kadish KM, Ou Z, Shao J, Gros CP, Barbe JM, Jérôme F, Bolze F, Burdet F, Guilard R (2002) Alkyl and aryl substituted corroles 3: reactions of cofacial cobalt biscorroles and porphyrin-corroles with pyridine and carbon monoxide. Inorg Chem 41:3990–4005
142. Barbe JM, Burdet F, Espinosa E, Gros CP, Guilard R (2003) New insights into the synthesis of porphyrin-corrole and biscorrole systems. J Porphyr Phthalocyanines 7:365–374
143. Guilard R, Barbe JM, Stern C, Kadish KM (2003) New developments in corrole chemistry: special emphasis on face-to-face bismacrocycles. In: Kadish KM, Smith KM, Guilard R (eds) The porphyrin handbook, Academic Press – San Diego, CA and London, UK
144. Guilard R, Burdet F, Barbe JM, Gros CP, Espinosa E, Shao J, Ou Z, Zhan R, Kadish KM (2005) Heterobimetallic complexes of cobalt(IV) porphyrin-corrole dyads: synthesis, physicochemical properties, and X-ray structural characterization. Inorg Chem 44:3972–3983
145. Kadish KM, Frémond L, Ou Z, Shao J, Shi C, Anson FC, Burdet F, Gros CP, Barbe JM, Guilard R (2005) Cobalt(III) corroles as electrocatalysts for the reduction of dioxygen: reactivity of a monocorrole, biscorroles, and porphyrin—corrole dyads. J Am Chem Soc 127:5625–5631
146. Kadish KM, Frémond L, Burdet F, Barbe JM, Gros CP, Guilard R (2006) Cobalt(IV) corroles as catalysts for the electroreduction of $O_2$: reactions of heterobimetallic dyads containing a face-to-face linked Mn(III) or Fe(III) porphyrin. J Inorg Biochem 100:858–868
147. Kadish KM, Shen J, Frémond L, Chen P, El Ojaimi M, Chkounda M, Gros CP, Barbe JM, Ohkubo K, Fukuzumi S, Guilard R (2008) Clarification of the oxidation state of cobalt corroles in heterogeneous and homogeneous catalytic reduction of dioxygen. Inorg Chem 47:6726–6737
148. Rosenthal J, Nocera DG (2007) Oxygen activation chemistry of pacman and hangman porphyrin architectures based on xanthene and dibenzofuran spacers. Prog Inorg Chem 55:483–544
149. Rosenthal J, Nocera DG (2007) Role of proton-coupled electron transfer in O-O bond activation. Acc Chem Res 40:543–553
150. Lee CH, Dogutan DK, Nocera DG (2011) Hydrogen generation by hangman metalloporphyrins. J Am Chem Soc 133:8775–8777
151. Schwalbe M, Dogutan DK, Stoian SA, Teets TS, Nocera DG (2011) Xanthene-modified and hangman iron corroles. Inorg Chem 50:1368–1377
152. Rosenthal J, Chng LL, Fried SD, Nocera DG (2007) Stereochemical control of $H_2O_2$ dismutation by hangman porphyrins. Chem Commun 25:2642–2644
153. Dogutan DK, Stoian SA, McGuire R Jr, Schwalbe M, Teets TS, Nocera DG (2011) Hangman corroles: efficient synthesis and oxygen reaction chemistry. J Am Chem Soc 133:131–140
154. Lindsey JS, Wagner RW (1989) Investigation of the synthesis of *ortho*-substituted tetraphenylporphyrins. J Org Chem 54:828–836
155. Dogutan DK, McGuire R Jr, Nocera DG (2011) Electrocatalytic water oxidation by cobalt (III) hangman β-octafluorocorroles. J Am Chem Soc. doi:10.1021/ja202138m
156. Grigg R, King TJ, Shelton G (1970) The structure of copper 2,3,7,13,17,18,21-heptamethylcorrole. J Chem Soc D Chem Commun 1:56

157. Broadhurst MJ, Grigg R, Shelton G, Woodworth AJ (1970) *N*-alkylation of porphins and related macrocycles. J Chem Soc D Chem Commun 4:231–233
158. Johnson AW, Grigg R, Shelton G (1971) Structures and thermal rearrangements of alkylated palladium and copper corroles. J Chem Soc C Organic 12:2287–2294
159. Murakami Y, Matsuda Y, Sakata K, Sunao Y, Tanaka Y, Aoyama Y (1981) Transition-metal complexes of pyrrole pigments, XVII: preparation and spectroscopic properties of corrole complexes. Bull Chem Soc Jpn 54:163–169
160. Will S, Lex J, Vogel E, Schmickler H, Gisselbrecht JP, Haubtmann C, Bernard M, Gross M (1997) Nickel and copper corroles: well-known complexes in a new light. Angew Chem Int Ed Engl 36:357–361
161. Brückner C, Briñas RP, Krause-Bauer JA (2003) X-ray structure and variable temperature NMR spectra of [*meso*-triarylcorrolato]copper(III). Inorg Chem 42:4495–4497
162. Ou Z, Shao J, Zhao H, Ohkubo K, Wasbotten IH, Fukuzumi S, Ghosh A, Kadish KM (2004) Spectroelectrochemical and ESR studies of highly substituted copper corroles. J Porphyr Phthalocyanines 8:1236–1247
163. Wasbotten IH, Wondimagegn T, Ghosh A (2002) Electronic absorption, resonance raman, and electrochemical studies of planar and saddled copper(III) *meso*-triarylcorroles: highly substituent-sensitive soret bands as a distinctive feature of high-valent transition-metal corroles. J Am Chem Soc 124:8104–8116
164. Luobeznova I, Simkhovich L, Goldberg I, Gross Z (2004) Electronic structures and reactivities of corrole-copper complexes. Eur J Inorg Chem 1724–1732
165. Bhattacharya D, Singh P, Sarkar S (2010) Synthesis, X-structure [sic] and solvent induced electronic states tuning of *meso*-tris(4-nitrophenyl)corrolato-copper complex. Inorg Chim Acta 363:4313–4318
166. Bröring M, Brégier F, Tejero EC, Hell C, Holthausen MC (2007) Revisiting the electronic ground state of copper corroles. Angew Chem Int Ed 46:445–448
167. Pierloot K, Zhao H, Vancoillie S (2010) Copper corroles: the question of non-innocence. Inorg Chem 49:10316–10329
168. Alemayehu A, Gonzalez E, Hansen LK, Ghosh A (2009) Copper corroles are inherently saddled. Inorg Chem 48:7794–7799
169. Steene E, Dey A, Ghosh A (2003) β-Octafluorocorroles. J Am Chem Soc 125:16300–16309
170. Thomas KE, Wasbotten IH, Ghosh A (2008) Copper β-octakis(trifluoromethyl)corroles: new paradigms for ligand substituent effects in transition metal complexes. Inorg Chem 47:10469–10478
171. Alemayehu A, Hansen LK, Ghosh A (2010) Nonplanar, noninnocent, and chiral: a strongly saddled metallocorrole. Inorg Chem 49:7608–7610
172. Sashuk V, Koszarna B, Winiarek P, Gryko DT, Grela K (2004) The simple synthesis of stable $A_3$- and *trans*-$A_2B$-molybdenum(V) corrolates. Inorg Chem Commun 7:871–875
173. Luobeznova I, Raizman M, Goldberg I, Gross Z (2006) Synthesis and full characterization of molybdenum and antimony corroles and utilization of the latter complexes as very efficient catalysts for highly selective aerobic oxygenation reactions. Inorg Chem 45:386–394
174. Czernuszewicz RS, Mody V, Zareba AA, Zaczek MB, Gałęzowski M, Sashuk V, Grela K, Gryko DT (2007) Solvent-dependent resonance raman spectra of high-valent oxomolybdenum (V) tris[3,5-bis(trifluoromethyl)phenyl]corrolate. Inorg Chem 46:5616–5624
175. Swain CG, Swain MS, Powell AL, Alunni S (1983) Solvent effects on chemical reactivity: evaluation of anion and cation solvation components. J Am Chem Soc 105:502–513
176. Gutmann V (1978) The donor—acceptor approach to molecular interactions. Plenum, New York
177. Su YO, Czernuczewicz RS, Miller LA, Spiro TG (1988) Effects of solvents, axial ligation, and radical cation formation on the V:O stretching raman frequency in vanadyl porphyrins: implications for peroxidase intermediates. J Am Chem Soc 110:4150–4157
178. Mody VV, Frtzpatrick MB, Zabaneh SS, Czernuszewicz RS, Gałęzowski M, Gryko DT (2009) Solvent effects on the electronic and vibrational properties of high-valent

oxomolybdenum(V) 5,10,15-triphenylcorrole probed by UV-visible and resonance raman spectroscopy. J Porphyr Phthalocyanines 13:1040–1052
179. Sinkhovich L, Luobeznova I, Goldberg I, Gross Z (2003) Mono- and binuclear ruthenium corroles: synthesis, spectroscopy, electrochemistry, and structural characterization. Chem Eur J 9:201–208
180. Collman JP, Arnold HJ (1993) Multiple metal-metal bonds in 4d and 5d metal-porphyrin dimers. Acc Chem Res 26:586–592
181. Collman JP, Barnes CE, Swepston PN, Ibers JA (1984) Synthesis, proton NMR spectroscopy, and structural characterization of binuclear ruthenium porphyrins. J Am Chem Soc 106:3500–3510
182. Collman JP, Hartford ST (1998) Structural and spectroscopic characterization of a metal-metal bonded ruthenium porphyrin dimer cation. Inorg Chem 37:4152–4153
183. Kadish KM, Burdet F, Jérôme F, Barbe JM, Ou Z, Shao J, Guilard R (2002) Synthesis, physicochemical and electrochemical properties of metal-metal bonded ruthenium corrole homodimers. J Organomet Chem 652:69–76
184. Hopf FR, O'Brien TP, Scheidt WR, Whitten DG (1975) Structure and Reactivity of Ruthenium(II) Porphyrin Complexes: Photochemical Ligand Ejection and Formation of Ruthenium Porphyrin Dimers. J Am Chem Soc 97:277–281
185. Boschi T, Licoccia S, Paolesse R, Tagliatesta P, Azarnia-Tehran M, Pelizzi G, Vitali F (1990) Synthesis and characterization of novel metal(III) complexes of corrole: crystal and molecular structure of (2,3,7,8,12,13,17,18-octamethylcorrolato)(triphenylarsine)rhodium(III). J Chem Soc Dalton Trans 2:463–468
186. Simkhovich L, Mahammed A, Goldberg I, Gross Z (2001) Synthesis and characterization of germanium, tin, phosphorous, iron and rhodium complexes of tris(pentafluorophenyl)corrole, and the utilization of the iron and rhodium corroles as cyclopropanation catalysts. Chem Eur J 7:1041–1055
187. Collman JP, Wang HJH, Decreau RA, Eberspacher TA, Sunderland CJ (2005) Synthesis and characterization of $Rh^{III}$ corroles: unusual reactivity patterns observed during metalation reactions. Chem Commun 2497–2499
188. Simkhovich L, Iyer P, Goldberg I, Gross Z (2002) Structure and chemistry of *N*-substituted corroles and their rhodium(I) and zinc(II) metal-ion complexes. Chem Eur J 8:2595–2601
189. Saltsman I, Simkhovich L, Balasz Y, Goldberg I, Gross Z (2004) Synthesis, spectroscopy, and structures of new rhodium(I) and rhodium(III) corroles and catalysis thereby. Inorg Chim Acta 357:3038–3046
190. Saltsman I, Balasz Y, Goldberg I, Gross Z (2006) Synthesis, spectroscopy, and structures of chiral rhodium(I) corrole complexes. J Mol Catal A 251:263–269
191. Rozenshtein V, Wagnert L, Berg A, Stavitski E, Berthold T, Kothe G, Saltsman I, Gross Z, Levanon H (2008) Probing the photoexcited states of rhodium corroles by time-resolved Q-Band EPR: observation of strong spin-orbit coupling effects. J Phys Chem A 112:5338–5343
192. Wagnert L, Berg A, Saltsman I, Gross Z, Rozenshtein V (2010) Time-resolved paramagnetic resonance study of rhodium(III) corrole excited states. J Phys Chem A 114:2059–2072
193. Wagnert L, Berg A, Stavitski E, Berthold T, Kothe G, Goldberg I, Mahammed A, Simkhovich L, Gross Z, Levanon H (2006) Exploring the photoexcited triplet states of aluminum and tin corroles by time-resolved Q-band EPR. Appl Magn Reson 30:591–604
194. Wagnert L, Rubin R, Berg A, Mahammed A, Gross Z, Levanon H (2010) Photoexcited triplet state properties of brominated and non-brominated Ga(III)-corroles as studied by time-resolved electron paramagnetic resonance. J Phys Chem B 114:14303–14308
195. Kadish KM, Lin XQ, Ding JQ, Araullo C (1986) A reinvestigation of silver porphyrin electrochemistry: reactions of Ag(III), Ag(II), and Ag(I). Inorg Chem 25:3236–3242
196. Brückner C (2004) The silver complexes of porphyrins, corroles, and carbaporphyrins: silver in the oxidation states II and III. J Chem Educ 81:1665–1669

197. Stefanelli M, Shen J, Zhu W, Mastroianni M, Mandoj F, Nardis S, Ou Z, Kadish KM, Fronczek FR, Smith KM, Paolesse R (2009) Demetalation of silver(III) corrolates. Inorg Chem 48:6879–6887
198. Capar C, Thomas KE, Ghosh A (2008) Reductive demetalation of copper corroles: first simple route to free-base β-octabromocorroles. J Porphyr Phthalocyanines 12:964–967
199. Capar C, Hansen LK, Conradie J, Ghosh A (2010) β-octabromo-*meso*-tris(pentafluorophenyl)corrole: reductive demetalation-based synthesis of a heretofore inaccessible, perhalogenated free-base corrole. J Porphyr Phthalocyanines 14:509–512
200. Palmer JH, Mahammed A, Lancaster KM, Gross Z, Gray HB (2009) Structures and reactivity patterns of group 9 metallocorroles. Inorg Chem 48:9308–9315
201. Dong SS, Nielsen RJ, Palmer JH, Gray HB, Gross Z, Dasgupta S, Goddard WA III (2011) Electronic structures of group 9 metallocorroles with axial ammines. Inorg Chem 50:764–770
202. Palmer JH, Durrell AC, Gross Z, Winkler JR, Gray HB (2010) Near-IR phosphorescence of iridium(III) corroles at ambient temperature. J Am Chem Soc 132:9230–9231
203. Palmer JH, Brock-Nannestad T, Mahammed A, Durrell AC, VanderVelde D, Virgil S, Gross Z, Gray HB (2011) Nitrogen insertion into a corrole ring: iridium monoazaporphyrins. Angew Chem Int Ed (Accepted Article)
204. Koren K, Borisov S, Saf R, Klimant I (2011) Strongly Phosphorescent iridium(III) porphyrins: new oxygen indicators with tuneable [sic] photophysical properties and functionalities. Eur J Inorg Chem 1531–1534

Struct Bond (2012) 142: 91–118
DOI: 10.1007/430_2011_59

Published online: 9 November 2011

# Chemical Sensing with Semiconducting Metal Phthalocyanines

**William C. Trogler**

**Abstract** Thin film conductivity sensors that use metal phthalocyanine semiconductors offer properties distinct from those of organic thin film transistor sensors.

**Keywords** Conductivity · OTFT · Phthalocyanine · Semiconductor · Sensor · Sensor array

## Contents



W.C. Trogler (✉)
Department of Chemistry and Biochemistry, University of California San Diego, La Jolla, CA 92093-0358, USA
e-mail: wtrogler@ucsd.edu

## Abbreviations

| | |
|---|---|
| AFM | Atomic force microscopy |
| DIMP | Diisopropyl methylphosphonate |
| DMMP | Dimethyl methylphosphonate |
| FET | Field effect transistor |
| LB | Langmuir–Blodgett |
| M | Metal |
| MEMS | Microelectromechanical systems |
| OTFT | Organic thin film transistor |
| OTMS | Octadecyltrimethoxysilane |
| Pc | Phthalocyanine |
| Por | Porphyrin |
| ppb | Part per billion |
| ppm | Part per million |
| ppth | Part per thousand |
| ROCs | Receiver operating characteristics |
| SAMs | Self-assembled monolayers |
| SAW | Surface acoustic wave |
| SCLC | Space charge limited conductivity |

## 1 Introduction

It is a pleasure to dedicate this review to honor the scientific contributions of Professor Carl Ballhausen, a pioneer in applying theoretical methods to the electronic structures and physical properties of transition metal complexes. His early interest in understanding the spectroscopic properties of metal complexes led [1, 2] to a classic text that remains the most succinctly rigorous introduction to ligand field theory [3]. Professor Ballhausen was among the first to adopt emerging molecular orbital methods and apply them to characterize the electronic structures of inorganic and organometallic compounds of broad interest to chemists [2, 4–6]. This philosophy of applying quantitative theories of electronic structure to describe the chemical and physical properties of transition metal compounds drives much current research in the fields of inorganic and materials chemistry.

Phthalocyanines (Pcs) are planar ligands that are closely related to porphyrins (Pors), as both can be deprotonated to form dianions that yield a wide variety of square planar metal(II) chelate complexes MPc and MPor. Porphyrins are crucial redox active metalloenzymes that play key roles in metabolism, oxygen storage, photosynthesis, and other biological processes. Phthalocyanines are important dye and paint pigments, as well as having electronic device applications in CD-recordable media, xerography, photovoltaics, and organic light emitting devices (OLEDs) [7–22]. Initial interest among physical chemists centered on the intense electronic absorptions associated with the ligand $\pi$ system in phthalocyanines

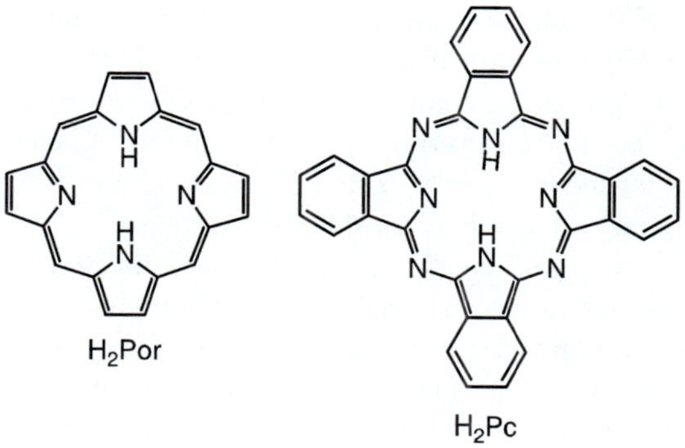


**Fig. 1** Chemical structures of the parent porphyrin ($H_2Por$) and phthalocyanine ($H_2Pc$) molecules

[23, 24], which adds a dominant contribution from low-lying empty ligand centered orbitals to the electronic structure characteristic of square planar metal complexes [4]. This was found to impart novel electronic materials properties, such as photoconductivity, to phthalocyanines [23, 25, 26]. Porphyrins and phthalocyanines have closely related electronic structures, as the phthalocyanine ligand can be viewed structurally as a tetrabenzoporphyrin where four methine groups have been replaced by nitrogen atoms (Fig. 1). One attractive feature of phthalocyanines is they are more easily synthesized than porphyrins, and exhibit high chemical and thermal stability [20, 27]. These properties make phthalocyanines attractive candidates as materials for chemical sensors.

Compact chemical sensors can be broadly classified as being based on electronic or optical readout mechanisms [28]. The electronic sensor types would include resistive, capacitive, surface acoustic wave (SAW), electrochemical, and mass (e.g., quartz crystal microbalance (QCM) and microelectromechanical systems (MEMSs)). Chemical specificity of most sensors relies critically on the materials designed either as part of the sensor readout itself (e.g., semiconducting metal oxides, nanoparticle films, or polymers in resistive sensors) or on a chemically sensitive coating (e.g., polymers used in MEMS, QCM, and SAW sensors). This review will focus on the mechanism of sensing in conductivity based chemical sensors that contain a semiconducting thin film of a phthalocyanine or metal phthalocyanine sensing layer.

Metal–organic materials offer several potential advantages over inorganic oxide semiconducting sensors. Fabrication of thin films generally occurs at much lower temperatures. The possibility of solution processing (e.g., spin coating or ink jet printing) offers inexpensive fabrication methods, which are well-suited to the fabrication of sensing films. Synthetic modifications are easier, more predictable, and offer the possibility of precise tailoring of electronic properties. Molecular recognition in the binding event can also be designed into the sensor film using specific binding interactions (e.g., hydrogen bonding, metal–ligand interactions, hydrophobic interactions, dipolar interactions, and electrostatic interactions), as well as by molecular imprinting as in shape selective polymer films [29–33].

One obstacle with application of these materials is the tendency of organic compounds to slowly oxidize in air, since they are thermodynamically unstable in the presence of atmospheric oxygen. Their oxidation proceeds by free radical chain processes initiated by trace atmospheric oxidants, such as ozone, hydrogen peroxide, and nitrogen dioxide [34]. Organic sensor films are generally more sensitive to aging effects than inorganic sensor films.

Sensors that convert a chemospecific sensing event into a voltage or current output are broadly classified as electronic sensors. Those that rely on conversion of an optical sensor signal into an electronic signal will not be considered here. Electronic sensors are valued for their low power requirements, sensitivity, and potential for mass fabrication in a compact form. The technology for measuring voltage and current is highly sensitive and of low cost, which makes such sensors attractive as compact inexpensive sensors. However, issues such as chemical sensitivity, selectivity, and stability pose challenging practical barriers to their application. One class of electronic sensors relies on capacitance changes for the sensing modality. These types of sensors will not be considered here. Their sensor response depends strongly on capacitance changes related to the bulk dielectric, surface conductivity, and swelling properties of the sensor film and adsorbed analyte [35].

## 2 Sensor Device Configuration

Many thin film electronic sensors based on metal phthalocyanines rely on a change in conductivity of a thin semiconducting film when a vapor phase analyte adsorbs on the surface or into the film porosity. There usually is a balance between sensitivity, selectivity, and the kinetics of reversibility in such sensors. While resistance may be measured directly for a material deposited between two sets of interdigitated electrodes (IDEs) [36], it is also possible to use an organic thin film transistor (OTFT) design, as shown in Fig. 2 [37–50]. An advantage of the OTFT device is that higher currents are obtained with thinner films (as thin as 5 nm) [51]. The use of ultrathin film OTFTs improves the stability, speed, and reversibility of the sensing process as equilibrium is more rapidly obtained between the film and vapor phase species, yet carrier mobilities remain reasonably high [42–45, 51–54]. Typical mobilities range from $2.6 \times 10^{-3}$ cm$^2$/Vs for $H_2Pc$, 0.15 cm$^2$/Vs for CuPc, and up to 10 cm$^2$/Vs for TiOPc [48]. The use of a bottom contact geometry shown in Fig. 2 permits either vapor deposition or solution deposition on the transistor after mass fabrication and offers improved analyte access [44]. The selectivity of an electronic sensor depends not only on binding between the analyte and sensor surface but also on electron or hole trapping or doping properties of the bound analyte. The connection between sensor response and material electronic characteristics is an area of active investigation.

Sensor measurements with OTFTs usually involve pulsed mode dosing of a vapor stream to a device held at a constant $V_{ds}$, which is high enough to lie on the flattening region of the current–voltage plot and at a $V_g$ that turns on the FET.

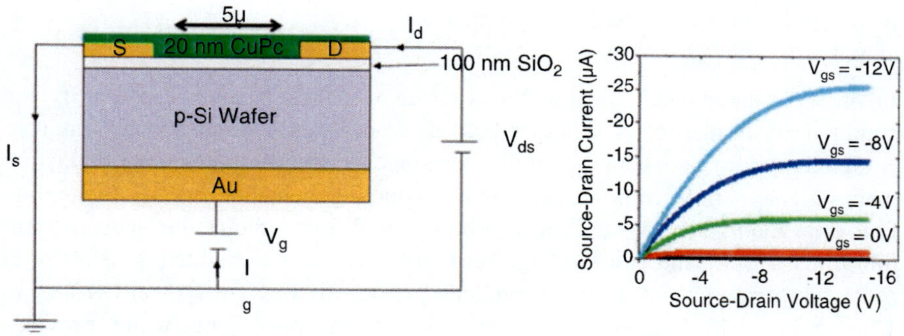


**Fig. 2** *Left* – bottom contact design of an OTFT sensor using copper(II) phthalocyanine as the channel material. Only one source (S)–drain (D) pair in the parallel array of 100 gold interdigitated electrodes (IDEs) is shown. $I_d$ and $I_s$ are the source and drain currents, respectively, $V_{ds}$ is the source drain voltage (typically −10 to +10 V range), and $V_g$ and $I_g$ are the gate voltage and current, respectively. Drawing not to scale. *Right* – FET characteristics of source drain current, which turns on at negative gate voltages for the hole-conducting CuPc

In practice, the sensitivity of an OTFT device is limited by the amount of gate leakage current, $I_g$, that occurs at a given voltage. While higher $V_g$ values lead to higher device currents, there is also a decrease in sensor lifetime and the generation of fixed charge at the gate dielectric interface, which degrades device characteristics and limits the sensor lifetime. The lifetime can be improved dramatically with the use of pulsed gate detection (a 10–100 μs on time) as long as the voltage pulse width is much shorter than the desired sensor response rate [55]. Mass transport for flowing vapor phase analytes usually limits detection response to a timescale of several seconds at best. A large current in the OTFT is not necessarily desirable since the sensor response is the ratio of the current in the presence of the analyte to the background current (e.g., in ambient air). The ratio $I/I_o$, where I is the current in the presence of analyte and $I_o$ the background current in its absence, may exhibit greater sensitivity at smaller percentage values of $I_o$, since at large values of $I_o$ the ratio will exhibit smaller percentage variations with analyte concentration. An advantage of pulsed mode dosing of analyte, where the sensor switches between a flow of background air (usually bottled dry air) and background air containing a known quantity of analyte, is that kinetics information about sensor response is obtained. Device current changes at the beginning of an analyte pulse are problematic due to mass transport limitations of gas switching, apparatus wall adsorption, and flow rate; however, the slower recovery after the analyte pulse ends has potential information content about analyte binding strength [56]. Even here, caution is needed due to the issue of analyte adsorption on the walls of the apparatus, sensor chamber, and any other surfaces present.

## 3 Mechanisms of Sensor Response

In order to understand the possible ways in which an analyte can modify the conductivity properties of a sensor film, it is necessary to consider the origin of conductivity in semiconducting molecular films. Conductivity in a semiconducting material may arise from intrinsic semiconducting behavior or due to doping. Phthalocyanines and most films composed of discrete molecular sensors form thin films with a band gap too large for high intrinsic semiconducting behavior at room temperature, and at 300 K undoped conductivities are typically $10^{-10}$ to $10^{-12}$ S cm$^{-1}$ [57]. Conductivity depends critically on doping, which provides the basis for use of these materials in vapor sensors. Both single crystal and thin film studies of conductivity in phthalocyanines and metal phthalocyanines suggest that impurity defects, especially those induced by molecular oxygen in air, are responsible for the extrinsic conductivity of phthalocyanine materials [25, 26, 58–68]. This not only leads to increased dark current in phthalocyanine materials but also enhanced photoconductivity as electron transfer to oxygen occurs, yielding detectable superoxide [69–71] and other oxygen radicals [72]. This process can be enhanced (Fig. 3) even with low optical energies in the visible and near IR spectral regions [73–75]. It should be noted that other chemical impurities present from the synthesis of metal phthalocyanines can contribute to their conductivity properties [76, 77]. Photoconductivity presents a practical problem in sensor applications,

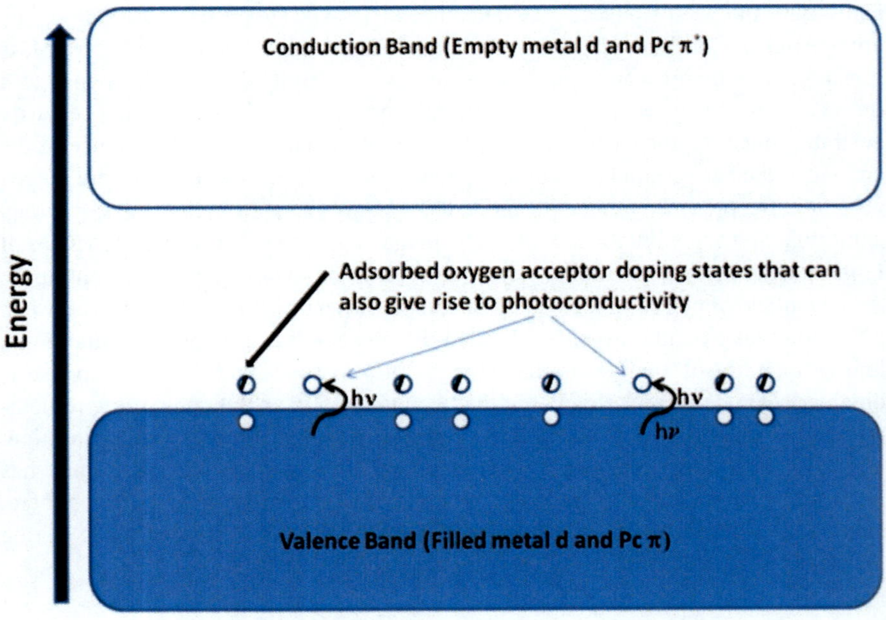


**Fig. 3** Schematic illustration of acceptor doping that gives rise to extrinsic hole conductivity and long-wavelength photoconductivity in metal phthalocyanines

since visible light must be excluded to determine the influence of vapor analytes on the dark current. While it is also possible to measure modulation of the sum of the dark current and photoconductivity by vapor phase analytes, this presents the additional problem of maintaining a constant light flux for meaningful results. The high photoconductivity of phthalocyanines is an advantage for other device applications, as TiOPc has a very high quantum yield for carrier generation at the long wavelengths easily accessible with diode lasers [75].

Like porphyrins, the phthalocyanines have low energy $\pi$–$\pi^*$ Q-bands in their electronic absorption spectra that can give rise to photoconductive behavior even in the absence of oxygen [24, 78–84]. This is an important consideration in the design of sensing experiments, as the oxygen doping level takes a day or more to stabilize in air, and it is desirable to conduct chemical sensing experiments with exclusion of light in order to separate the effects on the dark current from photoconductivity modulated sensing contributions [26, 85]. Free radical signals suggestive of phthalocyanine $\pi$ cation radicals have been detected in oxygen doped phthalocyanines in addition to oxidation at the metal [70, 71], and the materials generally behave as hole conducting semiconductors. Electron conductivity and n-doping have been observed in some fluorine and cyano-substituted phthalocyanine derivatives [86–90], and their conductivity can decrease in the presence of ambient oxygen. Some Langmuir–Blodgett films of phthalocyanines show conductivity that decreases in air and their behavior has been attributed to intrinsic conductivity that depends on the reduced HOMO–LUMO gap in the parent molecule [91].

## 3.1 *Charge Conduction Models*

The dark current density in phthalocyanine materials usually follows Ohm's law at low applied voltages (<1 V) and is limited by the electrode–semiconductor contact. It transitions to non-Ohmic space charged limited conductivity (SCLC) at higher applied voltages as charge becomes limited by the space charge that builds up in the relatively weakly conducting phthalocyanine semiconductor [92–98]. The SCLC expression for the current density for a discrete trap distribution is given as (1) [60].

$$\mathrm{J} = \mathrm{IA} = (9/8)(\varepsilon\,\theta\,\mu)\left(\mathrm{V}^2/\mathrm{d}^3\right), \tag{1}$$

where $J$ is the current density, $I$ is the current, $A$ is the electrode area, $\varepsilon$ is the permittivity of the material times the permittivity of free space, $\theta$ is the ratio between free and trapped charges, $\mu$ is the carrier mobility, $V$ is the voltage, and $d$ is the electrode spacing. At a constant temperature, voltage, and electrode spacing, the sensor response due to changes in material resistivity is expected to follow the proportionality of (2).

$$\text{Sensor Response} \propto (\mathrm{N_c} + \mathrm{FS_c})\left[(\mu)\left(1 + \mathrm{FS}_\mu\right)\right], \tag{2}$$

where $N_c$ is the number of charge carriers in the absence of analyte and $S_c$ are the added or subtracted carriers due to doping by the analyte S. Similarly, $S_\mu$ is the change in carrier mobility due to adsorption of analyte and F is the fraction of saturation coverage of the binding sites, which can be modeled using the Langmuir or Freundlich isotherms [99, 100]. For very sensitive detection, it is desirable for the analyte to interact strongly with the surface. Unfortunately, strong adsorptive interactions will preclude facile loss of the analyte from the surface. Sensors that strongly bind analytes should behave as integrating or dosiometric sensors. The tradeoff between sensitivity and reversibility is a vexing problem for most thin film sensors that rely on reversible adsorption of an analyte. It should be noted that there need be no correlation between the strength of adsorption, determined by F, and the values of $S_c$ and $S_\mu$ for an analyte; however, there may be a correlation if adsorption strength parallels charge donor or acceptor properties of the analyte.

The charge mobility term $[(\mu)(1 + FS_\mu)]$ should depend strongly on the energies of charge carrier trap states in the semiconducting material. The traps may be defect sites in the solid, interfacial trap sites, or new charge trapping sites created by the analyte. Conductivity can decrease on analyte adsorption either due to loss of charge carriers or increased stability and numbers of charge trapping sites. Interfacial properties between the gate and channel material are also important, as the trapping of fixed charge at the dielectric–channel material interface degrades OTFT performance and results in high turn-on voltages. Covalent modification of the dielectric gate (often $SiO_2$) with organic moieties has been shown to result in dramatic increases in device performance. An excellent review of chemical surface modification in OTFTs appeared recently [101]. In organic semiconductors, analytes are thought to primarily increase the average activation energy for carrier escape from charge trap states, where the mobility in the presence of analyte is given by $\mu = \mu_o e^{-(Ea/kT)}$. If the energy increase, $c\delta$, in $E_a$ is small and depends on the concentration, c, of the analyte, where $E_a'$ is the activation energy in the absence of analyte, then $\mu = \mu_o e^{-(E'a/kT)} e^{-(c\delta/kT)} \sim \mu_o e^{-(E'a/kT)}(1 - c\delta/kT)$ and the current change is approximately linear at low analyte concentrations as represented in (2), when the response is dominated by mobility changes.

Charge trapping may also occur at the gold–semiconductor channel contacts, but this is probably of lesser importance in micron scale devices [102–104]. For sensor response, the surface charge trap sites between adjacent grains in the channel material may play a key role as analyte adsorption is expected to enhance the stability of nearby trapped charge by dipolar interactions [44]. Additional complications that have yet to be explored fully are the differing contributions to current changes in an OTFT that arise from changes in carrier mobility as well as shifts in the threshold voltage for FET turn on. Often only device current changes are monitored at fixed $V_g$ and $V_{sd}$ during analyte dosing, without determining the complete operating characteristics for the FET device.

## *3.2 Analyte Induced Direct Doping*

Several studies of phthalocyanines in the presence of strong oxidants, such as chlorine, ozone, and nitrogen dioxide, attribute the primary electronic sensor response to the $S_c$ term (2) where the strong chemical oxidant gasses accept an electron as in (3)–(5) [38, 99, 105–125].

$$Cl_2 + 2MPc \rightarrow 2MPc^+ + 2Cl^- \tag{3}$$

$$2O_3 + MPc \rightarrow MPc^+ + O_2{}^- + 2O_2 \tag{4}$$

$$NO_2 + MPc \rightarrow MPc^+ + NO_2{}^- \tag{5}$$

$$H_2O_2 + 2MPc \rightarrow 2MPc^+ + 2OH^- \tag{6}$$

The $MPc^+$ hole carriers are created by trapping of an electron, or oxidation of MPc, by the chemical oxidant [126]. Strong chemical oxidants usually produce large conductivity increases in all p-type phthalocyanine chemical sensors. The oxidation reactions in (3)–(5) are also chemically reasonable, given the strong oxidizing nature of the analytes involved. In further support of this model are the highly conducting partly oxidized $MPcI_x$ materials that have been well-characterized, and whose conduction is primarily attributed to hole conduction between overlapping $\pi$ aromatic rings [127, 128]. For example, Cu(II)Pc is paramagnetic yet the unpaired metal d electron is not well-delocalized in the solid state and its intrinsic conductivity is low and similar to diamagnetic Ni(II)Pc and Zn(II) Pc [48]. Because oxidation processes are essentially irreversible chemical reactions, it has been suggested that at higher oxidant concentrations the conduction band should become nearly completely filled and conductivity could decrease. However, strong oxidants such as ozone, chlorine, and $NO_2$ also can attack and oxidize the organic phthalocyanine ring so that gross decomposition of the sensor film occurs at higher oxidant concentrations [129]. The decrease in current of phthalocyanine sensors that occurs with gaseous reductants, such as ammonia or hydrogen, has been attributed to destruction of hole carriers by the reducing analyte [130]; however, a similar behavior has been observed for a wide variety of molecular vapors, such as acetonitrile, acetone, trimethylphosphate, and DMSO, that are not normally considered good reducing agents [56, 131, 132].

The extreme sensitivity of phthalocyanine sensors to oxidants has raised the possibility that common urban pollutants ozone and $NO_x$ are responsible for much of the conductivity of phthalocyanine devices in ambient conditions [133]. Indeed, phthalocyanine OTFTs on prolonged exposure to ambient air often exhibit an increase in device current; however, their FET characteristics are also degraded so that they no longer can be switched off [34]. High humidity can reverse the process and restore FET characteristics. The high sensitivity of phthalocyanine OTFTs to pollutants may limit sensing applications and improved stability in ambient atmospheres is needed. Given these limitations, it is reasonable that, a primary focus of development for electronic sensors that use phthalocyanines has been detection of $NO_2$ [134].

## 3.3 *Analyte Induced Secondary Doping, Counter Doping, and Carrier Trapping*

While strong oxidants, such as those in (3)–(6), are capable of oxidation of $H_2Pc$ or MPc compounds to produce radical cation species as hole carriers, an alternative mechanism for conductivity changes could involve the formation of charge transfer complexes between the phthalocyanine $\pi$ system and acceptor or donor analyte. The direction of charge transfer would then determine the current response, which is similar to the explanation proposed to explain analyte induced current effects in conducting polymers. Formation of charge transfer complexes alters occupancy at the valence band edge and changes conductivity by this secondary doping process [42, 43, 45]. In this view, the analyte acts to change the bulk work function of the semiconducting sensor [43]. Direct and secondary doping that relies on analyte electron donor or acceptor properties would be expected to change sign depending on whether the semiconducting sensor material is a hole or electron conductor. Perfluorinated phthalocyanines provide an n-type semiconducting material, and their behavior toward analytes sheds some light on this issue [86, 89, 135–146]. While the conductivity of the normal p-type phthalocyanine complexes decreases under a nitrogen gas purge as dioxygen dopant is lost from the film, the n-type copper(II) perfluorophthalocyanine complex shows an increase in current under nitrogen purge. In the n-type material, adsorbed dioxygen acts as a counterdopant and scavenges electron carriers to decrease current [90, 141, 147]. The reductant ammonia, however, caused an increase in current and the differing changes seen with $NH_3$, $O_2$, and $NO_x$ were attributed to changes in carrier density rather than carrier mobility differences [141].

For a variety of analytes of varying basicity and binding ability (dimethyl methylphosphonate (DMMP), diisopropyl methylphosphonate (DIMP), nitrobenzene, methanol, and water) an opposite current response has been observed in n-type and p-type OTFTs made from copper(II) perfluorophthalocyanine and copper(II) phthalocyanine, respectively. The electron pair donors DMMP, DIMP, methanol, and water cause a strong current decrease in the p-type material but a current increase in the n-type OTFT. Nitrobenzene, which might be expected to form a $\pi$ charge transfer complex as an acceptor, showed a small current decrease in the p-type material and a small increase in the n-type OTFT. These results suggested that the analytes alter conductivity in these materials by different perturbation of the electron trap energies versus hole trap energies in the *n*-channel and p-channel OTFTs. The ease of oxidation and reduction of nitrogen substituted delocalized $\pi$ electron systems [148] allows ready formation of both $\pi$ cation and $\pi$ anion radicals that could serve as p-type or n-type charger carriers, respectively. The increase in electron conductivity with polar analytes in n-type metal phthalocyanines is difficult to reconcile with the postulate that charge trapping at grain boundaries is usually stabilized by adsorbed polar analytes.

### 3.4 Polar Analytes and Charge Trapping at Grain Boundaries

The conductivity changes obtained with the polar analytes of low redox activity (dimethyl methylphosphonate (DMMP), diisopropyl methylphosphonate (DIMP), nitrobenzene, methanol, and water) differ from the behavior observed for organic polymer OTFTs. Polar compounds are thought to adsorb and exert their effect primarily at grain boundaries in long channel length devices (i.e., where interfacial effects at contacts are of minor importance). Polar analytes are able to stabilize charges at the intergrain boundaries and serve as trap states that hinder charge mobility [44, 149]. In this model, which depends on analyte polarization around charge carriers, both electron- and hole-conductivity are expected to decrease in the presence of polar analytes. This is because the orientation of the molecular dipoles of the analyte molecules bound to the grain surface adjusts to stabilize the charge trap sites. The opposite behavior observed with n-type and p-type copper(II) phthalocyanines requires a different explanation for how the analytes perturb charge carrier trap states.

Figure 4 depicts a possible explanation for the effect of polar electron donor analytes on the conductivity in metal phthalocyanines. Electron donors would

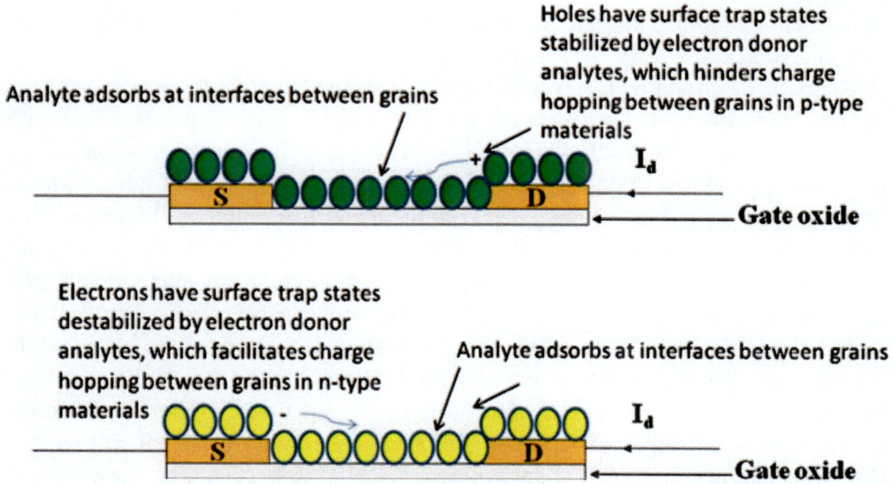


**Fig. 4** Schematic mechanism of the differing effect of electron donor analytes on the conductivity of hole conducting (*green*) and electron conducting (*yellow*) channel materials in an OTFT sensor where adsorption of analyte is directed by specific metal–ligand type coordination at the grain surface

reasonably be expected to stabilize positive holes and provide trap states that hinder charge hopping at the interface between grains in the channel material. Conversely, for electron conducting channel materials electron donors would destabilize (i.e., not trap) electron carriers at the interface and thereby facilitate charge transfer between grains. In order to obtain the opposite behavior, the analyte molecules would have to bind to the surface in a specific orientation so the dipoles do not alter their orientation in n- and p-type devices. This could occur if the primary interaction with the semiconductor involves coordination at the metal, which directs orientation of bound analyte.

## 4 Molecular Mechanism of Carrier Trapping in Phthalocyanines

The preceding discussion leaves open the question of how the polar donor analytes modulate carrier trap energies at the molecular level. The behavior differs from the adsorption of organic vapors by organic polymer film sensors where bulk material properties, such as polarity, hydrogen bonding, and electrostatic attraction determine analyte adsorption [36, 150–161]. In pentacene OTFTs, adsorption of polar analytes at grain boundaries is thought to stabilize surface trapped charge carriers by charge–dipole interactions and present a barrier to transfer charge at grain boundaries. Since charge hopping is thought to be the primary conduction mechanism in most phthalocyanines which are used as sensors, it is appropriate to view the process at a molecular level. Indeed the solid state optical spectra of phthalocyanines exhibit Davydov splittings [23] as observed in molecular crystals rather than exhibiting true ground state delocalization associated with band structure in a solid.

Study of the effect of 12 polar analytes of varying electron donor properties on the conductivity of a CoPc sensor 50 nm thick suggested a strong correlation with donor ligand properties of the analytes [132] rather than the linear solvation energy relationships (LSERs) that have been successfully used to model adsorption of polar analytes by polymer films [162]. For the metal free phthalocyanine sensor, a better correlation was obtained with the hydrogen bond donor ability of the polar analyte [132]. These observations suggest a specific chemical interaction of the metal phthalocyanine core with analytes that are capable of ligation and the stronger the electron donor nature of the analyte the stronger the decrease in conductivity due to charge trapping of the hole. For the metal free phthalocyanine, hydrogen bonding to the two central N–H groups appeared to dominate surface interactions with analyte. The linear sensor concentration dependent response at low coverages also can be fit to such a model [100]. From a chemical viewpoint, it would be expected that if $MPc^{+}$ radical cations represent the molecular nature of the hole carriers in these materials, then these species would be stabilized by coordination of an electron donor ligand to the metal center and that these holes would tend to accumulate and be trapped at surface sites where the donor analyte is bound to the metal. Consistent with this, polar substances, such as dichloromethane, which are poor ligands only caused very small changes in conductivity. Several theoretical studies have supported the role of metal coordination as the primary factor in determining the magnitude of the sensor response of metal phthalocyanine sensors [163–166]. Of particular interest is the observation (with the aid of surface sticking coefficient measurements) that the organic ring portion of the phthalocyanine can serve as a collector to initially weakly bind analytes and funnel them to the metal binding sites that represent a relatively small percentage of the sensor surface area [163].

With liquid crystal metal phthalocyanine compounds as mass sensors, the LSER approach has proven useful. Analyte uptake has been measured using QCM methods, and adsorption of volatile organic compounds (VOCs) into the liquid crystalline coating appears to follow similar trends as for organic polymer film sensors. It should be noted that the analytes examined (toluene, chloroform, carbon tetrachloride, benzene, hexane, and methanol) are volatile compounds that are very weak ligands toward metals [167]. Thus, the composite sensor response for metal phthalocyanine sensors based on conductivity is a complex property that depends on analyte redox properties, basicity, and sensor crystallinity.

## 5 Other Effects: Grain Size, Electrode Contacts, Fixed Charge at Interfaces

Several studies have examined the influence of film morphology on sensor behavior [109, 168–178]. Two main areas of interest are the influence of film thickness and grain size on the time scale for sensor response as well as the magnitude (sensitivity) of sensor response. Other factors that may influence the conductivity response are the nature of the semiconductor–metal device contacts, as well as charge accumulation at other device interfaces, such as the gate oxide in field effect sensor devices.

### *5.1 Grain Size*

One possible solution to minimizing the influence of grain size on electronic properties, as well as improving sensor response times and homogeneity, is through the use of Langmuir–Blodgett (LB) deposited films, liquid crystal films, and self-assembly approaches [7, 91, 128, 169, 179–197]. Surface adsorption of analyte, followed by diffusion through such films, has been modeled successfully, so the behavior appears to differ from polycrystalline solid state sensors [184]. Orientation of the MPc plane within the film does not appear to be very important for $NO_2$ sensing, and selectivity of sensing differs from that for sublimed MPc sensors. Thinner LB films appear to dramatically improve the response time of such sensors when compared to solid state MPc sensors [182, 186, 187]. The interdomain size in LB films has been proposed to correlate with sensitivity with fewer domains leading to higher sensitivity [198]. However, the suggestion that $N_2$ acts as an electron donor in these materials seems unlikely given its high ionization potential. It seems more likely that exposure to a nitrogen gas stream removes oxygen dopant from the film, as the experiments were done relative to air (20% oxygen) and oxygen could reversibly dope such films. The identical time constant observed for oxygen and nitrogen doses suggests a common species is responsible for the current changes observed. Another study of iodine adsorption by LB films suggested a stoichiometry of 2–4 iodine atoms occurred to induce large increases in conductivity. The stoichiometry was rigorously determined by simultaneous electrical and gravimetric measurements [128]. The reason for the high iodine stoichiometry probably arises from the proclivity of $I_2$ to form triiodide ion during oxidation reactions [199].

The morphology of vapor deposited MPc films can vary from highly crystalline thick films (100s nm thick, Fig. 5a) using conventional sublimation. Films as thin as several monolayers (4 nm thick) can be deposited using molecular beam epitaxy (MBE) deposition conditions under ultra high vacuum [53, 200–202]. Long range ordering enhances carrier mobility, since as noted earlier highly conducting partly oxidized $MPcI_x$ compounds exhibit conduction primarily attributed to the

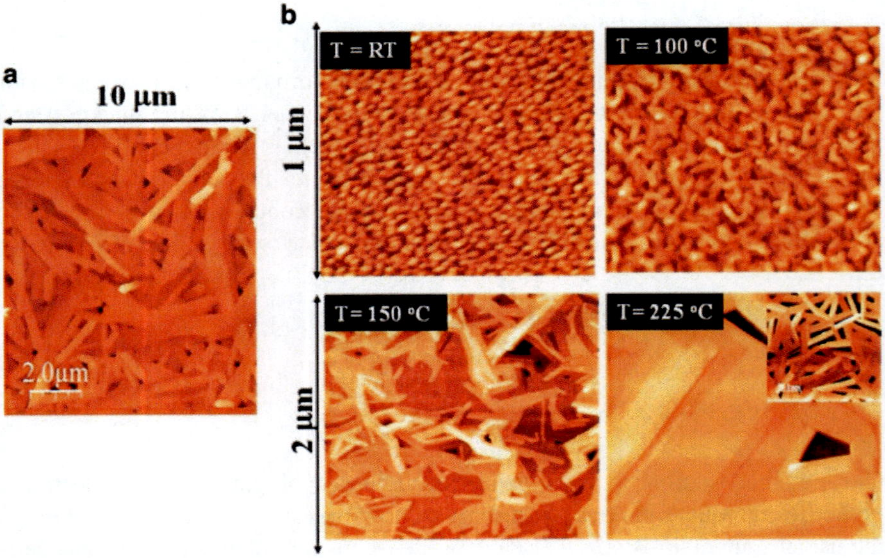


**Fig. 5** Atomic force microscope (AFM) images showing varying surface morphologies of CuPc thin films deposited on $Al_2O_3$ by (**a**) sublimed film several hundred nanometres thick; (**b**) MBE deposition of 10 nm thick films at temperatures indicated (Miller, Bohrer, Sharoni, Schuller, Kummel, and Trogler, unpublished work)

overlapping $\pi$ aromatic rings [127, 128]. The relative sensor response of MPc OTFTs to various analytes does not appear to depend primarily on film morphology, but is chiefly a function of the central metal atom. However, the ultrathin film OTFT sensors exhibit a net enhancement both in sensitivity, due to the close proximity of surface adsorbed analytes to the field effect carrier transport states, as well as an improved stability toward sensor drift due to depletion of surface trap states [202]. The mobility of CuPc OTFTs has been observed to saturate at only 6 monolayers in a top contact OTFT [203], and at 12 monolayers in a bottom contact OTFT [54], so it is possible to use extremely thin sensing films in such devices.

The grain structure of MBE deposited MPc films can vary from tens of nm in diameter up to the micron scale, depending on the substrate deposition temperature [204–206]. Above 120°C, CuPc deposits as the $\beta$-phase, but below that the $\alpha$-phase forms. Alignment with the substrate can also occur, similar to that in LB deposited films. The phthalocyanine plane lies with a near edge on alignment on $SiO_2$ and $Al_2O_3$ surfaces [204], but flat alignment of the MPc plane occurs on the surfaces of single crystal gold. However, polycrystalline and amorphous gold do not appear to favor flat alignment of phthalocyanines and it has been suggested that amorphous and polycrystalline films generally favor edge on alignment [207]. In addition, unsymmetrical substitution with alkyl groups on the MPc periphery destroys ordering and alignment on the molecular level [164].

## 5.2 *Electrode Contacts*

Several studies have shown that gold forms Ohmic contacts with ZnPc, while the lower work function metal aluminum produces blocking contacts because the thin layer of aluminum oxide between the metal and semiconductor adds an insulating junction [92, 208–212] that gives rise to a Schottky barrier for charge transport [92, 208, 213]. The sensor response of the DC conductivity of 50 nm thick films of CoPc at gold, platinum, and palladium electrodes has been characterized [98]. The transition to space-charge-limited conductivity (SCLC) at higher voltages depends on the specific metal electrode (Pt, Pd, and Au), but does not correlate with the work function of the electrode. AC impedance measurements show that the contribution of low-frequency traps associated with Ohmic behavior diminishes at higher applied voltage. Although device resistances are strongly influenced by the electrode material, and vary by a factor of over 300 for identical thickness films, the relative chemical sensor responses on exposure to DMMP, methanol, water, or toluene vapors were nearly identical for CoPc on Pt, Pd, and Au electrodes when operated in the SCLC regime. This was attributed to filling of the low-frequency interfacial trap states at room temperature. These considerations probably are restricted to micron scale devices, as it has been noted that the sensor contribution of the electrode contacts will increase when device sizes reach the nanoscale [102–104]. Band bending and perturbation of the energy levels of CuPc have been found to be limited to the first 2 nm at the interface with a gold metal electrode [207].

Lithographic sculpting of the shape of the metal electrodes also plays a role, since undercut electrodes do not form reliable contacts with vapor deposited MPcs in a bottom contact FET [54, 200].

### 5.3 *Fixed Charge at Interfaces*

Besides the electrode–phthalocyanine interface in an OTFT, the interface between the gate dielectric and the semiconducting material presents a problematic interface for carrier trapping. It is well-known that surface functionalization of inorganic oxide dielectrics can dramatically improve the device performance of OTFTs by reducing the gate leakage current [101]. However, there are additional beneficial changes that may arise from improved order and grain structure in the overlying hydrophobic organic semiconductor [214, 215]. For example, covalent surface functionalization of $SiO_2$ gate dielectric occurs by the reaction of surface hydroxyl and oxide groups with a reagent such as ( $CH_3O)_3SiR$, where R is one of many possible inert organic groups. Saturated R groups provide a σ-only conducting insulating layer that is only capable of nonresonant charge tunneling. In addition, the reactions to form the SAM appear to reduce the number of interfacial charge traps at the gate dielectric interface. For example, the use of octadecyltrimethoxysilane (OTMS) to functionalize $SiO_2$ imparts structural order to overlying films of $CuF_{16}Pc$ crystallites [87, 145]. The reduced and aligned grain boundaries observed with OTMS SAMs result in improvement of the field-effect mobility by almost an order of magnitude in this n-type OTFT. The number of grain boundaries also has been associated with aging effects in CuPc OTFTs due to the strong oxidants ozone and $NO_2$, which are present in ambient laboratory air [34]. In this case, thicker films with more grain surfaces exhibit increased rates of device degradation due to aging. Electron trap states also form at the junction between dissimilar p-type and n-type phthalocyanine semiconductors in multilayer structures [216].

## 6 Sensor Selectivity

Microsensors that rely on molecular recognition, adsorption, and chemical reactions for identification are, with few exceptions, of limited selectivity. Even with the use of array technology, as in an electronic nose, it is unreasonable to expect such sensors to be useful for identifying individual components in mixtures of similar compounds without a prior separation by chromatography or some other method [217]. Application of a single sensor in a compact electronic device will likely be limited to a sensor of high selectivity, or else the detection of an analyte in a mixture where the presence of interferents causing a response is known to be highly unlikely [218]. Films of p-type and n-type semiconducting films are potentially well-suited to electronic nose array sensing. Moment analysis of the pattern from an array of sensors, linear discriminant, and other statistical analysis schemes has been used to enhance the chemical selectivity information contained in the response from a cross-reactive array of microsensors. One sticking point is how to define selectivity of a sensor in the context of sensitivity. So far the fabrication of highly sensitive sensors has been a much easier task than solving the problem of sensor selectivity.

### *6.1 Sensitivity–Selectivity and Receiver Operating Characteristics*

A widely used method for quantifying the sensitivity–selectivity performance of a sensor uses a concept from signal theory used to evaluate diagnostic tests. The problem is to define the ability to make a yes/no decision about the presence of a particular diagnostic signal. The four possibilities are a true positive (TP), false positive (FP), false negative (FN), and true negative (TN). The sensitivity is defined as the true positive probability (TPP), which is given by:

$$\mathrm{TPP} = \mathrm{TP}/(\mathrm{TP} + \mathrm{FN}).$$

Accuracy (ACU) is defined by:

$$\mathrm{ACU} = (\mathrm{TP} + \mathrm{TN})/(\mathrm{TP} + \mathrm{FN} + \mathrm{FP} + \mathrm{TN}).$$

Specificity is defined as the true negative probability (TNP):

$$\mathrm{TNP} = \mathrm{TN}/(\mathrm{FP} + \mathrm{TN}).$$

The data are often plotted as a Receiver Operating Characteristics (ROC) curve, which plots the TPP vs. the false positive rate or probability (FPP), where:

$$\mathrm{FPP} = 1 - \mathrm{TNP}.$$

In general, when sensor sensitivity is increased, the fraction of false positives also increases. This presents the universal dilemma all types of chemical sensing, that increasing sensitivity alone does not improve system performance. One needs to simultaneously improve selectivity.

The typical types of behavior shown in the ROC curves of Fig. 6 illustrate the problem in developing a practical sensor system [219–221]. Usually as the sensitivity (TPP) of a sensor system is increased, for example, by signal amplification or preconcentration of analyte, the noise also increases, and the probability of a false positive increases (FPP). Characterizing a sensor system by this method is fairly time consuming, as there will be an ROC curve for each analyte concentration, and different background conditions may also have to be included for

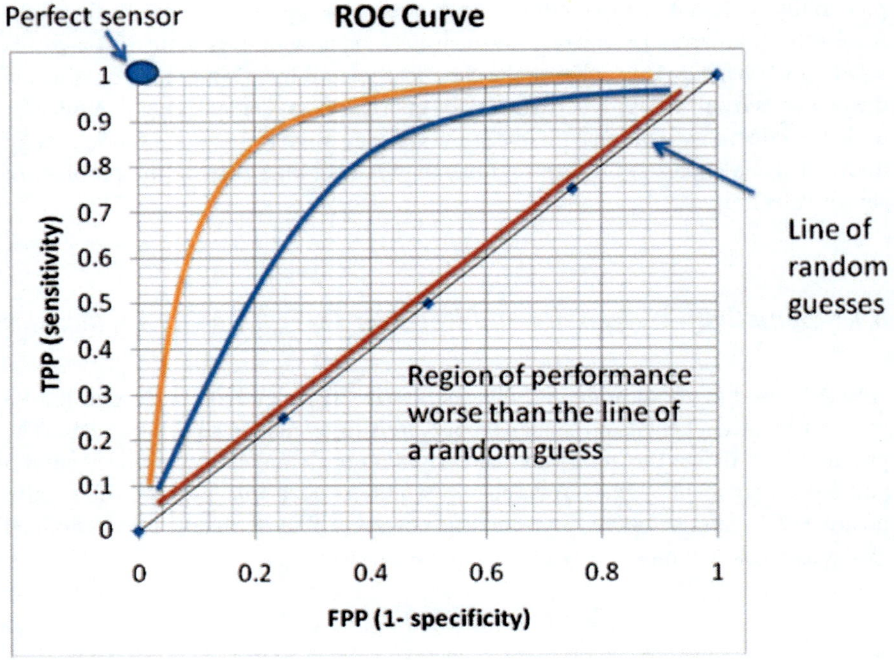


**Fig. 6** Receiver operating characteristics (ROCs) curve for typical sensor behavior. The diagonal line represents a sensor with no selectivity and the two upper curves improved sensor performance in both sensitivity and specificity

each concentration. Parameters of the instrument may have to be varied, and the time for obtaining the measurement is another possible variable to be included in the family of curves in an ROC plot. A discussion of practical sampling may be found in a DARPA report for critical chemical and biological detection problems that is available via the internet [222].

In evaluating a chemical sensor, most studies determine whether an individual analyte can be detected by a sensor system at a given concentration. The sensor system itself can usually be tuned by changing one or more variables to increase sensitivity at the cost of selectivity. This trade-off is what the ROC curve quantifies. Since the detection is posed as a detect/no detect decision, a curve must be constructed for each specific concentration level of analyte where a detection decision is to be determined.

A perfect sensor is defined by the point in the upper left of the ROC curve. The red line shows behavior typically observed when the analyte concentration lies below the detection limit of a sensor system. In this instance, attempts to increase sensitivity by amplifying the signal provide a yes/no detection answer near that of a random guess line. The intermediate blue curve displays typical behavior when analyte concentration approaches the useful range for a sensor. Increasing the sensitivity so that an analyte is reliably detected (increasing TPP) will also increase the probability of a false positive (FPP). The uppermost curve displays the behavior for a more reliable sensor system when the analyte concentration lies in the useful concentration range. Here, increasing the sensitivity of the system to obtain a TPP of 0.9 occurs with a FPP of 0.25. It is extremely unlikely that any single electronic sensor can exhibit the needed selectivity, so that arrays of sensors – the so-called electronic nose approach – are the likely solution with phthalocyanine based OTFT sensors [218]. An additional complication not considered is the presence of interferents. There are a large number of known chemicals; however, in a specific application, only a certain number is likely to be present. This introduces a complication that the experiments needed to construct Fig. 6 also need to include the possible mixtures of interfering chemicals present in the background for a specific application.

### 6.2 Phthalocyanine Sensor Arrays

The sensitivity of phthalocyanine OTFTs usually approaches ppb level (10–100 ppb) detection limits for strongly oxidizing analyte vapors such as $O_3$, $NO_2$, $H_2O_2$, and halogens. Good donor ligands, such as DMMP, DIMP, $NH_3$, and other amines generally have detection limits in the low ppm range. Weak donor ligands, such as methanol and ethanol, and VOCs such as toluene and dichloromethane, generally can be detected when concentrations are in the part per thousand range (ppth). It is difficult for any single phthalocyanine sensor to determine whether a response arises from a tiny amount of an oxidizing analyte, such as $O_3$, or a high concentration of a weakly detectable analyte, such as methanol. For electron donor ligands, which are thought to bind primarily to the metal center, it might be expected that the desorption kinetics of the analyte from the surface, when the analyte is exposed to the sensor in pulsed mode, would allow additional discrimination. Unfortunately, the relatively weak nature of the analyte-surface bonding only permits broad classification into strongly binding and weakly binding ligands by the desorption rate determined in a flow system [90, 132]. There are possibly mass transport issues as thin films exhibit significantly faster kinetics than thicker films [168, 202].

There are a few limited examples of array sensing with phthalocyanine OTFTs. Colorimetric array sensing with the use of metalloporphyrins as cross-reactive sensor arrays has been highly developed and proven successful [223–226]. Similar examples are known for phthalocyanines, but have not yet been developed extensively [227–230]. A QCM sensor array of eight chips coated with differing amounts of phthalocyanine has been used to determine the composition of acetone:trichloroethylene:ethanol vapors [231]. Voltammetric sensors using electrodes coated with lanthanide phthalocyanine complexes have been used in array sensing as an electronic tongue to identify different types of red wines [232]. Three types of phthalocyanine sensors have been used to discriminate among the aroma of four different types of olive oils [227]. An array of $H_2Pc$, ZnPc, CuPc, NiPc, and

CoPc chemiresistors has been used as a cross-reactive sensor array to discriminate among 12 different volatile analytes using linear discriminant analysis of the array current response. Due to the linear response of the sensors, it was possible to distinguish different analytes regardless of the concentration over the linear range of the sensor array, provided that the array response was normalized to one of the sensors (ZnPc) [56].

## 7 Summary and Outlook

A potential advantage of metal phthalocyanines is that compounds are known for over 70 different elements that could potentially be used in making dense lithographically patterned microsensor arrays [99, 233]. Below saturation coverage, the MPc OTFT sensors generally respond to analyte in a linear fashion, which facilitates mathematical analysis approaches. If one includes the soluble phthalocyanine derivatives that are potentially amenable to low-cost solution processing [234–238], then the potential for developing arrays containing varied sensing elements becomes even more appealing. The most likely applications for a stand-alone sensor array would be when one vapor phase analyte is expected to vary significantly against a relatively constant background. Given the high sensitivity of MPc sensors toward oxidants, application as a broad spectrum sensor for the release of toxic oxidants, such as ozone, halogens, and nitrogen oxides, could be envisioned. If multicomponent analyte variations are expected, then preseparation of the analyte stream is likely to be required. Application as a sensor in miniature gas chromatography devices seems possible.

Several challenges remain for the ultimate practical use of these sensors. The response time of the solid state sensors are short (seconds) for initial sensing, but recovery times range from minutes to hours at room temperature. The stability of the sensor to drift associated with accumulation of fixed charge at interfaces, as well as the high sensitivity to ubiquitous urban pollutants ozone and $NO_2$ are problematic. All MPc OTFTs show some response to moisture, and conductivity is also temperature sensitive so that humidity and temperature compensation are essential. On a basic research level, the detailed characterization of charge trapping states, electronic structure, and the interactions with analytes is not yet fully understood on a quantitative theoretical basis. The time response of sensor initiation and recovery is also not understood in a detailed manner. In spite of these limitations, the intrinsic chemical stability of MPc compounds and their compatibility with microsensor array fabrication make these candidate OTFTs for further research and development.

**Acknowledgments** Support of sensor research by the Air Force Office of Scientific Research is gratefully acknowledged, as are long-standing collaborations with Professors Andrew Kummel and Ivan Schuller on this research area.

## References


1. Bjerrum J, Ballhausen CJ, Jørgensen CK (1954) Acta Chem Scand 8:1275–1289
2. Dahl JP, Ballhausen CJ (1961) Kgl Danske Videnskab Selskab, Mat-Fys Medd 33(5):22 pp
3. Ballhausen CJ (1962) Introduction to ligand field theory. McGraw-Hill series in advanced chemistry. McGraw-Hill Book Company, New York
4. Gray HB, Ballhausen CJ (1963) J Am Chem Soc 85:260–265
5. Liehr AD, Ballhausen CJ (1957) Acta Chem Scand 11:207–218
6. Moffitt W, Ballhausen CJ (1956) Annu Rev Phys Chem 7:107–136
7. Armstrong NR, Xia W, Minch B, Simmonds A, Carter C, Donley CL, Zangmeister RAP, Drager A, Cherian SK, LaRussa L, Kippelen B, Yoo S, Domercq B, Mathine DL, O'Brien DF (2006) Proc – Electrochem Soc 2004-22 (Electron Transfer in Nanomaterials):376–384
8. Katz HE, Huang J (2009) Annu Rev Mater Res 39:71–92
9. Lee J-K, Jang S-I, Kim Y-G, Jang Y-W, Jeong B-H, Kim J-U, Jung K-S, Kim M-R (2008) Mol Cryst Liq Cryst 491:307–316
10. Zheng F, Shen Y, Gu F, Zhang J, Zhang J (2008) Proc SPIE 6984 (Sixth International Conference on Thin Film Physics and Applications, 2008):69843J/69841–69843J/69844
11. Ng TW, Lo MF, Zhou YC, Liu ZT, Lee CS, Kwon O, Lee ST (2009) Appl Phys Lett 94 (19):193304/193301-193304/193303
12. Sakurai T, Naito R, Toyoshima S, Ohashi T, Akimoto K (2009) Nanosci Nanotechnol Lett 1(1):23–27
13. Coffey DC, Ferguson AJ, Kopidakis N, Rumbles G (2010) ACS Nano 4(9):5437–5445
14. Burgdorf K (2008) Schweiz Lab-Z 65(2):54–57
15. Goes M, Verhoeven JW, Hofstraat H, Brunner K (2003) ChemPhysChem 4(4):349–358
16. Mori T, Ogawa T, Cho D-C, Mizutani T (2003) Appl Surf Sci 212–213:458–463
17. Cao J-S, Guan M, Cao G-H, Zeng Y-P, Li J-M, Qin D-S (2008) Chin Phys Lett 25(2):719–721
18. Fan Z, Cheng C, Yu S, Ye K, Sheng R, Xia D, Ma C, Wang X, Chang Y, Du G (2009) Opt Mater 31(6):889–894
19. Armstrong NR (2000) J Porphyrins Phthalocyanines 4(4):414–417
20. Leznoff CC, Lever ABP (eds) (1989) Phthalocyanines properties and applications, vol 1. Wiley, New York
21. Gregory P (2000) J Porphyrins Phthalocyanines 4(4):432–437
22. Van Biesen J (1987) Chem Mag 13(9):31–35
23. Day P, Williams RJP (1962) J Chem Phys 37(3):567–570
24. Gouterman M, Wagniere G, Snyder LC (1963) J Mol Spectrosc 11(2):108–127
25. Day P, Scregg G, Williams RJP (1963) Nature 197:589–590
26. Day P, Williams RJP (1965) J Chem Phys 42(11):4049–4050
27. Lawton EA (1958) J Phys Chem 62:384
28. Nylander C (1985) J Phys E: Sci Instrum 18(9):736–750
29. Dickert FL, Hayden O (1999) TrAC, Trends Anal Chem 18(3):192–199
30. Wu X, Shimizu KD (2008) Mol Recognit Polym 395–429
31. Shimizu KD, Stephenson CJ (2010) Curr Opin Chem Biol 14(6):743–750
32. BelBruno JJ (2009) Micro Nanosyst 1(3):163–180
33. Dickert FL, Hayden O (2001) Tech Instrum Anal Chem 23 (Molecularly Imprinted Polymers):503–525
34. Park J, Royer JE, Colesniuc CN, Bohrer FI, Sharoni A, Jin S, Schuller IK, Trogler WC, Kummel AC (2009) J Appl Phys 106(3):034505/034501–034505/034508
35. Igreja R, Dias CJ (2006) Sens Actuators, B B115(1):69–78
36. Grate JW, Klusty M, Barger WR, Snow AW (1990) Anal Chem 62(18):1927–1934
37. Bao Z, Lovinger AJ, Dodabalapur A (1996) Appl Phys Lett 69(20):3066–3068
38. Bouvet M (2006) Anal Bioanal Chem 384(2):366–373

39. Bouvet M, Pauly A (2006) Encycl Sens 6:227–269
40. Roberts ME, Sokolov AN, Bao Z (2009) J Mater Chem 19(21):3351–3363
41. Wang X, Someya T, Sekitani T, Kato Y, Iba S (2007) Mol Cryst Liq Cryst 462:29–36
42. Janata J, Josowicz M (2003) Nat Mater 2(1):19–24
43. Janata J, Josowicz M (2009) J Solid State Electrochem 13(1):41–49
44. Someya T, Dodabalapur A, Huang J, See KC, Katz HE (2010) Adv Mater 22(34):3799–3811
45. Janata J (2003) Phys Chem Chem Phys 5(23):5155–5158
46. Viricelle JP, Pauly A, Mazet L, Brunet J, Bouvet M, Varenne C, Pijolat C (2006) Mater Sci Eng, C 26(2–3):186–195
47. Bouvet M, Parra V, Locatelli C, Xiong H (2009) J Porphyrins Phthalocyanines 13(1):84–91
48. Li L, Tang Q, Li H, Hu W, Yang X, Shuai Z, Liu Y, Zhu D (2008) Pure Appl Chem 80 (11):2231–2240
49. Roberts ME, Queralto N, Mannsfeld SCB, Reinecke BN, Knoll W, Bao Z (2009) Chem Mater 21(11):2292–2299
50. Bao Z, Lovinger AJ, Dodabalapur A (1997) Adv Mater 9(1):42–44
51. Yang RD, Gredig T, Colesniuc CN, Schuller IK, Park J, Trogler WC, Kummel AC (2007) Appl Phys Lett 90:263506
52. Yang RD, Park J, Colesniuc CN, Schuller IK, Trogler WC, Kummel AC (2007) J Appl Phys 102:034515
53. Jiang H, Tan KJ, Zhang KK, Chen XD, Kloc C (2011) J Mater Chem 21(13):4771–4773
54. Royer JE, Park J, Colesniuc C, Lee JS, Gredig T, Lee S, Jin S, Schuller IK, Trogler WC, Kummel AC (2010) J Vac Sci Technol, B: Nanotechnol Microelectron: Mater, Process, Meas, Phenom 28(4):C5F22–C25F27
55. Yang RD, Park J, Colesniuc CN, Schuller IK, Trogler WC, Kummel AC (2007) J Appl Phys 102(3):034515/034511–034515/034517
56. Bohrer FI, Colesniuc CN, Park J, Ruidiaz ME, Schuller IK, Kummel AC, Trogler WC (2009) J Am Chem Soc 131(2):478–485
57. Snow AW, Barger WR (1989) Phthalocyanine films in chemical sensors. In: Lever ABP (ed) Phthalocyanines properties and applications, vol 1. Wiley, New York, pp 341–392
58. Day P, Price MG (1969) J Chem Soc A (2):236–240
59. Atkinson L, Day P, Price MG (1970) Phys Status Solidi A 2(3):K157–K159
60. Gould RD (1996) Coord Chem Rev 156:237–274
61. Hassan AK, Gould RD (1989) J Phys: Condens Matter 1(37):6679–6684
62. Zhou Q, Gould RD (1998) Thin Solid Films 317(1–2):432–435
63. Waite S, Pankow J, Collins G, Lee P, Armstrong NR (1989) Langmuir 5(3):797–805
64. Collins RA, Mohammed KA (1988) J Phys D: Appl Phys 21(1):154–161
65. Blochwitz J, Fritz T, Pfeiffer M, Leo K, Alloway DM, Lee PA, Armstrong NR (2001) Org Electron 2(2):97–104
66. Assour JM, Harrison SE (1965) Phys Chem Solids 26(3):670–672
67. Wright JD (1989) Prog Surf Sci 31(1–2):1–60
68. Maleysson C, Passard M, Blanc JP, Battut V, Germain JP, Pauly A, Demarne V, Grisel A, Tiret C et al (1995) Sens Actuators, B B26(1–3):144–149
69. Barbon A, Brustolon M, van Faassen EE (2001) Phys Chem Chem Phys 3(23):5342–5347
70. Yahiro H, Naka T, Kuramoto T, Kurohagi K, Okada G, Shiotani M (2005) Microporous Mesoporous Mater 79(1–3):291–297
71. Zwart J, Van Wolput JHMC (1979) J Mol Catal 5(1):51–64
72. Assour JM, Harrison SE (1964) J Phys Chem 68(4):872–876
73. Tezuka M, Ohkatsu Y, Osa T (1975) Bull Chem Soc Jpn 48(5):1471–1474
74. Maillard P, Krausz P, Giannotti C, Gaspard S (1980) J Organomet Chem 197(3):285–290
75. Law K-Y (1993) Chem Rev 93:449–486
76. Sadaoka Y, Sakai Y (1976) J Chem Soc, Faraday Trans 2 72(2):379–382
77. Salzman RF, Xue J, Rand BP, Alexander A, Thompson ME, Forrest SR (2005) Org Electron 6(5–6):242–246

78. Erk P, Hengelsberg H Phthalocyanine dyes and pigments, in Kadish KM, Smith KM, Guiland R (eds) (2003) The porphyrin handbook: phthalocyanines: spectroscopic and electrochemical characterization, 16:117–170
79. Schaffer AM, Gouterman M, Davidson ER (1973) Theor Chim Acta 30(1):9–30
80. Sayer P, Gouterman M, Connell CR (1982) Acc Chem Res 15(3):73–79
81. Minor PC, Gouterman M, Lever ABP (1985) Inorg Chem 24(12):1894–1900
82. Orti E, Bredas JL, Clarisse C (1990) J Chem Phys 92(2):1228–1235
83. Peralta GA, Seth M, Zhekova H, Ziegler T (2008) Inorg Chem 47(10):4185–4198
84. Mack J, Stillman MJ (2003) Porphyrin Handb 16:43–116
85. Hamann C (1972) Phys Status Solidi A 10:83–90
86. Bao Z, Lovinger AJ, Brown J (1998) J Am Chem Soc 120(1):207–208
87. de Oteyza DG, Barrena E, Osso JO, Dosch H, Meyer S, Pflaum J (2005) Appl Phys Lett 87(18):183504/183501–183504/183503
88. Karmann E, Meyer JP, Schlettwein D, Jaeger NI, Anderson M, Schmidt A, Armstrong NR (1996) Molecular crystals and liquid crystals science and technology, Section A: molecular crystals and liquid crystals 283 (Proceedings of the 7th International Conference on Unconventional Photoactive Systems, 1995):283–291
89. Michaelis W, Woehrle D, Schlettwein D (2004) J Mater Res 19(7):2040–2048
90. Yang RD, Park J, Colesniuc CN, Schuller IK, Royer JE, Trogler WC, Kummel AC (2009) J Chem Phys 130(16):164703/164701–164703/164708
91. Kutzler FW, Barger WR, Snow AW, Wohltjen H (1987) Thin Solid Films 155(1):1–16
92. Shafai TS, Gould RD (1992) Int J Electron 73(5):1043–1045
93. Abdel Malik TG, Abdel-Latif RM (1997) Thin Solid Films 305:336–340
94. Colesniuc CN, Biswas RR, Hevia SA, Balatsky AV, Schuller IK (2011) Phys Rev B: Condens Matter Mater Phys 83(8):085414/085411–085414/085417
95. Gould RD (1982) J Appl Phys 53(4):3353–3355
96. Hassan AK, Gould RD (1992) Int J Electronics 73(5):1047–1049
97. Amar N, Gould RD, Saleh AM (1998) Vacuum 50(1–2):53–56
98. Miller KA, Yang RD, Hale MJ, Park J, Fruhberger B, Colesniuc CN, Schuller IK, Kummel AC, Trogler WC (2006) J Phys Chem B 110(1):361–366
99. Rodriguez-Mendez ML (2006) Encycl Sens 9:111–133
100. Bora M, Schut D, Baldo MA (2007) Anal Chem 79(9):3298–3303
101. DiBenedetto SA, Facchetti A, Ratner MA, Marks TJ (2009) Adv Mater 21:1407–1433
102. Buttner WJ, Rieke PC, Armstrong NR (1985) J Phys Chem 89(7):1116–1121
103. Jimenez Tejada JA, Awawdeh KM, Villanueva JAL, Carceller JE, Deen MJ, Chaure NB, Basova T, Ray AK (2011) Org Electron 12(5):832–842
104. Jin Y, Yadav A, Sun K, Sun H, Pipe KP, Shtein M (2011) Appl Phys Lett 98(9):093305/093301–093305/093303
105. Abass AK, Krier A, Collins RA (1993) J Phys D: Appl Phys 26(7):1120–1124
106. Schuetze A, Pieper N, Zacheja J (1995) Sens Actuators, B B23(2–3):215–217
107. Bouvet M, Guillaud G, Leroy A, Maillard A, Spirkovitch S, Tournilhac F-G (2001) Sens Actuators, B 73:63–70
108. Chen Y, Bouvet M, Sizun T, Gao Y, Plassard C, Lesniewska E, Jiang J (2010) Phys Chem Chem Phys 12(39):12851–12861
109. Pizzini S, Timo GL, Beghi M, Butta N, Mari CM, Faltenmaier J (1989) Sens Actuators 17(3–4):481–491
110. Shu JH, Wikle HC, Chin BA (2010) Sens Actuators, B B148(2):498–503
111. Kolesar ES Jr, Wiseman JM (1989) Anal Chem 61:2355–2361
112. Mueller M, Bratke G, Gaertig J, Starke M, Hamann C (1989) Wiss Z Tech Univ Karl-Marx-Stadt 31(3):393–398
113. Sadaoka Y, Jones TA, Goepel W (1989) J Mater Sci Lett 8(9):1095–1097
114. Belghachi A, Collins RA (1990) J Phys D: Appl Phys 23(2):223–227

115. Starke M, Mrwa A, Mueller M, Hofmann A (1990) Wiss Z Tech Univ Karl-Marx-Stadt 32(1):79–84
116. Szczurek A, Lorenz K (1990) Int J Environ Anal Chem 41(1–2):57–63
117. Hamann C, Goepel W, Mrwa A, Mueller M, Rager A (1991) Sens Actuators, B 4(1–2):73–78
118. Gould RD, Ibrahim NA (2001) Thin Solid Films 398–399:432–437
119. Bott B, Jones TA (1984) Sens Actuators 5:43–53
120. Ray AK, Cook MJ, Thorpe SC, Mukhopadhyay S (1993) Phys Status Solidi A 140(2): K85–K88
121. Inagaki N, Tasaka S, Ikeda Y (1995) J Appl Polym Sci 55(10):1451–1454
122. Pakhomov GL, Pokhomov LG, Bagrov AM (1995) Khim Fiz 14(12):108–117
123. Ostrick B, Fleischer M, Meixner H (1998) Chem Sens, Tech Dig Int Meet, 7th:867–869
124. Bohrer FI, Colesniuc CN, Park J, Schuller IK, Kummel AC, Trogler WC (2008) J Am Chem Soc 130(12):3712–3713
125. Shilbayeh NF, Iskandarani MZ (2006) Am J Appl Sci 3(6):1879–1884
126. Basova TV, Kol'tsov EK, Igumenov IK (2005) Sens Actuators, B B105(2):259–265
127. Inabe T, Tajima H (2004) Chem Rev 104(11):5503–5533
128. Snow AW, Barger WR, Klusty M, Wohltjen H, Jarvis NL (1986) Langmuir 2(4):513–519
129. Kumar A, Singh A, Debnath AK, Samanta S, Aswal DK, Gupta SK, Yakhmi JV (2010) Talanta 82(4):1485–1489
130. Wu R, Jones TA (1990) Sens Actuators, B B2(1):33–42
131. Barger WR, Wohltjen H, Snow AW, Lint J, Jarvis NL (1986) ACS Symp Ser 309 (Fundam. Appl. Chem. Sens.):155–165
132. Bohrer FI, Sharoni A, Colesniuc C, Park J, Schuller IK, Kummel AC, Trogler WC (2007) J Am Chem Soc 129(17):5640–5646
133. de Haan A, Debilquy M, Decroly A (1999) Sens Actuators, B 57:69–74
134. Zhou R, Josse F, Gopel W, Oeztuerk ZZ, Bekaroglu O (1996) Appl Organomet Chem 10(8):557–577
135. Brinkmann H, Kelting C, Makarov S, Tsaryova O, Schnurpfeil G, Woehrle D, Schlettwein D (2008) Phys Status Solidi A 205(3):409–420
136. Brinkmann H, Kelting C, Makarov S, Tsaryova O, Schnurpfeil G, Woehrle D, Schlettwein D (2009) Org Electron 37–60
137. Liao M-S, Watts JD, Gorun SM, Scheiner S, Huang M-J (2008) J Theor Comput Chem 7(4):541–563
138. Liao M-S, Watts JD, Huang M-J, Gorun SM, Kar T, Scheiner S (2005) J Chem Theory Comput 1(6):1201–1210
139. Michaelis W, Kelting C, Hirth A, Woehrle D, Schlettwein D (2004) Macromol Symp 212 (Electronic Phenomena in Organic Solids):299–305
140. Muzikante I, Parra V, Dobulans R, Fonavs E, Latvels J, Bouvet M (2007) Sensors 7(11):2984–2996
141. Tada H, Touda H, Takada M, Matsushige K (1999) J Porphyrins Phthalocyanines 3(6/7):667–671
142. Karmann E, Meyer JP, Schlettwein D, Jaeger NI, Anderson M, Schmidt A, Armstrong NR (1996) Mol Cryst Liq Cryst Sci Technol, Sect A 283 (Proceedings of the 7th International Conference on Unconventional Photoactive Systems, 1995):283–291
143. Meyer JP, Schlettwein D (1996) Adv Mater Opt Electron 6(5&6):239–244
144. Hoshino S, Nagamatsu S, Chikamatsu M, Misaki M, Yoshida Y, Tanigaki N, Yase K (2002) Jpn J Appl Phys, Part 2 41(7A):L808–L810
145. Ling M-M, Bao Z (2006) Org Electron 7(6):568–575
146. Schon JH, Kloc C, Bao Z, Batlogg B (2000) Adv Mater 12(20):1539–1542
147. Nenon S, Kanehira D, Yoshimoto N, Fages F, Videlot-Ackermann C (2010) Thin Solid Films 518(19):5593–5598
148. Trogler WC, Johnson CE, Ellis DE (1981) Inorg Chem 20(4):980–986
149. Duarte D, Sharma D, Cobb B, Dodabalapur A (2011) Appl Phys Lett 98(13) 133302–133304

150. Grate JW, Klusty M (1991) Anal Chem 63(17):1719–1727
151. Grate JW, Martin SJ, White RM (1993) Anal Chem 65(21):940A
152. Grate JW, Patrash SJ, Abraham MH, Du CM (1996) Anal Chem 68(5):913–917
153. Grate JW, Abraham MH, McGill RA (1997) Handb Biosens Electron Noses 593–612
154. Finklea HO, Phillippi MA, Lompert E, Grate JW (1998) Anal Chem 70(7):1268–1276
155. Kaganove SN, Grate JW (1998) Polym Prepr (Am Chem Soc, Div Polym Chem) 39(1):556–557
156. Grate JW, Patrash SJ, Kaganove SN, Wise BM (1999) Anal Chem 71(5):1033–1040
157. Grate JW, Wise BM, Abraham MH (1999) Anal Chem 71(20):4544–4553
158. Grate JW, Zellers ET (2000) Anal Chem 72(13):2861–2868
159. Grate JW, Nelson DA (2003) Proc IEEE 91(6):881–889
160. Grate JW, Nelson DA, Skaggs R (2003) Anal Chem 75(8):1868–1879
161. Grate JW (2008) Chem Rev 108(2):726–745
162. Grate JW (2000) Chem Rev 100:2627–2648
163. Bishop SR, Tran NL, Poon GC, Kummel AC (2007) J Chem Phys 127(21):214702/214701–214702/214708
164. Tran NL, Bishop SR, Grassman TJ, Poon GC, Bohrer FI, Trogler WC, Kummel AC (2009) J Chem Phys 130(17):174305/174301–174305/174307
165. Tran NL, Bohrer FI, Trogler WC, Kummel AC (2009) J Chem Phys 130(20):204307/204301–204307/204307
166. Tran NL, Kummel AC (2007) J Chem Phys 127(21):214701/214701–214701/214707
167. Oeztuerk ZZ, Kilinc N, Atilla D, Guerek AG, Ahsen V (2009) J Porphyrins Phthalocyanines 13(11):1179–1187
168. Hsieh JC, Liu CJ, Ju YH (1998) Thin Solid Films 322(1–2):98–103
169. Armand F, Perez H, Fouriaux S, Araspin O, Pradeau JP, Claessens CG, Maya EM, Vazquez P, Torres T (1999) Synth Met 102(1–3):1476–1477
170. Generosi A, Paci B, Albertini VR, Generosi R, Paoletti AM, Pennesi G, Rossi G, Fosca M, Caminiti R (2008) Sens Actuators, B B134(2):396–402
171. Generosi A, Paci B, Albertini VR, Generosi R, Perfetti P, Paoletti AM, Pennesi G, Rossi G, Caminiti R (2007) J Phys Chem C 111(32):12045–12051
172. Liu CJ, Hou WC, Ju YH (2000) J Chin Inst Chem Eng 31(3):237–242
173. Liu CJ, Wang SY, Hsieh JC, Ju YH (2000) Sens Actuators, B B65(1–3):371–374
174. Takeuchi N, Nagasawa T, Murakami K, Watanabe K (1996) Chem Sens 12(suppl A):81–84
175. Volpati D, Alessio P, Zanfolim AA, Storti FC, Job AE, Ferreira M, Riul A Jr, Oliveira ON Jr, Constantino CJL (2008) J Phys Chem B 112(48):15275–15282
176. Wilson A, Rigby GP, Wright JD, Thorpe SC, Terui T, Maruyama Y (1992) J Mater Chem 2(3):303–308
177. Hsieh J-C (2009) Biomed Eng 21(6):467–470
178. Strelcov E, Kolmakov A (2008) J Nanosci Nanotechnol 8(1):212–221
179. Baker S, Petty MC, Roberts GG, Twigg MV (1983) Thin Solid Films 99:53–59
180. Moriizumi T (1988) Thin Solid Films 160:413–429
181. Kumaran N, Veneman PA, Minch BA, Mudalige A, Pemberton JE, O'Brien DF, Armstrong NR (2010) Chem Mater 22(8):2491–2501
182. Gu C, Sun L, Zhang T, Li T, Hirata M (1994) Thin Solid Films 244(1–2):909–912
183. Gutierrez N, Rodriguez-Mendez ML, de Saja JA (2001) Sens Actuators, B B77 (1–2):437–442
184. Lando JB, Wang HY, Ko WH (1991) Polym Prepr (Am Chem Soc, Div Polym Chem) 32(3):378–379
185. Lu A, Li Y, Pang X, Jiang D, Hua Y (1991) Gongneng Cailiao 22(1):41–45
186. Wang HY, Ko WH, Batzel DA, Kenney ME, Lando JB (1990) Sens Actuators, B B1 (1–6):138–141
187. Wohltjen H, Barger WR, Snow AW, Jarvis NL (1985) IEEE Trans Electron Devices ED-32 (7):1170–1174

188. Xu L, Krishnaswamy L, Burrows VA, George RD (1996) J Vac Sci Technol, B 14(5): 3386–3390
189. Crouch D, Thorpe SC, Cook MJ, Chambrier I, Ray AK (1994) Sens Actuators, B 19(1–3): 411–414
190. Hassan AK, Ray AK, Travis JR, Ghassemlooy Z, Cook MJ, Abass A, Collins RA (1998) Sens Actuators, B B49(3):235–239
191. Mukhopadhyay S, Hogarth CA, Thorpe SC, Cook MJ (1994) J Mater Sci: Mater Electron 5(6):321–323
192. Mukhopadhyay S, Mukherjee D, Cook MJ, Hogarth CA (1994) Int J Electron 76(5):757–761
193. Mukhopadhyay S, Ray AK, Cook MJ, Simmons JM, Hogarth CA (1992) J Mater Sci: Mater Electron 3(3):139–143
194. Nabok AV, Ray AK, Hassan AK, Travis JR, Cook MJ (1997) Supramol Sci 4(3–4):407–411
195. Ray AK, Hassan AK, Saatchi MR, Cook MJ (1997) Philos Mag B 76(6):961–971
196. Travis J, Ray AK, Thorpe SC, Cook MJ, James SA (1995) Meas Sci Technol 6(7):988–994
197. Velez M, Vieira S, Chambrier I, Cook MJ (1998) Langmuir 14(15):4227–4231
198. Rella R, Serra A, Siciliano P, Tepore A, Valli L, Zocco A (1997) Langmuir 13:6562–6567
199. Boschloo G, Hagfelt A (2009) Acc Chem Res 11:1819–1826
200. Park J, Yang RD, Colesniuc CN, Sharoni A, Jin S, Schuller IK, Trogler WC, Kummel AC (2008) Appl Phys Lett 92(19):193311/193311–193311/193313
201. Schmidt A, Chau LK, Back A, Armstrong NR (1996) Phthalocyanines 4:307–341
202. Yang RD, Gredig T, Colesniuc CN, Park J, Schuller IK, Trogler WC, Kummel AC (2007) Appl Phys Lett 90(26):263506/263501–263506/263503
203. Gao J, Xu JB, Zhu M, Ke N, Ma D (2007) J Phys D: Appl Phys 40(18):5666–5669
204. Miller CW, Sharoni A, Liu G, Colesniuc CN, Fruhberger B, Schuller IK (2005) Phys Rev B 72:104113-104111-104116
205. Gredig T, Bergman D, Gentry KP (2010) Nanotech Conf Expo 2010 1:119–121
206. Gentry KP, Gredig T, Schuller IK (2009) Phys Rev B: Condens Matter Mater Phys 80 (17):174118/174111–174118/174115
207. Knupfer M, Peisert H (2004) Phys Status Solidi A 201(6):1055–1074
208. Abu-Hilal AO, Saleh AM, Gould RD (2005) Mater Chem Phys 94(1):165–171
209. Saleh AM, Gould RD, Hassan AK (1993) Phys Status Solidi A 139(2):379–389
210. Saleh AM, Hassan AK, Gould RD (2003) J Phys Chem Solids 64:1297–1303
211. Amar NM, Saleh AM, Gould RD (2003) Appl Phys A: Mater Sci Process 76(1):77–82
212. Schlaf R, Crisafulli LA, Murata H, Merritt CD, Kafafi ZH, Schroeder PG, Nelson MW, Parkinson BA, Lee PA, Nebesny KW, Armstrong NR (1999) Proceedings of SPIE-The International Society for Optical Engineering 3797 (Organic Light-Emitting Materials and Devices III):189–197
213. Ahmad A, Collins RA (1991) Phys Stat Sol A 126:411–426
214. Klauk H, Zschieschang U, Pflaum J, Halik M (2007) Nature 445(7129):745–748
215. Weitz RT, Amsharov K, Zschieschang U, Burghard M, Jansen M, Kelsch M, Rhamati B, van Aken PA, Kern K, Klauk H (2009) Chem Mater 21(20):4949–4954
216. Schlettwein D, Oekermann T, Jaeger N, Armstrong NR, Wohrle D (2002) Chem Phys 285(1):103–112
217. Aernecke MJ, Guo J, Sonkusale S, Walt DR (2009) Anal Chem 81(13):5281–5290
218. Walt DR (2005) Anal Chem 77(3):45A
219. Fraga CG, Melville AM, Wright BW (2007) Analyst 132(3):230–236
220. Cotte-Rodriguez I, Justes Dina R, Nanita Sergio C, Noll Robert J, Mulligan Christopher C, Sanders Nathaniel L, Cooks RG (2006) Analyst 131(4):579–589
221. Lim S-H, Raorane D, Satyanarayana S, Majumdar A (2006) Sens Actuators, B B119(2):466–474
222. Carrano J (2004) Chemical and biological sensor study. DARPA Microsystems Technology Office, US Department of Defense, 31 pp

223. Drain CM, Hupp JT, Suslick KS, Wasielewski MR, Chen X (2002) J Porphyrins Phthalocyanines 6(4):243–258
224. Janzen MC, Ponder JB, Bailey DP, Ingison CK, Suslick KS (2006) Anal Chem 78(11): 3591–3600
225. Lim SH, Feng L, Kemling JW, Musto CJ, Suslick KS (2009) Nature Chem 1(7):562–567
226. Suslick KS, Rakow NA, Sen A (2004) Tetrahedron 60(49):11133–11138
227. Rella R, Siciliano P, Capone S, Spadavecchia J, Ciccarella G, Vasapollo G (2004) Sens Microsyst, Proc Ital Conf, 8th:115–120
228. Uttiya S, Kladsomboon S, Chamlek O, Suwannet W, Osotchan T, Kerdcharoen T, Brinkmann M, Pratontep S (2009) Springer Proc Phys 129 (Interface Controlled Organic Thin Films):211–215
229. Rella R, Capone S, Siciliano P, Spadavecchia J, Ciccarella G (2004) Proc SPIE-Int Soc Opt Eng 5502 (Second European Workshop on Optical Fibre Sensors, 2004):435–438
230. Uttiyh S, Pratontep S, Bhanthumnavin W, Buntem R, Kerdcharoen T (2008) Volatile organic compound sensor array based on zinc phthalocyanine and zinc porphyrin thin films. In: 2nd IEEE International Nanoelectronics Conference, 2008. pp 618–623
231. Oezmen A, Tekce F, Ebeoglu MA, Tasaltin C, Oeztuerk ZZ (2006) Sens Actuators, B B115(1):450–454
232. Parra V, Arrieta AA, Fernandez-Escudero JA, Garcia H, Apetrei C, Rodriguez-Mendez ML, de Saja JA (2006) Sens Actuators, B B115(1):54–61
233. LaFratta CN, Walt DR (2008) Chem Rev 108(2):614–637
234. Critchley SM, Willis MR, Maruyama Y, Bandow S, Cook MJ, McMurdo J (1993) Mol Cryst Liq Cryst Sci Technol, Sect A 229:47–51
235. Dick S, Peisert H, Dini D, Hanack M, Cook MJ, Chambrier I, Chasse T (2005) J Appl Phys 97(7):073715/073711–073715/073718
236. Haug A, Harbeck S, Dini D, Hanack M, Cook MJ, Peisert H, Chasse T (2005) Appl Surf Sci 252(1):139–142
237. Nicolau M, Del Rey B, Torres T, Mingotaud C, Delhaes P, Cook MJ, Thorpe SC (1999) Synth Met 102(1–3):1462–1463
238. Peisert H, Biswas I, Zhang L, Knupfer M, Hanack M, Dini D, Cook MJ, Chambrier I, Schmidt T, Batchelor D, Chasse T (2005) Chem Phys Lett 403(1–3):1–6

Struct Bond (2012) 142: 119–154
DOI: 10.1007/430_2011_49

Published online: 29 July 2011

# Biological Outer-Sphere Coordination

**Kyle M. Lancaster**

**Abstract** The concept of outer-sphere coordination (OSC) is surveyed in the context of bioinorganic chemistry. A distinction is made between electronic and structural OSC, both arising from the interaction of the protein matrix with inner sphere ligands. Electronic OSC entails the electronic interaction between the polypeptide and inner-sphere ligands. These effects principally arise from hydrogen-bonding interactions, though through-space dipolar interactions are also encountered. Structural OSC comprises primarily steric effects that do not necessarily impact ligand electronics but rather influence inner sphere topology. Additionally, the protein matrix can be envisioned as a local "solvent" whose bulk dielectric and point charges influence the metal center. Recurring themes are highlighted where OSC regulates the properties of various metalloproteins distinguished by cofactors and/or function. Finally, cases are presented where OSC has guided molecular design.

**Keywords** Bioinorganic chemistry · Outer sphere coordination · Protein design

## Contents



---

"With some exaggeration it might be said that the model builder who finally would succeed in mimicking a biological metal centre, stands a fair chance of finding that he has synthesized not only the metal centre but the protein matrix as well." G.W. Canters and G. Gilardi

K.M. Lancaster (✉)
Department of Chemistry and Chemical Biology, Cornell University, Ithaca, NY, USA
e-mail: kml236@cornell.edu

## 1 Introduction

Transition metal cofactors typically lose function when extracted from their host proteins. Related to this observation is the diminished, divergent, or nonexistent function of synthetic complexes modeled on metalloprotein active sites. Common to both of these phenomena is the loss of the tuning afforded by the protein matrix. The protein matrix surrounding such sites plays a vital role in the bioinorganic chemistry.

The interaction between the protein matrix and its nested coordination complexes is exemplary of outer sphere coordination (OSC). Recently, OSC has guided the design of both small molecules and metalloproteins to tune existing or imbue novel properties. While often cited to explain metalloprotein behavior, OSC is not typically treated in a descriptive manner. We will begin with Bjerrum's favored definition [1] of "coordination in the second sphere" (chosen from Werner's original postulation [2]): "a complex with a fully occupied first sphere has residual affinity to attach groups." This mode of OSC is involved in supramolecular chemistry [3, 4], but does not quite suit our present discussion.

Werner offers another possibility: "A group directly coordinated to a ligand in the first sphere" [2]. This class of OSC accords better with the definition as employed by bioinorganic chemists seeking to explain metalloprotein properties and reactivity. Exemplifying this class of OSC are hydrogen bond networks formed between the protein and ligands directly coordinated to the metal. However, this does not account for the entirety of "outer sphere" interactions. Groups interacting with inner-sphere ligands can impose steric restrictions, leading to energetically frustrated geometries in which coordination geometry is adopted by the transition metal that differs from "preferred" behavior. Additionally, consideration must be afforded to the explicit solvation environment conferred upon metal cofactors by their protein hosts. These factors are all vital to proper metalloprotein function. Establishment of a proper OSC environment is arguably as important an evolutionary selection criterion as adoption of a proper tertiary polypeptide fold.

Our formalism shall include three types of OSC. We will explore *electronic OSC*. Here, we will consider OSC where the electronics of a ligand are influenced by some species not directly bound to a metal; these perturbations manifest in altered spectroscopic and thermodynamic properties. *Electronic OSC* will also be considered where nonligating atoms participate in active site reactivity, as with hydrogen bonds that define electron transfer (ET) pathways. *Structural OSC* will be distinguished where noncoordinated groups perturb the metal geometry without themselves interacting electronically with the inner sphere. We will also encounter effects arising from the protein-defined *solvation environment*.

In searching for cases of these three types of OSC, we will survey a wide (but certainly not exhaustive) series of metalloproteins, distinguished principally by their metal cofactors. We will begin with metalloproteins whose cofactors are formed entirely by polypeptide–metal ion interactions. Protein-mediated tuning of cofactors assembled from metal ions bound by exogenous prosthetic groups will follow. The discussion will close with a brief survey of the efforts to exploit Nature's OSC motifs to the end of imparting new functions on metalloproteins as well as engineering functional small molecule models.

## 2 Polypeptide-Bound Metal Cofactors

Many metalloproteins directly incorporate transition metals without requiring prosthetic groups. Commonly encountered polypeptide-based ligands are cysteine thiolates, aspartate and glutamate carboxylates, asparagine and glutamine carboxamides, methionine thioethers, and histidine imidazoles. This side chain-based ligand library is supplemented by main-chain carbonyls. The capabilities of polypeptide-based ligation are extended by the protein matrix. The matrix may dictate ligand electronics through hydrogen-bonding interactions or local solvation. Furthermore, the protein fold can force topologies that would not be encountered in discrete coordination complexes possessing identical inner-sphere ligands.

### *2.1 Fe and Mn Superoxide Dismutases*

The superoxide dismutase (SOD) superfamily comprises a variety of metalloenzymes that deactivate superoxide ($O_2^{\bullet-}$). Because of the propensity for formation of $O_2^{\bullet-}$ during respiration, examples of these enzymes are found in nearly all aerobic organisms. The identities of the metal ion(s) incorporated in the active sites evolutionarily distinguish these proteins. The Fe and Mn SODs differ primarily in their cofactor, otherwise they are nearly homologous. Case in point, representative oxidized active sites are almost entirely superimposable (Fig. 1). Both active sites bind metals in trigonal bipyramidal geometry, with an equatorial inner sphere consisting of two histidine imidazoles and an aspartate carboxylate. The coordination sphere is completed by an axial histidine and either a water or hydroxide ligand. The protonation state of this axial ligand depends on the oxidation state of the metal, with the trivalent state binding hydroxide and the divalent state binding water. Despite this strong similarity, metal ion substitution (Fe to Mn and vice versa) obviates catalytic activity.

A proposal was offered by Miller: simply, the midpoint reduction potentials ($E_m$) of the two classes of SOD are tuned in opposite fashions [5]. As the $E_m$ values for

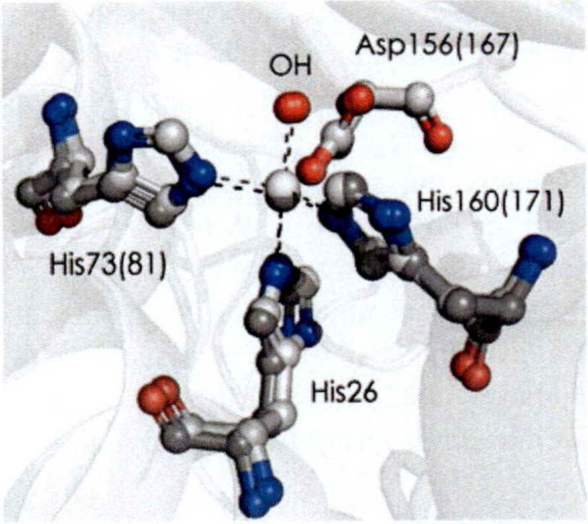


**Fig. 1** Overlaid $Fe^{III}$SOD (*white*, PDBID: 1ISB) and $Mn^{III}$SOD (*gray*, PDBID: 1VEW) inner coordination spheres, showing a high degree of structural similarity. Figures are numbered according to *Escherichia coli* FeSOD, with MnSOD numbering indicated in parentheses if different

$Mn^{III/II}$ are much more positive in small molecules than those of $Fe^{III/II}$ [6], the MnSOD must lower the $Mn^{III/II}$ couple to a range suitable for superoxide dismutation while FeSOD raises the $Fe^{III/II}$ couple accordingly. Early support for this explanation came from measuring the $E_m$ of Fe-substituted MnSOD [Fe(Mn) SOD] in comparison to that of FeSOD. Indeed, the reduction potential of Fe(Mn) SOD drops from the wild-type value of approximately 200 mV to −240 mV (vs. NHE). This report launched extensive study by Miller, Brunold, and others of the mechanisms by which these highly similar proteins tune their active site $E_m$ [7, 8].

In the $Fe^{II}$, $H_2O$-bound state, FeSOD, and Fe(Mn)SOD are very similar spectroscopically [9]. Electrostatics are insufficiently altered to explain the ~400 mV difference in reduction potential. However, EPR spectra of the $Fe^{III}$, OH-bound sites revealed striking differences in solvent-binding mode and active site $pK_a$ [9–11]. As redox in FeSOD is coupled to the protonation/deprotonation of the axial solvent molecule, differing interactions of the proteins with this axial ligand were implicated in the observed redox behavior.

Research focused on a conserved glutamine at position 69 in FeSOD (146 in MnSOD). This residue acts as a hydrogen bond donor to the axial solvent via its side chain amide proton. The residue is oriented by participation in a network of hydrogen bonds between Tyr34 and Trp122 (Trp128 in MnSOD) (Fig. 2). Studies of Gln69X variants of FeSOD generated by site-directed mutagenesis demonstrated that this residue plays a critical role in tuning the $E_m(Fe^{III/II})$. Gln69His and Gln69Glu have $E_m$ values raised by 220 and 660 mV, respectively [12]. Miller and coworkers concluded from detailed study that Gln69 stabilizes the $Fe^{III}$ state by ~220 mV owing to hydrogen bond donation to $OH^-$. In Gln69Glu, the hydrogen bond is now accepted from the solvent ligand, stabilizing the $Fe^{II}$, $H_2O$-bound state by ~200 mV. Moreover, Glu69 undergoes deprotonation during redox cycling, providing a proton transfer relay to the axial ligand. This facile proton transfer from this nearby residue to the axial solvent molecule confers another ~300 mV of stability on the $Fe^{II}$ state.

This OSC tuning is also operative in MnSOD. As with the Fe protein, MnSOD and Mn(Fe)SOD are spectroscopically similar, indicating that Mn electronic structure differences are a minor contribution to the observed changes in $E_m$ [7].

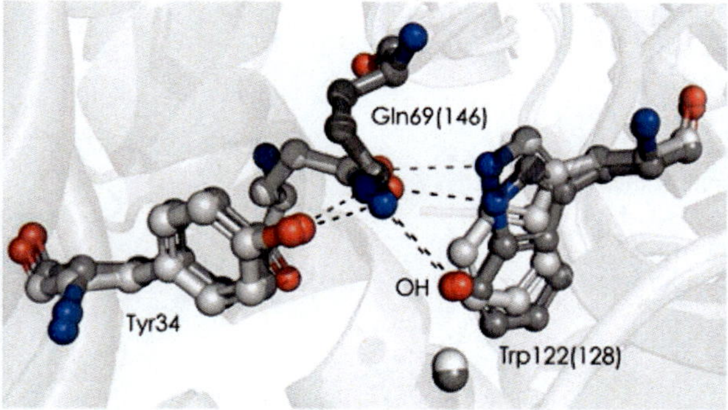


**Fig. 2** Overlaid hydrogen bond OSC network of $Fe^{III}$SOD (*white*, PDBID: 1ISB) and $Mn^{III}$SOD (*gray*, PDBID: 1VEW). Gln69(146) donates a hydrogen bond to the axial solvent. Hydrogen bonds from Tyr34 and Trp122(128) enforce this interaction by orienting the carboxamide side chain. Figures are numbered according to *Escherichia coli* FeSOD, with MnSOD numbering indicated in parentheses if different

The requirement for a $Mn^{III/II}$ couple with $E_m$ dropped sufficiently to allow superoxide dismutation suggests a stronger Gln-axial solvent hydrogen bond in MnSOD. This was observed by $^{15}N$ NMR spectroscopy. Paramagnetic shifting of Gln146 by 79 ppm in $Fe^{II}$(Mn)SOD exceeds that of Gln69 in $Fe^{II}$SOD, which is shifted 45 ppm [13, 14]. This stronger bond tunes the axial solvent to lower p$K_a$, favoring $OH^-$ and thus the oxidized trivalent state.

The Fe and Mn SODs demonstrate clearly the dramatic effects of electronic OSC over active site properties. Differences between these two sites arise purely due to the hydrogen-bonding networks influencing the acidity of the axial solvent ligand. These variations in OSC are sufficient to tune $E_m$ across a large potential window in sites with essentially identical inner-sphere coordination environments.

## 2.2 α-Ketoglutarate-Dependent Fe Dioxygenases

$Fe^{II}$/α-ketoglutarate (α-KG)-dependent dioxygenases hydroxylate a variety of substrates via the oxidative decarboxylation of α-KG [15]. They share an active site structure wherein $Fe^{II}$ is coordinated by a facial triad of two His and one Asp. The sites are primed for reactivity with $O_2$ by the binding of α-KG, which serves as a bidentate ligand to Fe. A recent study [16] examined the role of OSC in the reactivity of one member of this family, "factor inhibiting HIF-1" (FIH). FIH participates in human $O_2$ sensing by catalyzing hydroxylation of $C_\beta$ of Asn803 in the C-terminal transactivation domain (CTAD) of hypoxia-inducible factor (HIF) [17]. The active site in the α-KG and HIF-1 bound form of FIH features an extensive hydrogen bonding network (Fig. 3). Knapp and coworkers identified Arg238, Asn205, and Asn294 as important determinants of the reactivity behavior of the enzyme. The effects of their interactions with the $Fe^{II}$ ligands could be

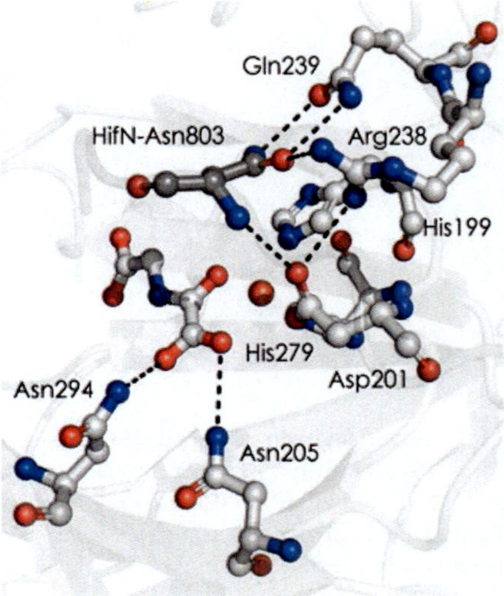


**Fig. 3** Active site of FIH with α-KG and the HIF CTAD (indicated in *gray*) bound (PDBID: 1H2K). Asn294 and Asn205 have been shown to control the electronics of the α-KG-$Fe^{II}$ interaction, while Arg238 orients substrate with respect to $Fe^{II}$

distinguished by electronic spectroscopy and activity measurements of point mutants where the hydrogen bond network was perturbed. EPR spectra of $Cu^{II}$-substituted FIH variants are reported as a claim of minimal structural perturbation by these mutants, but such evidence is circumstantial as numerous competing effects giving rise to EPR spectra can mask subtle structural perturbations.

Asn205 and Asn294 participate in a largely electronic OSC network. Asn205Ala and Asn294Ala mutations each led to a blue shift of a metal-to-ligand charge-transfer (MLCT) band in the absorption spectra of $Fe^{II}$/α-KG FIH. This was attributed to a destabilization of α-KG π* orbitals owing to increased electron density following removal of hydrogen bonds to the carboxylate oxygens. This increased electron density destabilizes the buildup of negative charge that occurs during catalysis; this is reflected in attenuated $k_{cat}$ values for these two mutants.

The isosteric substitution Arg238Met obviates hydrogen bonds to Asp201 and the HIF-Asn803 substrate. This mutation renders the enzyme inactive. Knapp and coworkers attribute this to incorrect positioning of the CTAD substrate. However, here the electronics are slightly perturbed as well, evidenced by a blue-shifted MLCT. This likely arises due to reorientation of Asp201 in the absence of Arg238 hydrogen bonding. Moreover, the removal of the positively charged guanidinium side chain would influence charge distribution over the carboxylate, resulting in a more electron-rich coordinating oxygen.

## 2.3 Type 1 Copper Proteins

Type 1 or "blue" copper (T1Cu) proteins comprise a rich case where the effects of electronic OSC, structural OSC, and the solvation sphere have all been explored.

T1Cu proteins incorporate Cu ions into active sites with $E_m$ values elevated relative to aqueous $Cu^{II}$ (~ 1 V > $E_m$ > 200 mV, vs. ~ 150 mV). These sites mediate efficient electron transfer (ET) reactions [18]. T1Cu domains participate in intermolecular ET mediated by small electron carriers (the cupredoxins). They also serve as sites for intramolecular ET to catalytic domains in larger proteins such as multicopper oxidases (MCOs) and nitrite reductases (NiRs). T1Cu is readily identified by an intense blue color and by a small axial $^{63/65}Cu$ hyperfine splitting ($25 < A_{||} < 95 \times 10^{-4}$ $cm^{-1}$) [19]. Their molecular structures consist of a conserved His-His-Cys equatorial inner-sphere and weaker axial interactions that vary among different proteins [18]. Solomon and coworkers have intensively studied the electronic structure of T1Cu [20]. The Cys thiolate forms a highly π-covalent interaction with the $Cu^{II}$ that delocalizes ~40% of the unpaired spin onto sulfur and shrinks $A_{||}$. This strong π overlap also gives rise to the intense blue absorption band, assigned as an S π → Cu $3d_{x2\text{-}y2}$ ligand-to-metal charge transfer (LMCT) transition. Site-directed mutagenesis (SDM), protein engineering, and synthetic modeling studies have demonstrated that the His-His-Cys ligand set is necessary but not sufficient for the generation of a T1Cu site [21, 22]. OSC tuning of the electronics and geometry of this ligand set completes the formation of a T1Cu site.

### 2.3.1 Thermodynamic and Spectroscopic Properties

$Cu^{II/I}$ $E_m$ values in naturally occurring T1Cu sites range from 180 to >1,000 mV vs. NHE [23]. Values for four cupredoxins are presented in Table 1. Comparison of the active site structures of these proteins (Fig. 4) differentiates stellacyanin by a tightly coordinated axial glutamine, relative to the weakly interacting methionine of plastocyanin and rusticyanin [24–26]. Inner-sphere coordination readily explains the low $E_m$ of stellacyanin; SDM mutagenesis studies on a variety of cupredoxins have demonstrated significant sensitivity of $E_m$ to the nature of this axial ligand. Azurin's elevated $E_m$ has been attributed to the substitution of this ligand for a weakly interacting methionine thioether. The lower azurin $E_m$ relative to plastocyanin is rationalized by stabilization of $Cu^{II}$ by the Gly45 carbonyl at 2.6 Å [27].

Plastocyanin and rusticyanin possess practically identical inner coordination spheres, yet differ markedly in their $E_m$ values. This discrepancy in $E_m$ has been attributed to the greater site hydrophobicity of rusticyanin. A SDM study by Lu and coworkers of the azurin from *Pseudomonas aeruginosa* has explicitly examined the

**Table 1** Reduction potentials of T1Cu in cupredoxins spanning the natural range

| Protein | $E_m$ (mV vs. NHE) | pH | Reference |
|---|---|---|---|
| *R. vernicifera* stellacyanin | 184 | 7.1 | [28] |
| Cucumber stellacyanin (non-glycosylated) | 260 | 7 | [29] |
| *P. aeruginosa* azurin | 304 | 7 | [30] |
| Poplar plastocyanin | 375 | 7 | [23] |
| *T. ferooxidans* rusticyanin | 680 | 3.2 | [31] |

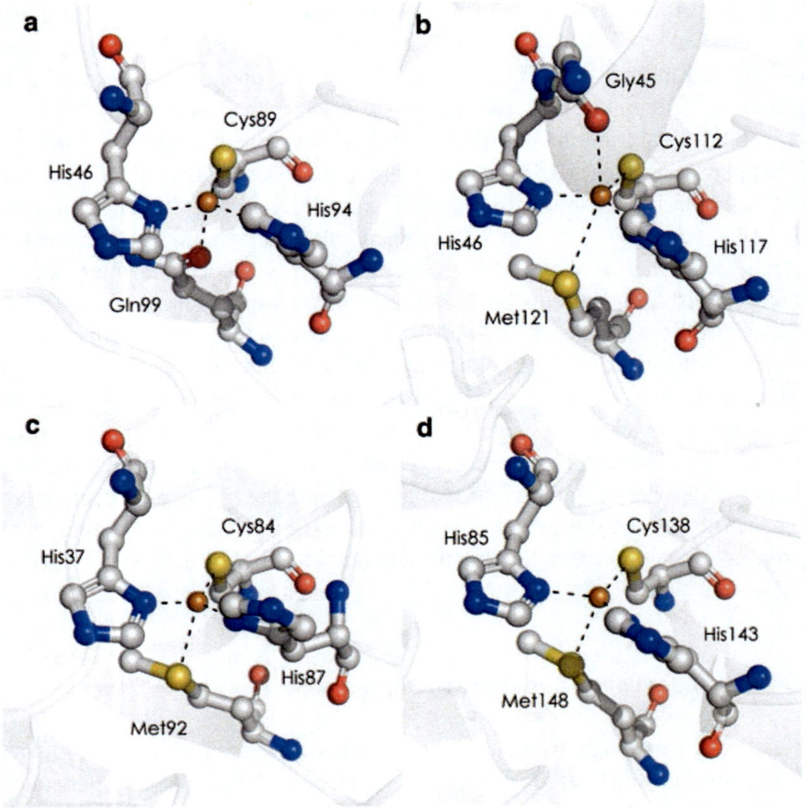


**Fig. 4** T1Cu sites of cucumber stellacyanin (**a**, PDBID: 1JER), *P. aeruginosa* azurin (**b**, PDBID: 4AZU), poplar plastocyanin (**c**, PDBID: 1PLC), and *T. ferooxidans* rusticyanin (**d**, PDBID: 2CAK), in order of increasing $E_m$. Stellacyanin features coordination by an axial glutamine, which increases ligand field strength and consequently lowers $E_m$. In the other three proteins, this interaction is replaced by a weakly coordinated methionine thioether. A weakly interacting axial carbonyl in azurin lowers $E_m$ relative to plastocyanin, while high site hydrophobicity raises $E_m$ in rusticyanin relative to plastocyanin

effects of site hydrophobicity on $E_m$ [32]. The early observation of an increased $E_m$ in the Met121Leu azurin mutant [33] was extended by the substitution of Met121 with isostructural nonstandard amino acids norleucine (thioether substituted by methylene) and selenomethionine. $E_m$ was found to vary linearly with $\log_P$, a measure of the hydrophobicity of the amino acid at position 121. This work was extended by Berry and coworkers, who introduced phenylalanine residues surrounding the metal-binding site [34]. The added hydrophobicity of these residues elevated $E_m$ by approximately 30 mV per phenylalanine with minimal perturbations to the $Cu^{II}$ electronic structure.

The influence of solvation extends beyond bulk active site hydrophobicity. A quantum-mechanics/molecular mechanics (QM/MM) study by Sinnecker and

Neese has demonstrated that T1Cu spectroscopic properties experience substantial influence from backbone charges as well as the protein-solvating water molecules [35]. Surface charges have also been shown to modulate $E_m$. Sola and coworkers investigated variants of plastocyanin where neutral (polar and nonpolar) surface residues were substituted with charged residues. These substitutions demonstrated appreciable effects on $E_m$ [36]. They observed intuitive effects, i.e., adding positive charge raises $E_m$, adding negative charge lowers $E_m$. However, the overall electrostatic potential distribution around the metal site was found to heavily influence $E_m$ as well.

The electronic effects of OSC on metal–ligand bonds have been explored in T1Cu proteins. The cysteine thiolates are tuned by hydrogen bonds donated from backbone amide residues; these networks are often referred to as "the rack" [37]. Early work by Spiro and coworkers showed that removal of hydrogen bonds from Asn38 in plastocyanin resulted in spectroscopic perturbations and decreased protein stability, though 3D structural characterization was not reported for any of the mutants discussed [38]. Thorough investigation has been performed with the hydrogen-bonding network in azurin. In azurin, this network comprises Asn47 and Phe114 (Fig. 5a). Dennison and coworkers removed one of these hydrogen bonds via the Phe114Pro mutation [39]. $E_m$ in this variant decreases 90 mV concomitant with an 800 $cm^{-1}$ blue shift and decreased intensity of the S$\pi$ LMCT band. Structural characterization (Fig. 5b) reveals a 0.06–0.09 Å decrease in the Cu-S(Cys112) bond length, along with a ~0.35 Å decrease in Cu-S(Met121). Loss of the Phe114 hydrogen bond causes Cys112 to rotate about $C_\beta$, raising the Cu-S-$C_\beta$-$C_\alpha$ dihedral from 7° to 35°. The bond length decrease toward Met121 results in a ~1 Å longer Cu-O(Gly45) distance. Thus, decoupling inner-sphere effects from OSC in tuning $E_m$ is difficult. Moreover, here both structural and electronic OSC effects are at play.

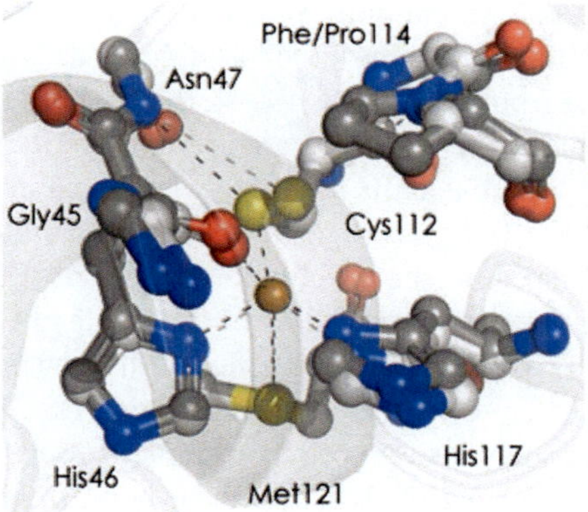


**Fig. 5** T1Cu sites of wild-type (*white*, PDBID: 4AZU) and Phe114Pro (*gray*, PDBID: 2GHZ) *P. aeruginosa* azurin. Wild-type azurin features a hydrogen-bond OSC network, with backbone amide protons from Asn47 and Phe114 donated to the Cys112 thiolate. In Phe114Pro, this network is disrupted, with only the Asn47-Cys112 hydrogen bond remaining. This results in a 90 mV drop in $E_m$, concomitant with dramatically perturbed spectroscopic features and decreased ET activity (See Sect. 2.3.2)

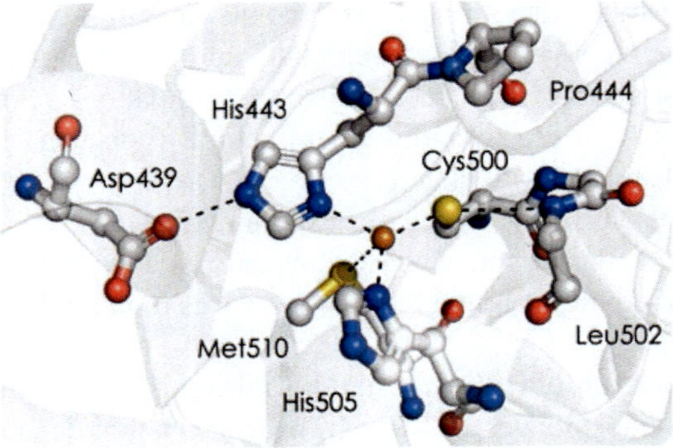


**Fig. 6** T1Cu site of CuEO (PDBID: 1N68). Electronic OSC comprises hydrogen bonds to Cys500 from the Leu502 backbone amide and to His443 from the Asp439 side chain. Installation of a hydrogen bond at position 444 by replacement of Pro with aliphatic amino acids removes electron density from the Cys500 thiolate, raising $E_m$. Removal of the Asp439 hydrogen bond removes electron density from His443, also raising $E_m$

Removal of the hydrophobic Phe114 side chain should lower the reduction potential, as should the removal of electron density from the cysteine thiolate via hydrogen bonding. A stronger axial interaction would also be expected to stabilize $Cu^{II}$.

Sakurai and coworkers present another scenario with mutants of the MCO copper efflux oxidase (CuEO) from *E. coli* [40]. CuEO participates in metal ion homeostasis by coupling the four electron reduction of dioxygen to water by oxidation of $Cu^{I}$ ions to $Cu^{II}$ [41]. Common to MCOs is the ET coupling of a T1Cu site to a trinuclear cluster consisting of a T2Cu and a binuclear type 3 copper (T3Cu) site [42]. In CuEO, the analog to the "rack" comprises only a hydrogen bond to the Cys500 thiolate from the backbone amide of Leu502 (Fig. 6). A proline is found at position 444, prohibiting a second hydrogen bond to the thiolate. Reversing Dennison's approach, the Pro444X (X = Ala, Leu, Ile) mutants raise the reduction potential from 360 mV to 400–430 mV. While structural characterization is not available, UV/vis, EPR, and resonance Raman (RR) data were presented. Here, no substantial shifts in the π LMCT band maxima were observed. EPR spectra were qualitatively similar, though spin Hamiltonian parameters are unreported. The RR spectra show ~2 $cm^{-1}$ shifts of Cu-S modes to lower frequencies. Together, the data suggest that the T1Cu site of CuEO in these mutants is largely unperturbed by the addition of a second hydrogen bond to Cys500 and thus shifts in $E_m$ can be largely attributed to electronic OSC tuning of the thiolate.

The Sakurai CuEO study also probes the effects of tuning histidine electronics. Common to metal oxidase MCOs is a hydrogen bond from a carboxylate at the substrate-binding site to a T1Cu histidine. The Asp439Ala mutation removes this interaction with His443, resulting in a 70 mV increase in $E_m$. This hydrogen bond donates electron density to the imidazole ring; thus its removal destabilizes $Cu^{II}$. A similar observation has been made in the Asp409Ala mutation of the MCO Fet3p, where $E_m$ increases by 120 mV [43].

### 2.3.2 Reactivity

The "rack" system of hydrogen bonds not only tunes the electronics of the cysteine thiolate but also affects protein reactivity by enhancing ET. In some cases, OSC in T1Cu prevents adventitious chemistry. In the former case, we again consider the Phe114Pro mutant of azurin [39]. By NMR studies, the activation-controlled electron self-exchange (ESE) rate constant for Phe114Pro is found to be lower than wild type by an order of magnitude. The rate constant for ESE is given as $k_{ese} = K_a k_{ET}$, where $K_a$ is the association constant for formation of the ET encounter complex, and $k_{ET}$ is given by the semi-classical Marcus expression [44] (1):

$$k_{ET} = \sqrt{\frac{4\pi^3}{h^2 \lambda k_B T}} H_{AB}^2 exp\left\{\frac{-(\Delta G^\circ + \lambda)^2}{4\lambda k_B T}\right\}. \tag{1}$$

Here, $k_B$ is the Boltzmann constant, $\lambda$ is the reorganization energy, $\Delta G^\circ$ is the driving force for the reaction, and $H_{AB}$ is the electronic coupling matrix element between the donor and the acceptor. $\lambda$ contains contributions from the electronics of the metal site, the solvent dielectric, and oxidation-state dependent metal-ligand vibrations and bond distances. Thus, high site rigidity during redox cycling is a critical factor in rapid ET. Thus, both structural and electronic OSC can have an influence over ET reactivity by constraining active-site topology such that it is either intermediate between oxidation states or simply cannot change during redox cycling. This is analogous to enforcement of transition state geometries in enzymatic catalysis to impose low activation barriers.

ESE in azurin has been shown to be mediated by the hydrophobic patch surrounding His117 [45, 46], suggesting that ET occurs through this residue with participation from solvent molecules. Thus, the $H_{AB}$ term in the ESE reaction is not likely to be affected by alterations to the cysteine thiolate electronics. The removal of Phe114 from the hydrophobic patch was implicated in decreasing the electron transfer rate [39], but this would necessitate an order of magnitude decrease in $K_a$. Crystal structures of both $Cu^I$ and $Cu^{II}$ Phe114Pro azurin show significant structural differences from one another. Analogous wild-type structures are superimposable (Fig. 7) [47]. These structural changes likely also account for the observed ET behavior. A precise value of $\lambda$ has not been reported for the Phe114Pro azurin.

The "rack" networks in T1Cu sites have also been shown to confer a conserved orientation upon the cysteine, wherein the S-$C_\beta$-$C_\alpha$-N dihedrals are ~0°. In multidomain T1Cu proteins, ET occurs through the cysteine, which is positioned adjacent to a histidine coordinated to the downstream active site. The conservation of this side chain orientation is thought to promote efficient ET, likely by conferring a large $H_{AB}$. In Phe114Pro azurin, this dihedral increases to 20°. The consequences of this dihedral on ET through Cys112 merit investigation, although some indirect evidence has been provided in "non-blue" variants of azurin (*vide infra*).

The structural constraints on the cysteine in T1Cu have also been proposed to prevent adventitious chemistry such as thiolate oxidation reactions. In the study of

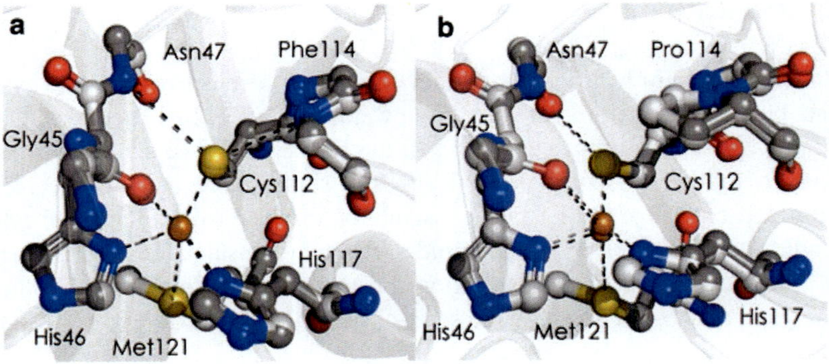


**Fig. 7** (**a**) $Cu^{II}$ (*white*, PDBID: 4AZU) and $Cu^{I}$ (*gray*, PDBID: 1E5Y) *P. aeruginosa* azurin active sites are superimposable. Removal of the Phe114 amide hydrogen bond to Cys112 attenuates structural rigidity, in this mutant $Cu^{II}$ (*white*, PDBID: 2GHZ) and $Cu^{I}$ (*gray*, PDBID: 2GI0) forms show structural differences. Most notably, the Cu-S(Met121) distance shortens by 0.3 Å

Asn38 mutants by Spiro, Asn38Thr, Asn38Gln, and Asn38Leu all demonstrated instability of the T1Cu chromophore [38]. Typical of "rack" residues, the Asn38 residue donates a hydrogen bond from its amide to the Cys84 thiolate, but it also forms a "clasp" with Ser85 that is thought to rigidify the site. Loss of this rigidity is thought to allow Cys84 conformational freedom to form intermolecular disulfides. While evidence for the formation of disulfide-linked plastocyanin dimers was not directly provided, irreversible color loss was observed that is indicative of thiolate oxidation. The OSC in plastocyanin not only tunes site electronics and ET reactivity but also provides a defense mechanism against reactions that would compromise site integrity.

OSC hydrogen-bond networks have been implicated in forming ET paths. Electrochemical ET studies of azurin mutants immobilized on self-assembled monolayer (SAM)-coated gold electrodes suggest that the Asn47-Cys112 hydrogen-bond provides the main route for electron flow in these experiments. The pathways for ET from pulse-radiolytically generated disulfide radicals at Cys3/Cys26 in azurin are proposed to involve several noncovalent contacts (Fig. 8). Notably, electron entry points involve a hydrogen bond from the Asn10 carbonyl to His46 as determined by ET pathway calculations [48].

### 2.3.3 Hard-Ligand Azurins

The Cys112Asp mutant of *P. aeruginosa* azurin transforms the active site from T1Cu to type 2 copper (T2Cu, analogous to copper aquo species) with a marked drop in reduction potential from 304 mV to 180 mV [49]. A study of Met121X (X = Leu, His, Glu) variants of the Cys112Asp mutant [50] gave confusing results: the Cys112Asp/Met121Glu mutant had a higher $E_m$ at pH 7.0 than the single mutant. Cys112Asp/Met121Leu showed a predicted reduction potential increase,

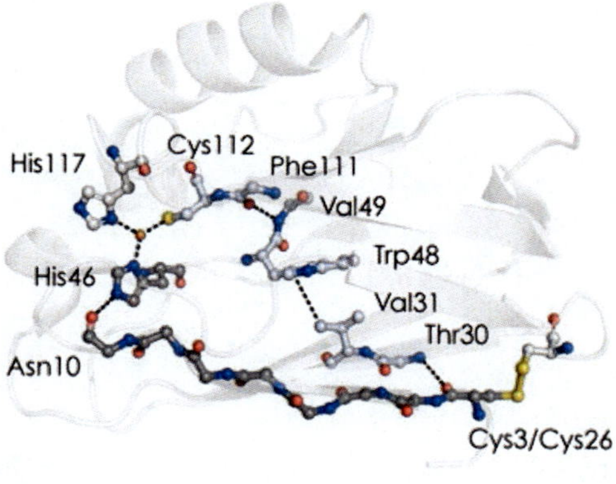


**Fig. 8** Electron transfer pathways from pulse-radiolytically generated Cys3/Cys26 disulfide radical to the T1Cu site in azurin. The *gray route* largely passes through covalent bonds, though it passes through a hydrogen bond between Asn10 and His46. The *white path* involves many non-bonded contacts, though is favored by the highly covalent Cu-S(Cys112) bond. Adapted from [48]

but exhibited an unexpected, dramatic perturbation of its electronic structure where spectroscopic features were intermediate between T1Cu and T2Cu.

The pH-dependence of the Cys112Asp/Met121Glu $E_m$ was studied by redox titrations [51]. A nearly 200 mV drop in $E_m$ was observed as the pH was raised from 5.5 to 10.0. Structural and spectroscopic characterization of low- and high-pH forms of the protein revealed that structural OSC governed this behavior. As predicted by EPR, UV/Vis, and X-ray absorption spectroscopy (XAS), at low pH Glu121 is not coordinated to $Cu^{II}$. At high pH, Glu121 coordinates, stabilizing the Cu in the divalent oxidation state. However, Glu121 protonation state does not control this coordination.

Azurin exhibits pH-dependent ET reactivity that has been attributed to the p$K_a$ of His35. His35 either accepts a hydrogen bond from Pro36 or donates a hydrogen bond to Gly37 depending on the imidazole protonation state. In Cys112Asp/Met121Glu azurin, the protonated form of His35 sterically clashes with His46 (Fig. 9). This rigidifies the metal-binding site and effectively locks Cu away from Glu121, enforcing unfavorable 3-coordinate ligation on $Cu^{II}$. Upon deprotonation, H35 shifts outward, alleviating this constraint on H46, which allows the Cu to sink toward Glu121, establishing a 4-coordinate site. The inner-sphere topology is constrained sterically by residues in the outer sphere.

The Cys112Asp/Met121Leu mutant, in addition to an elevated $E_m$ compared to the single mutant, displays an $A_{\|}$ of $100 \times 10^{-4}$ cm$^{-1}$ that is intermediate between T1Cu and T2Cu. Moreover, it exhibits increased ET reactivity over the single mutant. Structural characterization of this "type zero" copper protein revealed a tetrahedral geometry wherein the Asn47-X112-Phe114 "rack" is reconstituted and forces monodentate coordination by Asp112 (Fig. 10) [52]. Moreover, the Cu-$O_\varepsilon$-$C_\gamma$-$C_\beta$ dihedral decreases from 40° to 3°. This monodentate coordination has been proposed to give rise to the type zero electronic structure and is the subject of an upcoming report [53].

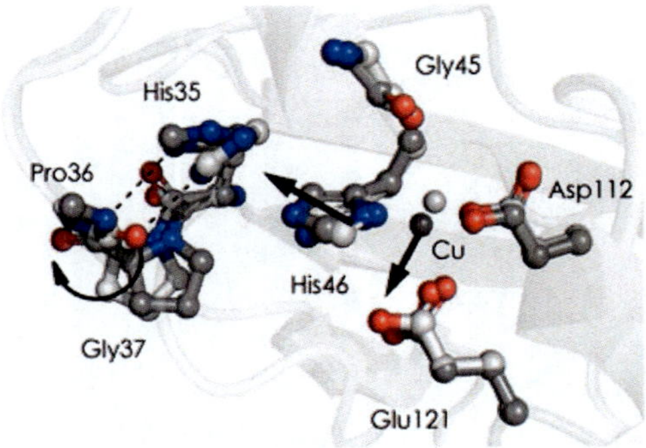


**Fig. 9** Overlaid Cys112Asp/M121E azurin active sites at pH 7.0 (*white*, PDBID: 3NP3) and pH 10.0 (*gray*, PDBID: 3OQR). Upon deprotonation of His35, the Pro36/Gly37 peptide bond flips, substituting a hydrogen bond with the Pro36 carbonyl to one with the Gly37 amide. This affords His46 flexibility, which in turn allows Cu to move toward Glu121. His117 is omitted for clarity

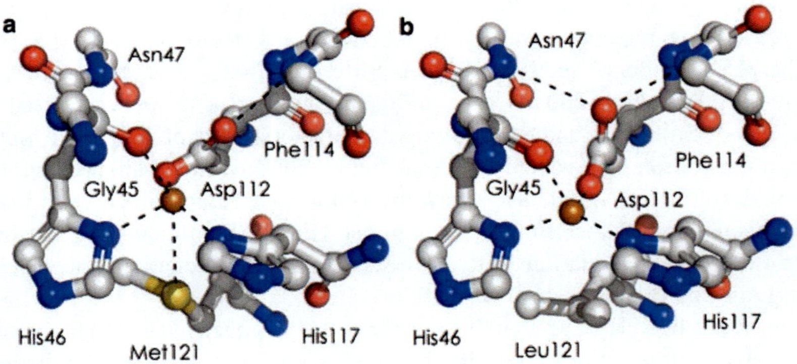


**Fig. 10** T2Cu site of Cys112Asp azurin (**a**, PDBID: 3FQY) and type zero copper site of Cys112Asp/Met121Leu azurin (**b**, PDBID: 3FPY)

As with T1Cu, the rack also impacts site reactivity of type zero copper. Cys112Asp azurin ET is ~4 orders of magnitude slower than WT. This has been attributed to elevated reorganization energy, as a 0.2 Å expansion is observed in the Cys112Asp active site upon reduction [54]. The Cys112Asp/Met121Leu recovers over an order of magnitude in ET efficiency as measured by voltammetry and stopped-flow kinetics. Structural and activation studies attribute this to site rigidification with the restoration of structural OSC from the rack [55]. However, measurement of ET rates from a pulse-radiolytically generated disulfide radical ~24 Å from the Cu-binding site showed slower ET in Cys112Asp/Met121Leu relative to Cys112Asp azurin [55]. The dihedral of the aspartate may confer an increased $H_{AB}$ to the ET pathway involving this residue. This would result in ET pathway interference with the route through His46, consequently slowing ET.

## 2.4 *Metallo-β-Lactamases*

The β-lactamase enzyme family confers antibiotic resistance to bacteria and as such has garnered substantial attention [56]. β-lactams are structural motifs common to antibiotics; hydrolysis leads to their deactivation. Hydrolysis of β-lactams proceeds entirely via protein side chains in the class A, C, and D enzymes. However, class B proteins employ one or two $Zn^{II}$ ions in their reaction pathways. These so-called metallo-β-lactamases (MβLs) are notable in their promiscuity toward β-lactams (with the exception of MβL subtype B2, which is strictly reactive toward the antibiotic carbapenem) [57]. Substrate promiscuity has been posited to be achieved at the expense of catalytic activity, suggesting evolutionary immaturity in MβLs [57]. Directed evolution of the B1 MβL BcII from *Bacillus cereus* has produced a variant, dubbed M5, with enhanced activity toward cephalosporins [58]. This enhancement did not restrict substrate scope. However, individually effecting the mutations of the M5 mutant did not necessarily improve activity, in some cases it was reduced. The dependence of individual mutations on the entire genetic background is referred to in genetics as sign epistasis [59]. The sign epistasis of the mutations in M5, which in combination give rise to an enzyme with both enhanced catalytic activity and broad substrate scope, indicate the importance of OSC as an evolutionary selection criterion.

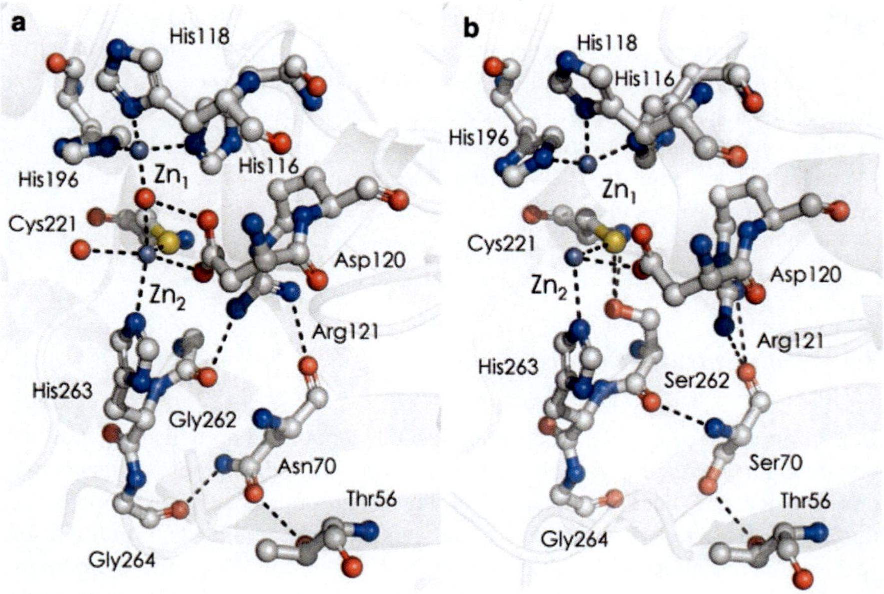


**Fig. 11** Active sites of wild-type (**a**, PBDIB: 1BC2) and M5 variant (**b**, PDBID: 3FCZ) *B. cereus* BcII MβL. The Gly262Ser mutation directly adjusts the coordination of $Zn_2$, leading to greater solvent exposure. The more distant Asn70Ser mutation confers backbone flexibility (and enhancement of substrate scope) by removal of hydrogen bonding interactions

The active form of BcII binds two $Zn^{II}$ ions in vitro; both are thought to be required in vivo [60] (Fig. 11a). $Zn_1$ is bound by three histidines and a solvent molecule ($OH^-$ or $H_2O$) that bridges to $Zn_2$. $Zn_2$ is ligated by a cysteine, a histidine, an aspartate, and an additional $H_2O$. $Zn_2$ is considered essential for orientation of substrate and intermediate stabilization [61]. The M5 variant features a Gly262Ser mutation that introduces a hydrogen bond to the Zn-binding Cys221 (Fig. 11b). This structural OSC repositions $Zn_2$, conferring greater solvent and thus substrate accessibility. Consequently, catalytic activity toward the antibiotic cephalexin is enhanced. However, this enhancement comes at the penalty of reduced substrate scope, attributed by molecular dynamics studies to restriction of backbone flexibility [58].

M5 exhibited the additional mutation Asn70Ser. Position 70 is distant from the active site; consequently its effects on the active site geometry are expected to be minimal (though confirmation requires structural characterization). Asn70Ser BcII exhibits a broader substrate scope, though at the expense of catalytic activity toward cephalexin. Molecular dynamics studies here show greater backbone mobility due to the obviation of two hydrogen bonds from Asn70 to Gly264 and Gly262.

## 3 Cofactors with Prosthetic Groups

Some metalloproteins bind cofactors whose inner coordination spheres contain a substantial number of exogenous ligands. These exogenous ligands range in complexity from monoatomic species such as sulfide to larger frameworks such as porphyrin macrocycles. We will consider both ends of this spectrum, noting that these cofactors exhibit properties that are lost when isolated from their host protein. In other words, simply nesting a coordination complex within a protein is not sufficient to confer function. OSC from the protein matrix strongly influences the properties of these metalloprotein active sites.

### *3.1 Iron Sulfur Proteins*

Iron sulfur clusters are often encountered as ET domains in proteins [62]. As such, precise control of the $E_m$ values of these cofactors is crucial to the establishment of physiological ET chains. A variety of these clusters occur, with complexities ranging from a single iron bound by endogenous cysteine ligands to multinuclear species incorporating sulfide. The former case is represented by the rubredoxins, which still merit mention despite their lack of exogenous ligands. Larger examples include $Fe_4S_4$ clusters, which are not only found in ET proteins such as bacterial ferredoxins and high potential iron proteins (HiPIPs) but also as constituents of intramolecular ET chains in metalloenzymes such as Fe-only hydrogenases.

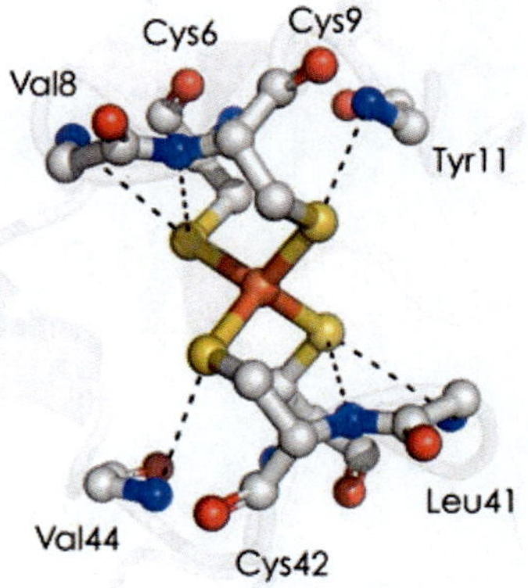


**Fig. 12** $FeS_4$ cluster of oxidized *C. pasteurianum* rubredoxin (PDBID: 1IR0). The OSC consists of multiple hydrogen-bonds donated from backbone amides directly to cysteine thiolates

### 3.1.1 Rubredoxins

Rubredoxins are ET proteins that exhibit values for the $Fe^{III/II}$ $E_m$ from −80 to 40 mV vs. NHE [62]. XAS studies failed to correlate differences in $E_m$ with Fe–S covalency [63]. However, the cysteine thiolate ligands are involved in hydrogen bond OSC networks, with these bonds donated by proximal backbone amides (Fig. 12) [64]. These hydrogen bonds are purported to serve a dual purpose: they tune the cofactor $E_m$ as well as define ET pathways [65–67]. Markley and coworkers reported a $^{15}N$ NMR study of mutants of Val44 in *Clostridium pasteurianum* rubredoxin in which the hydrogen bond from the backbone amide at position 44 to the iron-ligating Cys42 varies systematically [68, 69]. In these variants, no significant changes are noticed in the lengths of other hydrogen bonds. Here, the $E_m$ was found to decrease ~15 mV per 0.1 Å increase in hydrogen bond length. A follow-up study [70] showed that in mutants where multiple hydrogen bonds are perturbed, the effects are cooperative, though not necessarily additive.

### 3.1.2 $Fe_4S_4$ Clusters

Higher-complexity iron–sulfur clusters include $Fe_4S_4$ cofactors [62]. Here, the cluster is a roughly cubic arrangement of 4 iron atoms in a tetrahedral arrangement, with the remaining vertices of the cube occupied by sulfides. The clusters are bound to the protein at iron by cysteine thiolates. $Fe_4S_4$-binding proteins exhibit far more dramatic differences in their reduction potentials than the rubredoxins. In bacterial ferredoxins, the $Fe_4S_4$ clusters exhibit physiological redox cycles with $E_m$ values ranging from −400 to −600 mV vs. NHE for the $Fe_2^{III}Fe_2^{II}/Fe_1^{III}Fe_3^{II}$ or "2+/1+" (total charge) couple. High-potential iron proteins (HiPIPs), exhibit higher physiological $E_m$ values for their clusters of +50 to +500 mV [71], though this corresponds to the $Fe_3^{III}Fe^{II}/Fe_2^{III}Fe_2^{II}$ or cluster "3+/2+" couple. The topologies of the $Fe_4S_4$ clusters in these two protein families are nearly identical [72]. Thus, differences in $E_m$ fall to OSC. Structural characterizations of representative sites revealed a larger conserved hydrogen-bonding network in ferredoxins as compared to HiPIPs (Fig. 13) [65, 71, 73].

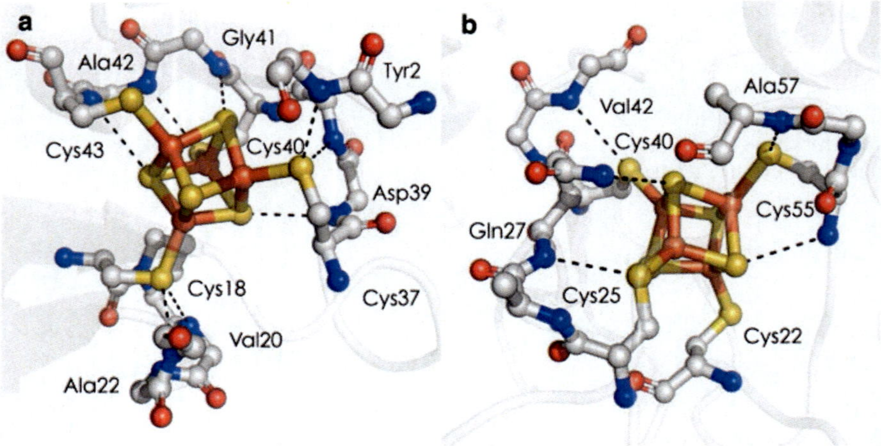


**Fig. 13** One $Fe_4S_4$ cluster from *Clostridium acideruci* ferredoxin (**a**, PDBID: 1FDN) and the $Fe_4S_4$ cluster of *Rhodocyclus tenuis* (**b**, PDBID: 1ISU). Hydrogen bonds to cysteine thiolates and inorganic sulfides are shown. For $Fe_4S_4$ clusters in ferredoxins, there are a total of 8 such hydrogen bonds conserved across species, while for the HiPIPs only five hydrogen bonds are conserved. Some side chains are omitted for clarity

Comparison of several ferredoxins and HiPIPs with synthetic clusters allowed Carter to quantify the tuning by these networks at +80 mV/hydrogen bond [74]. These hydrogen bonds have been proposed to shift $E_m$ values such that the $[Fe_4S_4]^{3+/2+}$ couple is accessible for HiPIP, while for ferredoxins this couple is tuned too high to be physiologically useful [71]. Low and Hill chemically generated variants of *Rhodocyclus tenuis* HiPIP with ester substitutions for amide linkages at positions 42 and 57, removing hydrogen bond donors to $Fe_4S_4$-binding cysteine residues [75]. The engineered holoproteins were shown to have lower reduction potentials, with values of 200 and 240 mV vs. NHE compared to the wild-type 325 mV, in agreement with Carter's prediction.

Differences between ferredoxin and HiPIP $E_m$ values have also been attributed to greater solvent exposure of the $Fe_4S_4$ cluster in the HiPIPs relative to the ferredoxins [76]. Sulfur K-edge XAS studies by Solomon and coworkers showed that, unlike with rubredoxins, the impact of hydrogen bonding on sulfur covalency can be correlated to differences in $E_m$ [77]. Moreover, comparison of sulfur K-edge XAS of HiPIP, ferredoxin, and model complexes in different solvents shows that the higher accessibility of the ferredoxin active site allows for more hydrogen bonds with solvent, decreasing Fe–S covalency and lowering $E_m$.

### 3.1.3 Fe-Only Hydrogenases

Fe-only hydrogenases contain multiple Fe–S clusters that either serve as relay points along ET chains or as reaction centers [78, 79]. The latter case is represented

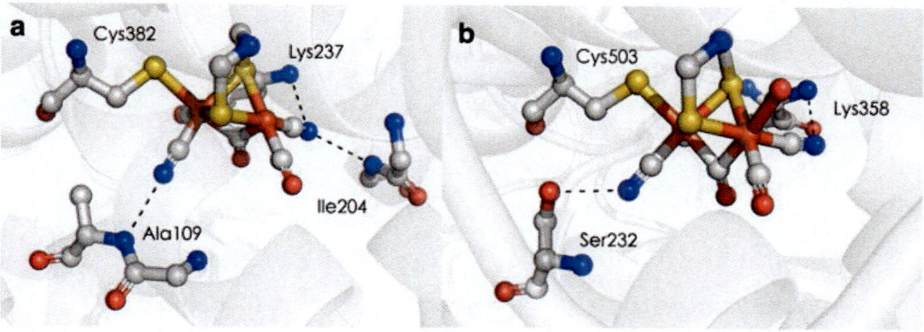


**Fig. 14** Active site models of the H-clusters of *D. desulfuricans* (**a**, based on PDBID: 1HFE) and *C. pasteurianum* (**b**, based on PDBID: 38CY) Fe-only hydrogenases. The catalytic centers are distinguished by an extra hydrogen bond to a $CN^-$ ligand in the *D. desulfuricans* enzyme, which feasibly could direct the $H_2$ chemistry toward oxidation. $Fe_4S_4$ clusters have been omitted for clarity

by the so-called "H-cluster," an $Fe_4S_4$ cluster that is covalently bridged by a cysteine thiolate to a binuclear iron center that catalyzes either the reduction of protons to dihydrogen or the reverse reaction of $H_2$ oxidation to protons. Both of these reactions are carried out rapidly and at low (100–500 mV) overpotentials. The direction of the reaction is species dependent; the periplasmic Fe-only hydrogenase from *Desulfovibrio desulfuricans* oxidizes $H_2$ while the cytoplasmic *Clostridium pasteurianum* enzyme reduces protons [80]. The binuclear components of the H-clusters from both proteins are structurally identical (Fig. 14). In the oxidized, mixed-valent $Fe^{I}$-$Fe^{II}$ form [80], the two metals are bridged by a CO and a dithiomethylamine [81]. The remaining coordination sites are populated by CO, $CN^-$, and $OH^-$. Directionality of catalysis is controlled by OSC. The $H_2$ oxidizing hydrogenase features 3 hydrogen bonds to its $CN^-$ ligands, while only 2 are found in the $H_2$-producing hydrogenase. Given the ~80 mV/hydrogen bond tuning seen for $Fe_4S_4$ clusters, this simple perturbation should be sufficient to reverse the direction of electron flow. Recombinant expression systems for these proteins merit development to test this hypothesis.

## 3.2 *Heme Proteins*

Myriad metalloproteins bind iron-protoporphyrin IX, known as heme (Fig. 15). Heme protein properties are determined by a variety of factors within the inner coordination sphere and without. These include chemical modifications to the porphyrin macrocycle, different axial ligation, perturbations to conformation, and protein dynamics surrounding the cofactor. Because of the extensive proliferation of heme proteins, we will limit ourselves to a small subset. These will include the cytochromes *c*, myoglobins, heme oxygenases and peroxidases, and a heme-based chemical sensor.

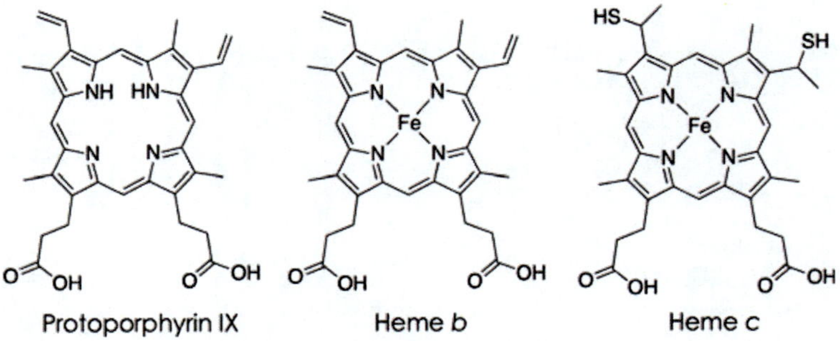


**Fig. 15** Protoporphyrin IX is a common cofactor in metalloproteins. When it incorporates iron, it is known as heme *b*. Further modifications to the macrocycle are encountered; modification of the vinyl groups to secondary thiols gives heme *c*, which may be covalently incorporated into proteins via disulfide linkages

### 3.2.1 Cytochromes *c*

Cytochromes *c* (cyt*c*) are ET proteins that use heme as the redox-active cofactor. They are distinguished by covalent attachment of *c*-type heme via a conserved Cys-X-X-Cys-His binding domain. They exhibit a wide $E_m$ range owing to variety in heme type as well as axial Fe ligation. Cyt*c* heme axial ligands consist of endogenous amino acid side chains as well as exogenous ligands such as $H_2O$. The endogenous ligands most commonly encountered for cyt*c* are histidines and methionines, either in bis-His or His-Met configurations. This coordinative saturation suits these proteins to ET, as it prevents adventitious chemistry at open coordination sites. Within conserved inner-sphere coordination environments, wide ranges in $E_m$ are observed. The bis-His coordinated cytochromes exhibit a ~800 mV variation [82]. OSC arising from hydrogen-bonding to heme as well as to its axial ligands has been implicated in tuning $E_m$ in heme proteins [83]. In the case of histidine-ligated cyts*c*, the degree of anionic character of the imidazole has been correlated to $E_m$ and reactivity. Tuning of the imidazole electrostatics has been attributed to hydrogen-bonding at the noncoordinated nitrogen.

A recent NMR study by Bowman and Bren directly probed the influence of imidazolate character on the $E_m$ of *Hydrogenobacter thermophilus* cyt$c_{552}$ [84]. The active site of this cyt*c* contains a heme *c* bound by an axial methionine (Met61) and an axial histidine (His16) (Fig. 16). The OSC network comprises a hydrogen bond from His16 to the backbone carbonyl of Pro25. The Cys-X-X-Cys-His heme-binding domain also participates in tuning His16; the backbone carbonyl of Cys15 accepts a hydrogen bond from the His16 backbone amide. Adjustment of these hydrogen bond strengths was effected by mutations at Met13 and Lys22. Increased His16-Pro25 hydrogen bond strength reflects greater imidazolate character, with concomitant decrease in $E_m$ and increased electron delocalization onto Met61.

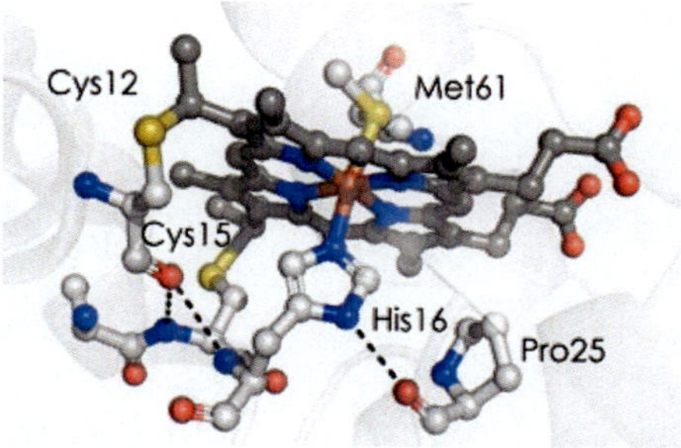


**Fig. 16** Heme binding site of *H. thermophilus* cyt$c_{552}$ (PDBID: 1YNR). The His16 imidazolate character is tuned by hydrogen bonds from the backbone carbonyls of Pro25 and Cys15. Greater imidazolate character stabilizes $Fe^{III}$, resulting in lower values of $E_m$

### 3.2.2 Myoglobins

OSC can also modulate substrate-binding properties; here, the myoglobins are exemplary. Myoglobins are $O_2$ transport proteins, where $O_2$ is bound as an axial ligand to a noncovalently incorporated heme *b* cofactor [85]. Deoxygenated myoglobin bears a water at this position. An axial histidine trans to the $O_2/H_2O$ completes the coordination sphere in both cases. Unlike with the cyts*c*, OSC to the axial histidine has minimal effect on the properties of the myoglobins. Myoglobins display high affinity for and tight binding to oxygen. Heme groups typically bind CO $10^3$–$10^4$ times more tightly than $O_2$; this is not the case for myoglobin hemes *b* [86]. Moreover, myoglobin-bound hemes autooxidize to an $Fe^{III}$ species that cannot bind $O_2$ at rates that are orders of magnitude slower than free heme.

A conserved distal histidine proximal to the $O_2$-binding site has been implicated in defining the functional traits of the myoglobins [87, 88]. An X-ray structure of $O_2$-bound myoglobin (oxymyoglobin) suggests that there is a hydrogen bond from the distal histidine to the bound oxygen (Fig. 17) [89]; this has been confirmed by neutron diffraction structure [90]. The conformation of the histidine, His64, favors the $O_2$-preferred bent end-on binding mode [91]; meanwhile, it sterically disfavors the linear end-on binding preference of CO [92]. Val68 is thought to sterically orient His68, defining in turn the orientation of the $O_2$. Replacement of this histidine by glycine results in a 100-fold increase in the rate of $O_2$ dissociation [93]. Only a ~2-fold increase is seen in the rate of CO dissociation. The histidine positioning inhibits a proper CO-binding conformation, whereas the hydrogen bond from this residue traps the $O_2$. This OSC effect offsets the lower intrinsic heme affinity for $O_2$ compared to CO.

An additional 9 point mutants at the distal histidine were generated by Sligar, Olson, and coworkers [94]. Replacement of the distal His64 in sperm whale myoglobin by valine or phenylalanine further compromised $O_2$ binding, presumably due to the local dielectric of the $O_2$-binding site. The rate of autooxidation was also measured in these mutants. In all cases, autooxidation was at least 40-fold faster than for wild type. This result seemed paradoxical, in that mutants with less proclivity toward $O_2$ binding exhibited greater susceptibility to autooxidation. Perutz proposed that the histidine acts as a "proton trap" at neutral pH. A mechanistic study of myoglobin autooxidation by Olson and coworkers revealed this to be the case [95]; pH dependence of autooxidation was found to reflect Fe–$O_2$

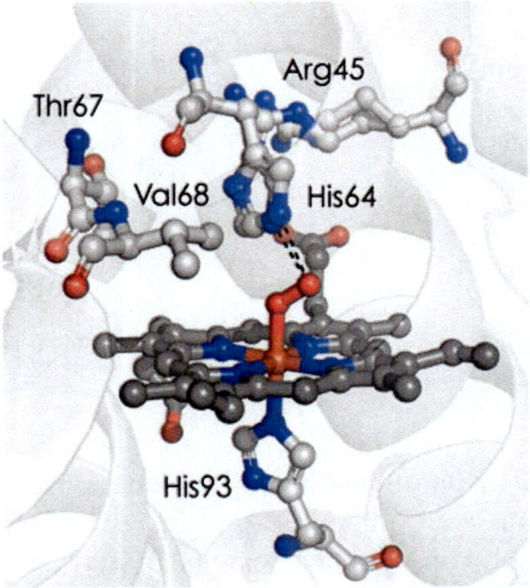


**Fig. 17** $O_2$ binding pocket of sperm whale oxymyoglobin (PDBID: 2Z6S). The distal histidine, His64, acts as a steric block against CO binding, stabilizes the Fe–$O_2$ interaction, and inhibits autooxidation by suppressing protonation of the bound $O_2$. Mutations to Arg45 and Thr67 have been shown to affect rates of autooxidation by modulating the p$K_a$ of His64 and/or the Fe–$O_2$ adduct. Val68 has been implicated in sterically positioning H64, defining the orientation of $O_2$ with respect to the Fe-pyrole axes. Alternative crystallographic conformations of Arg45 and His64 are displayed

protonation. The neutral His64 was proposed to inhibit this protonation; this assertion was supported by the inhibition and enhancement of autooxidation rates after installing positively and negatively charged residues, respectively, proximal to the distal histidine at positions 45 and 67.

There is ongoing debate concerning exact nature of the oxymyoglobin Fe–$O_2$ bond. The available structures of $O_2$-heme model complexes and of oxymyoglobin, combined with the established diamagnetism of oxymyoglobin, have been explained via three models (Fig. 18). Pauling's proposal invokes distal histidine-mediated polarization of $O_2$ that ligates via an end-on double bond to low-spin $Fe^{II}$. Weiss' model invokes ET from $Fe^{II}$ to $O_2$, resulting in S=1/2 superoxide bound end-on to $S = 1/2$ $Fe^{III}$ [96]. The $O_2^-$ and $Fe^{III}$ are then proposed to be antiferromagnetically coupled, accounting for the observed diamagnetic behavior. One further model, the McClure-Goddard "ozone model," features S=1 oxygen bound in a 4 electron, 3 center $\pi$ fashion and antiferromagnetically coupled to $S = 1$ $Fe^{II}$ [97, 98]. Extensive theoretical studies have provided an array of electronic structure solutions, with no clearly favored bonding scheme.

A recent study by Shaik and coworkers explored the influence of OSC on the calculated Fe–$O_2$ bonding interaction [99]. Here, DFT and complete active space self-consistent field (CASSCF) calculations were performed either in the gas-phase or in combination with MM, thus explicitly including the influence of the distal histidine as well as the effect of the protein environment on the polarization of the oxyheme cofactor. DFT/MM and CASSCF/MM calculations reproduce experimentally determined myoglobin oxyheme structures acceptably with the O–O bond eclipsing an

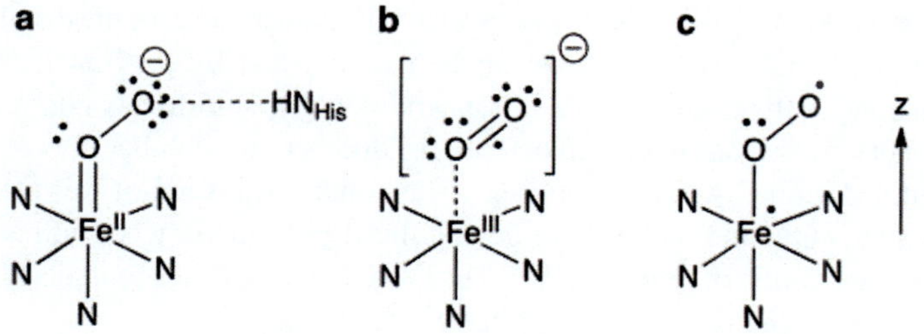


**Fig. 18** Three proposed electronic structures for oxyheme in myoglobin. The Pauling model (**a**) consists of $O_2$ doubly-bound end-on to S=1 $Fe^{II}$. The Fe-bound O is neutral, while the other O exhibits negative charge, and is stabilized by the distal histidine. The Weiss model (**b**) involves ET from $O_2$ to $Fe^{II}$ after binding, resulting S = ½ $Fe^{III}$ bound to S = ½ $O_2^-$. These two radicals are antiferromagnetically coupled. Finally, the Goddard–McClure model (**c**) consists of $S = 1$ $Fe^{II}$ bound to S=1 $O_2$ via a $d_{z2}$-$\pi^*$ interaction, leading to a 4-electron, 3-center bond akin to that of ozone. Adapted from [99]

Fe–N bond. Gas-phase calculations result in structures with the O–O bond bisecting an N–Fe–N angle.

Shaik and coworkers found that a more significant influence of the protein in the calculations was the polarization of orbitals involved in Fe–$O_2$ bonding. The group melded the three aforementioned bonding models in a valence bond framework in with the terms of a bonding wavefunction given by (2):

$$\Psi = \sqrt{(1 + S_1^2)/2}\Phi_{\text{Weiss}} + \lambda\Phi_{\text{Pauling}} + \delta\sqrt{(1 + S_1^2)(1 + S_2^2)}\Phi_{\text{McClure-Goddard}}. \quad (2)$$

Here, $S_1^2$ and $S_2^2$ are overlap integrals between Fe 3d-orbitals and O $\pi^*$ orbitals, the terms $\Phi_i$ are the configurations corresponding to the named model, and $\lambda$ and $\delta$ represent the mixing parameters for the Pauling and McClure–Goddard bonding schemes, respectively. Their relative magnitudes are defined by the protein environment and the axial ligand trans to $O_2$. In the gas phase, large orbital mixing coefficients lead to greater mixtures of the Pauling and McClure–Goddard configurations. The protein environment polarizes the valence bond orbitals, diminishing the mixing coefficients and thus favoring the Weiss model. Shaik and coworkers thus provide a potentially useful theoretical framework for rationalizing heme-$O_2$ interactions in metalloproteins in terms of relative contributions from axial ligation as well as OSC arising from some combination of residues directly interacting with oxyheme ligands as well as the protein-defined oxyheme solvation environment. This approach would benefit from decisive experimental characterization of the Fe–$O_2$ species.

### 3.2.3 Heme Peroxidases

Heme peroxidases such as horseradish peroxidase and cytochrome *c* peroxidase effect peroxide-mediated oxidations [11]. Like myoglobin, they feature a histidine-ligated heme with an open coordination site for peroxide. Near to the peroxide-binding site is a histidine residue that forms hydrogen bonds akin to those of the distal histidine of myoglobin. The coordination sphere is distinguished by a ~0.1–0.2 Å shorter hydrogen bond to the axial histidine [100]. Additionally, a positively charged residue is found near to the distal histidine. This residue acts to stabilize the charge that accumulates during O–O bond cleavage. The concerted efforts of this strong axial histidine interaction and the proton-relaying distal histidine/Lewis acid pair effect a "push-pull" activation of peroxide [11]. Thus, OSC distinguishes proteins binding the same cofactor, giving rise to entirely different function.

### 3.2.4 Cytochrome P450

The cytochrome P450 superfamily oxygenates a wide range of substrates via $O_2$ activation [101]. P450 enzymes feature an axial thiolate ligand that has been proposed to act as an endogenous electron donor for the activation and cleavage of $O_2$. The thiolate provides sufficient electron donation ("push") to obviate the need for an OSC network near to the substrate responsible for "pull" to cleave O–O bonds as in the peroxidases [102, 103]. However, the axial ligand is itself tuned by OSC. P450s feature a network comprised of multiple amide protons proximal to this axial thiolate in a region referred to as the "cysteine pocket" (Fig. 19) [104]. These amides have been shown to influence $E_m$. In *Pseudomonas putida* P450$_{cam}$, removal of a backbone amide near Cys357 by the Leu358Pro mutation lowers $E_m$ by 35 mV via electrostatics [105]. Meanwhile, Gln360 donates a side chain amide proton to the Cys357 backbone carbonyl. The backbone amide of Gln360 is positioned near to the cysteine thiolate. Removal of the hydrogen bond via Gln360L lowers $E_m$ by 45 mV, while removal of this interaction and the thiolate-proximal proton lowers $E_m$ by 70 mV [106].

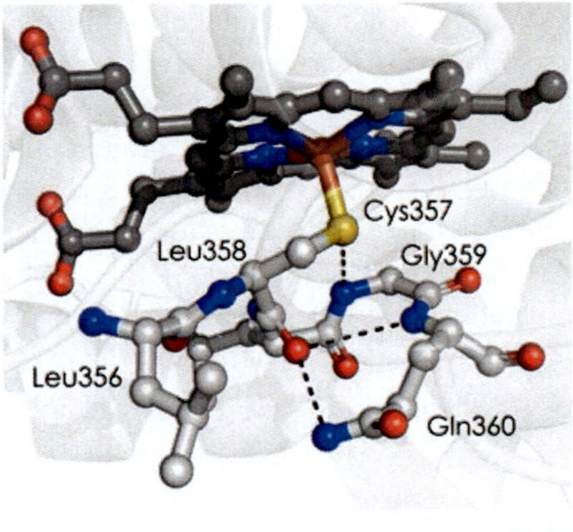


**Fig. 19** The "cysteine pocket" of P450 cam from *P. putida* (PDBID: 2CPP) comprises amides from Leu358, Gly359, and Gln360. The Gly359 amide donates a hydrogen bond to the Cys357 thiolate, while the backbone and side chain amides donate hydrogen bonds to the Cys357 carbonyl. The Leu358 amide does not participate in hydrogen bonding, but is thought to polarize the bonding electrons of the Cys357 thiolate

Morishima and coworkers proposed, based on NMR and RR results, that the Leu358 interaction tunes the thiolate $\pi$-electrons, while the end-on Gln360 proton tunes $\sigma$-electrons [106]. Increased $\pi$-electron donation from the axial thiolate was shown to correlate to accelerated $O_2$ bond heterolysis. Subsequent work by Leinert and coworkers disputes the notion that these mutations have any significant consequence on Fe–S bond strength and covalency, suggesting that they merely serve to orient the axial thiolate and lower its $pK_a$ to prevent protonation and site inactivation [107].

Confounding the discussion, in both studies the Leu358 and Gln360 interactions with Cys357 are incorrectly described as hydrogen bonds. The crystal structure of $P450_{cam}$ shows the backbone amides pointed decidedly away from the thiolate. The Gln360 amide does hydrogen bond to the Cys357 carbonyl. Without direct electronic interaction with the thiolate, mutations at these positions are not expected to influence bond strengths. Their effects on $E_m$ and reactivity likely arise due to polarization of the thiolate electrons. The Morishima conclusion of differential $\sigma$- or $\pi$-overlap with iron is not incongruous with such a proposal. This then exemplifies the influence of electronic OSC on reactivity by modifying the type rather than simply the strength of a metal–ligand interaction.

Schelvis and coworkers have demonstrated structural OSC at play on the properties of P450 hemes [108]. Multiple investigators have proposed that the orientation of heme substituents influences the electronic structure and $E_m$ [109–112]. The Schelvis team generated Phe393 mutants of P450 BM3 from *Bacillus megaterium*. RR data showed this position to affect vinyl orientations while leaving propionate groups largely unperturbed. The vinyl orientations were shown to correlate to $E_m$, with in-plane orientations having higher $E_m$. This is consistent with a proposal by Walker and coworkers that such an orientation confers greater conjugation on the porphyrin ring [113]. This conjugation leads to increased delocalization of electron density away from the iron.

### 3.2.5 *T. tengcongensis* H-NOX

Some heme proteins function as biological sensors for small molecules such as $O_2$, CO, and NO. One such protein, the heme nitric oxide/oxygen binding (H-NOX) protein from *Thermoanaerobacter tengcongensis*, binds a remarkably distorted heme *b* [114]. Marletta and coworkers relaxed the heme toward planarity via the Pro115Ala mutation, resulting in a 171 mV decrease in $E_m$ from 167 to −4 mV vs. NHE [115]. In a subsequent study, mutants were generated that exhibited variably distorted hemes (Fig. 20). These perturbations were effected entirely via van der Waals contacts from the protein; binding pocket hydrophobicity was left largely unperturbed. Heme distortion and $E_m$ correlated; increased planarity conferred a lower $E_m$. Optical pH titrations demonstrated that decreasing planarity leads to a lower $pK_a$ of axial $H_2O$ ligands. Thus, distortion removes electron density from the heme iron.

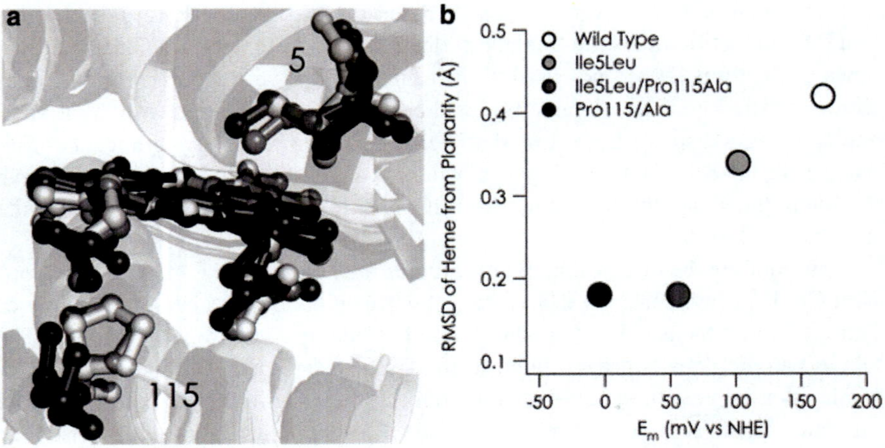


**Fig. 20** (**a**) Structural overlay of *T. tengcongensis* H-NOX (PDBID: 1U4H) with Ile5Leu (PDBID: 3NVR), Ile5Leu/Pro115Ala (PDBID: 3NVU), and Pro115Ala (PDBID: 3EEE) mutants. (**b**) Plot of heme distortion (RMSD from planarity, Å) versus $E_m$. Heme distortion is primarily influenced by van der Waals contacts; relaxation of this distortion appears to increase electron density at the iron, lowering $E_m$. Adapted from [116]

# 4 Strategies for Molecular Engineering

Taking inspiration from nature, chemists have sought to translate important chemical reactions from biology to the benchtop. Two general strategies exist. Existing protein scaffolds can be remodeled to enhance or confer novel properties. Additionally, bioinspired small molecules can be synthesized to the end of mimicking protein reactivity. Chemists employing either approach have recognized the vital importance of OSC. By employing motifs such as those discussed above, they have reported remarkable achievements, a few of which we will now briefly survey.

## 4.1 *Metalloproteins*

Both *de novo* and "remodeling" approaches have been applied to the design of novel metalloproteins [117]. In the former case, both the metal site and the protein scaffold are rationally built. "Remodeling" a protein involves the insertion of a metal-binding site in a previously "empty" space, or changes a pre-existing site to some desired end. While the *de novo* design of novel folds certainly poses a significant intellectual challenge, Lu has noted that nature commonly recycles and repurposes polypeptide scaffolds [118]. This observation indicates that metalloprotein design is not limited by scaffold diversity. Furthermore, the rational design of metal-binding sites in proteins alone represents a daunting challenge.

The "remodeling" approach thus seems to promise more immediate results. In either case, rationally designing the ISC has dominated efforts to date.

Nevertheless, protein designers have recognized the importance of precisely defined OSC. Watanabe and coworkers have exploited the importance of the distal heme in $O_2$-binding heme proteins to repurpose myoglobin for peroxidase activity [119]. Myoglobin reactivity with $H_2O_2$ is very slow compared to peroxidases. The origin of this poor reactivity was thought to be due to the distal heme orientation [120]. In myoglobin, the distal histidine can donate a hydrogen bond to either oxygen of bound $O_2/H_2O_2$. However, the hydrogen bond is donated to a single oxygen in the peroxidases. Relocation of the distal histidine via the Phe43His/His64Leu mutant of sperm whale myoglobin dramatically increases peroxidase activity in myoglobin. Increasing the distance of this histidine further via the Leu29His/His64Leu variant diminishes peroxidase activity relative to wild-type myoglobin.

The impact of the OSC has allowed remarkable control over $E_m$. Modulating charge distribution in *de novo* designed four-helix-bundle proteins has contributed to the ~400 mV tunability in the $E_m$ of the heme cofactors [121]. Even greater control has been demonstrated in the T1Cu site of azurin. Lu and coworkers demonstrated a 700 mV $E_m$ range effected entirely by modulating site hydrophobicity and the Asn47-Cys112-Phe114 hydrogen bonding network [122].

## 4.2 Coordination Complexes

Biomimetic inorganic synthesis has played an invaluable role in understanding the reactivity of metalloproteins, with the ultimate goal of producing robust, efficient catalysts for processes such as water splitting and small molecule activation. In the last decade, biological OSC motifs have guided synthetic efforts. The substantial progress in this arena has been reviewed with an emphasis on dioxygen activation in a recent forum article by Shook and Borovik [123]. Examples will be presented with an emphasis on strategy, rather than a quantitative discussion of reactivity.

Thermodynamics favor the conversion of iron-oxygen compounds to μ-oxo $Fe^{III}$ dimers [124]. However, as seen with myoglobin, OSC allows proteins to bind dioxygen without such dimerization. Appending steric bulk to iron porphyrins produced the "picket-fence" family of porphyrins [125], which has allowed the characterization of monomeric Fe–$O_2$ binding modes (Fig. 21). This use of structural OSC has extended to forcing other disfavored metal-binding modes. A complex of uranium(III) with a sterically bulky 1,4,7-triazacyclononane (tacn) derivative has been synthesized that binds $CO_2$ in the unprecedented linear $\eta^1$-OCO fashion [126] (Fig. 21).

Ligand scaffolds with hydrogen-bonding functionalities have been reported by a number of groups. Nocera and coworkers designed "hangman" porphyrins and corroles with pendant hydrogen-bonding groups that confer proton-coupled dioxygen activation activity [127, 128]. This iron-mediated dioxygen activation has been extended to nonheme complexes. Chang and coworkers reported

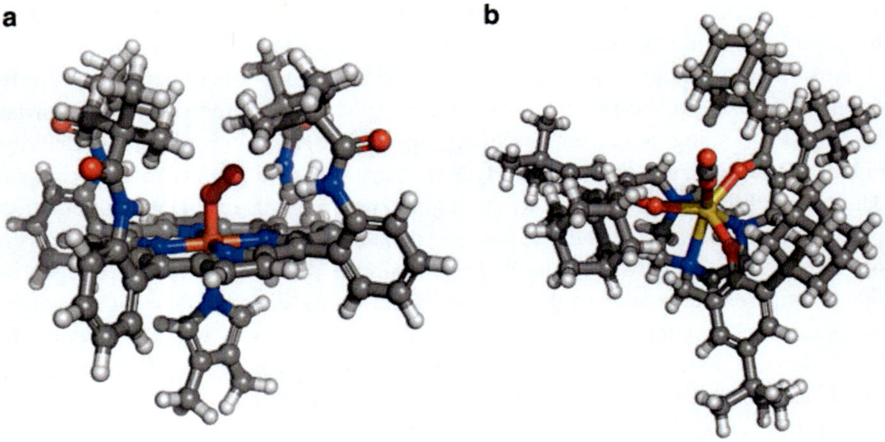


**Fig. 21** Defining binding site sterics by OSC has permitted isolation of thermodynamically unfavorable monomeric Fe–$O_2$ species, as exemplified by the "picket fence porphyrins" (**a**). This strategy has also been employed to force non-preferential binding modes, such as end-on $\eta^1$ coordination of $CO_2$ by $U^{III}$ coordinated by a bulky tacn scaffold (**b**)

hydrogen-bond functionalized derivatives of *N*,*N*-bis(2-pyridylmethyl)-bis (2-pyridyl)methylamine (N4Py) that provide an acid- and base-stable platform for dioxygen reduction via isolable $Fe^{II}$- and $Fe^{III}$-OH intermediates [129]. Borovik and coworkers have explored the effects of varying the numbers of hydrogen-bonding groups in tripodal metal complexes [123]. Variable efficacies of dioxygen and C–H activation have been demonstrated among families of complexes distinguished principally by the number of hydrogen bonds available to axial oxygen species. Paralleling Fe and MnSOD, the origins of the differences in reactivity arise due to the influences of OSC on axial ligand p$K_a$, which in turn modulates $E_m$.

## 5 Conclusions

Precisely defined OSC environments represent the principal feature distinguishing metalloproteins from synthetic coordination complexes. Moreover, the sign epistasis of mutations perturbing the OSC in MβLs demonstrates that optimizing OSC is a vital evolutionary criterion. We have distinguished OSC within several disparate protein families. We have indicated cases where the protein electronically tunes the inner-sphere ligands, where it enforces normally unfavorable geometries, and where it defines the local dielectric environment. Electronic OSC generally dictates the redox behavior of metallocofactors; this is achieved by hydrogen bond interactions between the protein matrix and inner-sphere ligands. This tuning profoundly affects the reactivity of active sites; in the case of Fe and Mn SODs, it represents the primary mode of metal ion selectivity. In myoglobin, OSC imparts substrate selectivity; moreover, it influences the electronic structure of the cofactor-substrate interaction. Variations within very similar electronic OSC motifs can be used to transform an $O_2$-carrying protein such as myoglobin into an $O_2$-activating enzyme such as the peroxidases or cytochromes P450.

Meanwhile, structural effects arising through both hydrogen bonding and steric interactions can force protein-bound metal ions into geometries that are unfavorable relative to isolated solution geometries. In T1Cu proteins, this results in constrained active sites with low reorganization barriers for ET. Moreover, these interactions can precisely position metal ions for substrate reactivity, as seen in FIH and the BCII MβL. Finally, the protein imparts specific solvation on nested coordination complexes beyond simply sheathing these species in a hydrophobic environment. Through-space interactions with backbone and side chain dipoles perturb the electronic structure, shown in studies on T1Cu as well as cytochrome $P450_{cam}$.

Within the last couple of decades, investigators have incorporated OSC into molecular design. Remarkable small-molecule activating complexes have been reported by many groups. These systems incorporate hydrogen bonds to tune ligand electronics as well as bulky groups to force geometries and prevent adventitious chemistry. Moreover, metalloproteins have undergone OSC environment redesign to the ends of tuning properties and changing reactivity. These approaches remain immature; further exploitation of motifs from biological OSC will extend the capabilities of synthetic chemists and metalloprotein engineers. Ultimately, building molecules with precisely defined OSC will allow chemists to realize goals such as biomimetic $H_2O$ splitting and $N_2$ reduction.

**Acknowledgments** I have my thesis advisor Harry B. Gray to thank for my participation in this memorial to C.J. Ballhausen. Among other things, Harry guided me through the formulation of the ligand field theory, which required careful and thorough reading of Ballhausen's opus *Introduction to Ligand Field Theory*. Ultimately, I largely failed to find a place for LFT in my graduate work, which in the end concerned the above topic of OSC. However, my efforts were rewarded with the opportunity to meet Ballhausen, who regarded me as a "poor chap" for having endured *Introduction* and made amends with Carlsberg lager and ginger cookies. Jeffrey Warren, Joshua Palmer, and Stephen Sproules provided useful comments on this manuscript. Finally, my postdoctoral advisor Serena DeBeer is gratefully acknowledged for reading this manuscript and for financial support.

# References

1. Bjerrum J (1967) Coordination in the second sphere. In: Werner centennial. Advances in chemistry, vol 62. American Chemical Society, Washington, DC, pp 178–186
2. Werner A (1913) Neuere anschauungen auf dem gebiete der anorganischen chemie, 3rd edn. Vieweg und Sohn, Braunschweig
3. Colquhoun HM, Stoddart JF, Williams DJ (1986) Second-sphere coordination–a novel rǒle for molecular receptors. Angew Chem Int Ed 25:487–507
4. Lehn J-M (1995) Supramolecular chemistry: concepts and perspectives. Wiley-VCH, Weinheim

5. Vance CK, Miller AF (1998) Simple proposal that can explain the inactivity of metal-substituted superoxide dismutases. J Am Chem Soc 120:461–467
6. Stein J, Fackler JP, Mcclune GJ, Fee JA, Chan LT (1979) Superoxide and manganese(III) - reactions of Mn-EDTA and Mn-CYDTA complexes with $O_2^-$ - X-ray structure of KMnEDTA•$2H_2O$. Inorg Chem 18:3511–3519
7. Miller AF (2008) Redox tuning over almost 1 V in a structurally conserved active site: lessons from Fe-containing superoxide dismutase. Acc Chem Res 41(4):501–510
8. Jackson TA, Brunold TC (2004) Combined spectroscopic/computational studies on Fe- and Mn-dependent superoxide dismutases: insights into second-sphere tuning of active site properties. Acc Chem Res 37:461–470
9. Grove LE, Xie J, Yikilmaz E, Miller AF, Brunold TC (2008) Spectroscopic and computational investigation of second-sphere contributions to redox tuning in Escherichia coli iron superoxide dismutase. Inorg Chem 47:3978–3992
10. Vance CK, Miller AF (1998) Spectroscopic comparisons of the pH dependencies of Fe-substituted (Mn)superoxide dismutase and Fe-superoxide dismutase. Biochemistry 37:5518–5527
11. Dawson JH (1988) Probing structure-function relations in heme-containing oxygenases and peroxidases. Science 240:433–439
12. Yikilmaz E, Porta J, Grove LE, Vahedi-Faridi A, Bronshteyn Y, Brunold TC, Borgstahl GEO, Miller AF (2007) How can a single second sphere amino acid substitution cause reduction midpoint potential changes of hundreds of millivolts? J Am Chem Soc 129:9927–9940
13. Vance CK, Kang YM, Miller AF (1997) Selective N-15 labeling and direct observation by NMR of the active-site glutamine of Fe-containing superoxide dismutase. J Biomol NMR 9:201–206
14. Costa-Neto CM, Mikakawa AA, Oliveira L, Hjorth SA, Schwartz TW, Paiva ACM (2000) Mutational analysis of the interaction of the N- and C-terminal ends of angiotensin II with the rat AT(1A) receptor. Br J Pharmacol 130:1263–1268
15. Price JC, Barr EW, Tirupati B, Bollinger JM, Krebs C (2003) The first direct characterization of a high-valent iron intermediate in the reaction of an alpha-ketoglutarate-dependent dioxygenase: a high-spin Fe(IV) complex in taurine/alpha-ketoglutarate dioxygenase (TauD) from Escherichia coli. Biochemistry 42:7497–7508
16. Saban E, Chen Y-H, Hangasky J, Taabazuing C, Holmes BE, Knapp MJ (2011) The second coordination sphere of FIH controls hydroxylation. Biochemistry. doi:10.1021/bi102042t
17. McNeill LA, Hewitson KS, Claridge TD, Seibel JF, Horsfall LE, Schofield CJ (2002) Hypoxia-inducible factor asparaginyl hydroxylase (FIH-1) catalyses hydroxylation at the beta-carbon of asparagine-803. Biochem J 367:571–575
18. Gray HB, Malmstrom BG, Williams RJP (2000) Copper coordination in blue proteins. J Biol Inorg Chem 5(5):551–559
19. Malkin R, Malmström BG (1970) The state and function of copper in biological systems. Adv Enzymol Relat Areas Mol Biol 33:177–244
20. Solomon EI (2006) Spectroscopic methods in bioinorganic chemistry: blue to green to red copper sites. Inorg Chem 45:8012–8025
21. Canters GW, Gilardi G (1993) Engineering type 1 copper sites in proteins. FEBS Lett 325:39–48
22. Holland PL, Tolman WB (1999) Three-coordinate Cu(II) complexes: structural models of trigonal-planar type 1 copper protein active sites. J Am Chem Soc 121(31):7270–7271
23. Li H, Webb SP, Ivanic J, Jensen JH (2004) Determinants of the relative reduction potentials of type-1 copper sites in proteins. J Am Chem Soc 126(25):8010–8019
24. Hart PJ, Eisenberg D, Nersissian AM, Valentine JS, Herrmann RG, Nalbandyan RM (1996) A missing link in cupredoxins: crystal structure of cucumber stellacyanin at 1.6 Å resolution. Protein Sci 5:2175–2183
25. Guss JM, Bartunik HD, Freeman HC (1992) Accuracy and precision in protein structure analysis: restrained least-squares refinement of the structure of poplar plastocyanin at 1.33 A resolution. Acta Crystallogr B 48:790–811

26. Barrett ML, Harvey I, Sundararajan M, Surendran R, Hall JF, Ellis MJ, Hough MA, Strange RW, Hillier IH, Hasnain SS (2006) Atomic resolution crystal structures, EXAFS, and quantum chemical studies of rusticyanin and its two mutants provide insight into its unusual properties. Biochemistry 45:2927–2939
27. Nar H, Messerschmidt A, Huber R, Vandekamp M, Canters GW (1991) Crystal-structure analysis of oxidized pseudomonas aeruginosa azurin at pH 5.5 and oH 9.0 - a pH-induced conformational transition involves a peptide-bond flip. J Mol Biol 221:765–772
28. Reinhammar BRM (1972) Oxidation-reduction potentials of electron acceptors in laccases and stellacyanin. Biochim Biophys Acta 275:245
29. Nersissian AM, Immoos C, Hill MG, Hart PJ, Williams G, Herrmann RG, Valentine JS (1998) Uclacyanins, stellacyanins, and plantacyanins are distinct subfamilies of phytocyanins: plant-specific mononuclear blue copper proteins. Protein Sci 7:1915–1929
30. Rosen P, Pecht I (1976) Conformational equilibria accompanying electron-transfer between cytochrome-c (P551) and azurin from pseudomonas aeruginosa. Biochemistry 15:775–786
31. Ingledew WJ, Cobley JG (1980) A potentiometric and kinetic study on the respiratory chain of ferrous iron grown thiobacillus ferrooxidans. Biochim Biophys Acta 590:141–158
32. Berry SM, Ralle M, Low DW, Blackburn NJ, Lu Y (2003) Probing the role of axial methionine in the blue copper center of azurin with unnatural amino acids. J Am Chem Soc 125:8760–8768
33. Pascher T, Karlsson BG, Nordling M, Malmstrom BG, Vanngard T (1993) Reduction potentials and their pH-dependence in site-directed-mutant forms of azurin from Pseudomonas-aeruginosa. Eur J Biochem 212(2):289–296
34. Berry SM, Baker MH, Reardon NJ (2010) Reduction potential variations in azurin through secondary coordination sphere phenylalanine incorporations. J Inorg Biochem 104 (10):1071–1078
35. Sinnecker S, Neese F (2006) QM/MM calculations with DFT for taking into account protein effects on the EPR and optical spectra of metalloproteins. Plastocyanin as a case study. J Comput Chem 27:1463–1475
36. Battistuzzi G, Borsari M, Loschi L, Menziani MC, De Rienzo F, Sola M (2001) Control of metalloprotein reduction potential: the role of electrostatic and solvation effects probed on plastocyanin mutants. Biochemistry 40:6422–6430
37. Malmstrom BG (1994) Rack-induced bonding in blue-copper proteins. Eur J Biochem 223:711–718
38. Dong SL, Ybe JA, Hecht MH, Spiro TG (1999) H-bonding maintains the active site of type 1 copper proteins: mutagenesis of Asn38 in poplar plastocyanin. Biochemistry 38:3379–3385
39. Yanagisawa S, Banfield MJ, Dennison C (2006) The role of hydrogen bonding at the active site of a cupredoxin: the Phe114Pro azurin variant. Biochemistry 45:8812–8822
40. Kataoka K, Hirota S, Maeda Y, Kogi H, Shinohara N, Sekimoto M, Sakurai T (2011) Enhancement of laccase activity through the construction and breakdown of a hydrogen bond at the type I copper center in Escherichia coli CueO and the deletion mutant delta alpha 5-7 CueO. Biochemistry 50:558–565
41. Roberts SA, Weichsel A, Grass G, Thakali K, Hazzard JT, Tollin G, Rensing C, Montfort WR (2002) Crystal structure and electron transfer kinetics of CueO, a multicopper oxidase required for copper homeostasis in Escherichia coli. Proc Natl Acad Sci USA 99:2766–2771
42. Solomon EI, Sundaram UM, Machonkin TE (1996) Multicopper oxidases and oxygenases. Chem Rev 96:2563–2605
43. Stoj CS, Augustine AJ, Zeigler L, Solomon EI, Kosman DJ (2006) Structural basis of the ferrous iron specificity of the yeast ferroxidase, Fet3p. Biochemistry 45:12741–12749
44. Marcus RA, Sutin N (1985) Electron transfers in chemistry and biology. Biochim Biophys Acta 811:265–322
45. Vanpouderoyen G, Mazumdar S, Hunt NI, Hill HAO, Canters GW (1994) The introduction of a negative charge into the hydrophobic patch of Pseudomonas aeruginosa azurin affects the alectron self-exchange rate and the electrochemistry. Eur J Biochem 222:583–588

46. van Amsterdam IMC, Ubbink M, Einsle O, Messerschmidt A, Merli A, Cavazzini D, Rossi GL, Canters GW (2002) Dramatic modulation of electron transfer in protein complexes by crosslinking. Nat Struct Biol 9:48–52
47. Shepard WEB, Anderson BF, Lewandoski DA, Norris GE, Baker EN (1990) Copper coordination geometry in azurin undergoes minimal change on reduction of copper(II) to copper(I). J Am Chem Soc 112:7817–7819
48. Farver O, Pecht I (1997) The role of the medium in long-range electron transfer. J Biol Inorg Chem 2:387–392
49. Mizoguchi TJ, Dibilio AJ, Gray HB, Richards JH (1992) Blue to type-2 binding - copper(II) and cobalt(II) derivatives of a Cys112Asp Mutant of Pseudomonas aeruginosa azurin. J Am Chem Soc 114:10076–10078
50. Lancaster KM, Yokoyama K, Richards JH, Winkler JR, Gray HB (2009) High-potential C112D/M121X (X = M, E H, L) Pseudomonas aeruginosa azurins. Inorg Chem 48:1278–1280
51. Lancaster KM, Sproules S, Palmer JH, Richards JH, Gray HB (2010) Outer-sphere effects on reduction potentials of copper sites in proteins: the curious case of high potential type 2 C112D/M121E Pseudomonas aeruginosa azurin. J Am Chem Soc 132:14590–14595
52. Lancaster KM, George SD, Yokoyama K, Richards JH, Gray HB (2009) Type-zero copper proteins. Nat Chem 1:711–715
53. Lancaster KM, Zaballa M-E, Sproules S, Sundararajan M, DeBeer S, Neese F, Vila AJ, Richards JH, Gray HB (In Preparation) The type zero copper site: Type 1 copper without thiolate ligation
54. DeBeer S, Kiser CN, Mines GA, Richards JH, Gray HB, Solomon EI, Hedman B, Hodgson KO (1999) X-ray absorption spectra of the oxidized and reduced forms of C112D azurin from Pseudomonas aeruginosa. Inorg Chem 38:433–438
55. Lancaster KM, Farver O, Wherland S, Crane EJ, Richards JH, Pecht I, Gray HB (2011) Electron transfer reactivity of type zero Pseudomonas aeruginosa azurin. J Am Chem Soc 133:4865–4873
56. Fisher JF, Meroueh SO, Mobashery S (2005) Bacterial resistance to beta-lactam antibiotics: compelling opportunism, compelling opportunity. Chem Rev 105:395–424
57. Crowder MW, Spencer J, Vila AJ (2006) Metallo-beta-lactamases: novel weaponry for antibiotic resistance in bacteria. Acc Chem Res 39:721–728
58. Tomatis PE, Fabiane SM, Simona F, Carloni P, Sutton BJ, Vila AJ (2008) Adaptive protein evolution grants organismal fitness by improving catalysis and flexibility. Proc Natl Acad Sci USA 105:20605–20610
59. Weinreich DM, Watson RA, Chao L (2005) Perspective: sign epistasis and genetic constraint on evolutionary trajectories. Evolution 59:1165–1174
60. Abriata LA, Gonzalez LJ, Llarrull LI, Tomatis PE, Myers WK, Costello AL, Tierney DL, Vila AJ (2008) Engineered mononuclear variants in Bacillus cereus metallo-beta-lactamase BcII are inactive. Biochemistry 47:8590–8599
61. Gonzalez JM, Martin FJM, Costello AL, Tierney DL, Vila AJ (2007) The zn2 position in metallo-beta-lactamases is critical for activity: a study on chimeric metal sites on a conserved protein scaffold. J Mol Biol 373:1141–1156
62. Cammack R (1992) Iron–sulfur clusters in enzymes: themes and variations. In: Richard C (ed) Advances in inorganic chemistry, vol 38. Academic Press, San Diego, pp 281–322
63. Rose K, Shadle SE, Eidsness MK, Kurtz DM, Scott RA, Hedman B, Hodgson KO, Solomon EI (1998) Investigation of iron-sulfur covalency in rubredoxins and a model system using sulfur K-edge X-ray absorption spectroscopy. J Am Chem Soc 120:10743–10747
64. Dauter Z, Wilson KS, Sieker LC, Moulis JM, Meyer J (1996) Zinc- and iron-rubredoxins from Clostridium pasteurianum at atomic resolution: A high-precision model of a ZnS4 coordination unit in a protein. Proc Natl Acad Sci USA 93:8836–8840
65. Adman E, Watenpaugh KD, Jensen LH (1975) NH•••S hydrogen-bonds in Peptococcus-aerogenes ferredoxin, Clostridium pasteurianum rubredoxin, and Chromatium high potential iron protein. Proc Natl Acad Sci USA 72:4854–4858

66. Kümmerle R, Zhuang-Jackson H, Gaillard J, Moulis J-M (1997) Site-directed mutagenesis of rubredoxin reveals the molecular basis of its electron transfer properties. Biochemistry 36:15983–15991
67. Okamura T-A, Takamizawa S, Ueyama N, Nakamura A (1998) Novel rubredoxin model tetrathiolato iron(II) and cobalt(II) complexes containing intramolecular single and double NH•••S hydrogen bonds. Inorg Chem 37:18–28
68. Lin IJ, Gebel EB, Machonkin TE, Westler WM, Markley JL (2003) Correlation between hydrogen bond lengths and reduction potentials in Clostridium pasteurianum rubredoxin. J Am Chem Soc 125:1464–1465
69. Xiao ZG, Maher MJ, Cross M, Bond CS, Guss JM, Wedd AG (2000) Mutation of the surface valine residues 8 and 44 in the rubredoxin from Clostridium pasteurianum: solvent access versus structural changes as determinants of reversible potential. J Biol Inorg Chem 5:75–84
70. Lin IJ, Gebel EB, Machonkin TE, Westler WM, Markley JL (2005) Changes in hydrogen-bond strengths explain reduction potentials in 10 rubredoxin variants. Proc Natl Acad Sci USA 102:14581–14586
71. Heering HA, Bulsink YBM, Hagen WR, Meyer TE (1995) Influence of charge and polarity on the redox potentials of high-potential iron-sulfur proteins - evidence for the existence of 2 groups. Biochemistry 34:14675–14686
72. Adman ET (1979) Comparison of the structures of electron-transfer Proteins. Biochim Biophys Acta 549:107–144
73. Backes G, Mino Y, Loehr TM, Meyer TE, Cusanovich MA, Sweeney WV, Adman ET, Sanders-Loehr J (1991) The environment of Fe4S4 clusters in ferredoxins and high-potential iron proteins – new information from X-ray crystallography and resonance Raman-spectroscopy. J Am Chem Soc 113:2055–2064
74. Carter CW (1977) New stereochemical analogies between iron-sulfur electron-transport proteins. J Biol Chem 252(21):7802–7811
75. Low DW, Hill MG (2000) Backbone-engineered high-potential iron proteins: effects of active-site hydrogen bonding on reduction potential. J Am Chem Soc 122:11039–11040
76. Kassner RJ, Yang W (1977) Theoretical-model for effects of solvent and protein dielectric on redox potentials of iron-sulfur clusters. J Am Chem Soc 99(13):4351–4355
77. Dey A, Francis EJ, Adams MWW, Babini E, Takahashi Y, Fukuyama K, Hodgson KO, Hedman B, Solomon EI (2007) Solvent tuning of electrochemical potentials in the active sites of HiPIP versus ferredoxin. Science 318:1464–1468
78. Peters JW, Lanzilotta WN, Lemon BJ, Seefeldt LC (1998) X-ray crystal structure of the Fe-only hydrogenase (Cpl) from Clostridium pasteurianum to 1.8 Ångström resolution. Science 282:1853–1858
79. Nicolet Y, Piras C, Legrand P, Hatchikian CE, Fontecilla-Camps JC (1999) Desulfovibrio desulfuricans iron hydrogenase: the structure shows unusual coordination to an active site Fe binuclear center. Structure 7:13–23
80. Tard C, Pickett CJ (2009) Structural and functional analogues of the active Sites of the [Fe]-, [NiFe]-, and [FeFe]-hydrogenases. Chem Rev 109:2245–2274
81. Silakov A, Wenk B, Reijerse E, Lubitz W (2009) N-14 HYSCORE investigation of the H-cluster of [FeFe] hydrogenase: evidence for a nitrogen in the dithiol bridge. Phys Chem Chem Phys 11:6592–6599
82. Walker FA (2004) Models of the bis-histidine-ligated electron-transferring cytochromes. Comparative geometric and electronic structure of low-spin ferro- and ferrihemes. Chem Rev 104:589–616
83. Banci L, Bertini I, Kuan IC, Tien M, Turano P, Vila AJ (1993) NMR investigation of isotopically labeled cyanide derivatives of lignin peroxidase and manganese peroxidase. Biochemistry 32:13483–13489
84. Bowman SEJ, Bren KL (2010) Variation and analysis of second-sphere interactions and axial histidinate character in c-type cytochromes. Inorg Chem 49:7890–7897

85. Springer BA, Sligar SG, Olson JS, Phillips GN (1994) Mechanisms of ligand recognition in myoglobin. Chem Rev 94:699–714
86. Collman JP, Brauman JI, Halbert TR, Suslick KS (1976) Nature of $O_2$ and CO Binding to metalloporphyrins and heme proteins. Proc Natl Acad Sci USA 73:3333–3337
87. Pauling L, Weiss JJ (1964) Nature of iron-oxygen bond in oxyhaemoglobin. Nature 203:182–183
88. Perutz MF (1989) Myoglobin and hemoglobin – role of distal residues in reactions with heme ligands. Trends Biochem Sci 14:42–44
89. Phillips SEV (1980) Structure and refinement of oxymyoglobin at 1.6 Å resolution. J Mol Biol 142:531–554
90. Phillips SEV, Schoenborn BP (1981) Neutron-diffraction reveals oxygen-histidine hydrogen-Bond in oxymyoglobin. Nature 292:81–82
91. Jameson GB, Molinaro FS, Ibers JA, Collman JP, Brauman JI, Rose E, Suslick KS (1978) Structural changes upon oxygenation of an iron(II)(porphyrinato)(imidazole) complex. J Am Chem Soc 100:6769–6770
92. Peng SM, Ibers JA (1976) Stereochemistry of carbonylmetalloporphyrins – structure of (pyridine)(carbonyl)(5,10,15,20-tetraphenylporphinato)Iron(Ii). J Am Chem Soc 98:8032–8036
93. Olson JS, Mathews AJ, Rohlfs RJ, Springer BA, Egeberg KD, Sligar SG, Tame J, Renaud JP, Nagai K (1988) The role of the distal histidine in myoglobin and hemoglobin. Nature 336:265–266
94. Springer BA, Egeberg KD, Sligar SG, Rohlfs RJ, Mathews AJ, Olson JS (1989) Discrimination between oxygen and carbon monoxide and inhibition of autooxidation by myoglobin - site-directed mutagenesis of the distal histidine. J Biol Chem 264:3057–3060
95. Brantley RE, Smerdon SJ, Wilkinson AJ, Singleton EW, Olson JS (1993) The mechanism of autooxidation of myoglobin. J Biol Chem 268:6995–7010
96. Weiss JJ (1964) Nature of iron-oxygen bond in oxyhaemoglobin. Nature 202:83–84
97. McClure DS (1960) Electronic structure of transition metal complex ions. Radiat Res Suppl 2:218–242
98. Goddard WA, Olafson BD (1975) Ozone model for bonding of an $O_2$ to heme in oxyhemoglobin. Proc Natl Acad Sci USA 72:2335–2339
99. Chen H, Ikeda-Saito M, Shaik S (2008) Nature of the Fe-$O_2$ bonding in oxy-myoglobin: effect of the Protein. J Am Chem Soc 130:14778–14790
100. Berglund GI, Carlsson GH, Smith AT, Szöke H, Henriksen A, Hajdu J (2002) The catalytic pathway of horseradish peroxidase at high resolution. Nature 417:463–468
101. Denisov IG, Makris TM, Sligar SG, Schlichting I (2005) Structure and chemistry of cytochrome P450. Chem Rev 105:2253–2278
102. Dawson JH, Holm RH, Trudell JR, Barth G, Linder RE, Bunnenberg E, Djerassi C, Tang SC (1976) Oxidized cytochrome-P450 – magnetic circular dichroism evidence for thiolate ligation in substrate-bound form – implications for catalytic mechanism. J Am Chem Soc 98:3707–3709
103. Sono M, Andersson LA, Dawson JH (1982) Sulfur donor ligand-binding to ferric cytochrome-$P450_{cam}$ and myoglobin – ultraviolet-visible absorption, magnetic circular-dichroism, and electron-paramagnetic resonance spectroscopic investigation of the complexes. J Biol Chem 257:8308–8320
104. Poulos TL, Finzel BC, Howard AJ (1987) High-resolution crystal structure of cytochrome-$P450_{cam}$. J Mol Biol 195:687–700
105. Yoshioka S, Takahashi S, Ishimori K, Morishima I (2000) Roles of the axial push effect in cytochrome P450cam studied with the site-directed mutagenesis at the heme proximal site. J Inorg Biochem 81:141–151
106. Yoshioka S, Tosha T, Takahashi S, Ishimori K, Hori H, Morishima I (2002) Roles of the proximal hydrogen bonding network in cytochrome P450(cam)-catalyzed oxygenation. J Am Chem Soc 124:14571–14579
107. Galinato MGI, Spolitak T, Ballou DP, Lehnert N (2011) Elucidating the role of the proximal cysteine hydrogen-bonding network in ferric cytochrome $P450_{cam}$ and corresponding mutants using magnetic circular dichroism spectroscopy. Biochemistry 50:1053–1069

108. Chen ZC, Ost TWB, Schelvis JPM (2004) Phe393 mutants of cytochrome P450BM3 with modified heme redox potentials have altered heme vinyl and propionate conformations. Biochemistry 43:1798–1808
109. Argos P, Mathews FS (1975) Structure of ferrocytochrome-B5 at 2.8 Å resolution. J Biol Chem 250:747–751
110. Reid LS, Taniguchi VT, Gray HB, Mauk AG (1982) Oxidation reduction equilibrium of cytochrome-B5. J Am Chem Soc 104:7516–7519
111. Reid LS, Mauk MR, Mauk AG (1984) Role of heme propionate groups in cytochrome-B5 electron transfer. J Am Chem Soc 106:2182–2185
112. Reid LS, Lim AR, Mauk AG (1986) Role of heme vinyl groups in cytochrome-B5 electron-transfer. J Am Chem Soc 108:8197–8201
113. Lee KB, Jun ES, Lamar GN, Rezzano IN, Pandey RK, Smith KM, Walker FA, Buttlaire DH (1991) Influence of heme vinyl-protein and carboxylate protein contacts on structure and redox properties of bovine cytochrome-B5. J Am Chem Soc 113:3576–3583
114. Pellicena P, Karow DS, Boon EM, Marletta MA, Kuriyan J (2004) Crystal structure of an oxygen-binding heme domain related to soluble guanylate cyclases. Proc Natl Acad Sci USA 101:12854–12859
115. Olea C, Boon EM, Pellicena P, Kuriyan J, Marletta MA (2008) Probing the function of heme distortion in the H-NOX family. ACS Chem Biol 3(11):703–710
116. Olea C, Kuriyan J, Marletta MA (2010) Modulating heme redox potential through protein-induced porphyrin distortion. J Am Chem Soc 132:12794–12795
117. Lu Y, Yeung N, Sieracki N, Marshall NM (2009) Design of functional metalloproteins. Nature 460:855–862
118. Lu Y, Berry SM, Pfister TD (2001) Engineering novel metalloproteins: design of metal-binding sites into native protein scaffolds. Chem Rev 101:3047–3080
119. Ozaki SI, Roach MP, Matsui T, Watanabe Y (2001) Investigations of the roles of the distal heme environment and the proximal heme iron ligand in peroxide activation by heme enzymes via molecular engineering of myoglobin. Acc Chem Res 34:818–825
120. Matsui T, Ozaki S, Liong E, Phillips GN, Watanabe Y (1999) Effects of the location of distal histidine in the reaction of myoglobin with hydrogen peroxide. J Biol Chem 274:2838–2844
121. Shifman JM, Gibney BR, Sharp RE, Dutton PL (2000) Heme redox potential control in de novo designed four-alpha-helix bundle proteins. Biochemistry 39:14813–14821
122. Marshall NM, Garner DK, Wilson TD, Gao YG, Robinson H, Nilges MJ, Lu Y (2009) Rationally tuning the reduction potential of a single cupredoxin beyond the natural range. Nature 462:113–127
123. Shook RL, Borovik AS (2010) Role of the secondary coordination sphere in metal-mediated dioxygen activation. Inorg Chem 49:3646–3660
124. Kurtz DM (1990) Oxo- and hydroxo-bridged diiron complexes: a chemical perspective on a biological unit. Chem Rev 90:585–606
125. Collman JP, Gagne RR, Reed C, Halbert TR, Lang G, Robinson WT (1975) Picket fence porphyrins. Synthetic models for oxygen binding hemoproteins. J Am Chem Soc 97:1427–1439
126. Castro-Rodriguez I, Nakai H, Zakharov LN, Rheingold AL, Meyer K (2004) A linear, O-coordinated $\eta^1$-$CO_2$ bound to uranium. Science 305:1757–1759
127. Rosenthal J, Nocera DG (2008) Oxygen activation chemistry of pacman and hangman porphyrin architectures based on xanthene and dibenzofuran spacers. Progress in Inorganic Chemistry. Wiley, Hoboken, NJ
128. Dogutan DK, Stoian SA, McGuire R, Schwalbe M, Teets TS, Nocera DG (2010) Hangman corroles: efficient synthesis and oxygen reaction chemistry. J Am Chem Soc 133:131–140
129. Sen Soo H, Komor AC, Iavarone AT, Chang CJ (2009) A hydrogen-bond facilitated cycle for oxygen reduction by an acid- and base-compatible iron Pplatform. Inorg Chem 48:10024–10035

Struct Bond (2012) 142: 155–184
DOI: 10.1007/430_2011_60

Published online: 15 November 2011

# Ligand Field and Molecular Orbital Theories of Transition Metal X-ray Absorption Edge Transitions

**Rosalie K. Hocking and Edward I. Solomon**

**Abstract** Carl Ballhausen made a wide range of seminal contributions to ligand field theory and its application to ground state and ligand field excited state spectroscopies. These provided a fundamental basis for probing the nature of transition metal complexes using their visible spectra and a range of magnetic spectroscopies. The advent of synchrotrons provided access to high flux electromagnetic radiation that could be tuned across a wide range of energies including X-ray. This expanded the scope of spectroscopic techniques available to inculde X-ray Absorption Edge Spectroscopies. Paralleling a visible absorption experiment, X-ray spectra (metal K-edge, i.e. 1s→3d and metal L-edge, i.e. 2p→3d) taken at a synchrotron are dominated by ligand field splittings, electron repulsion effects and covalency. These can be used to obtain important insight into the properties of a diverse range of materials from solar cells to the catalytic centers of metalloenzymes. Herein we systematically consider applications of ligand field theory to X-ray absorption edge transitions.

**Keywords** Ligand-field theory · Spectroscopy · Magnetism · X-ray Absorption Spectroscopy · Synchrotron · Molecular Orbital Theory

---

R.K. Hocking
Monash Centre for Synchrotron Science, The Australian Centre for Electromaterials Science and School of Chemistry, Monash University, Melbourne 3800, Australia

E.I. Solomon (✉)
Department of Chemistry, Stanford University, Stanford, CA 94305, USA

Stanford Synchrotron Radiation Lightsource, SLAC, Stanford University, Stanford, CA 94309, USA
e-mail: Edward.Solomon@stanford.edu

**Contents**



## 1 Ligand Field Theory: From d → d to Core → d Transitions

Carl Ballhausen made a wide range of seminal contributions to ligand field theory and its application to ground state and ligand field excited state spectroscopies [1–7]. These provided a fundamental basis for probing the nature of transition metal complexes using their visible spectra and a range of magnetic spectroscopies. Today, this field has expanded in many directions, including applications developed in bioinorganic chemistry [8–11], where ligand field theory has enabled researchers to probe metalloprotein active sites in their functional states.

The advent of synchrotrons provided access to high flux electromagnetic radiation that could be tuned across a wide range of energies including X-rays. This expanded the scope of spectroscopic techniques available to include the X-ray absorption edge spectroscopies [12, 13]. As for a laboratory-based UV-Vis-NIR absorption spectroscopy experiment, an X-ray absorption edge experiment fundamentally involves varying the incoming wavelength of light (i.e., X-rays) and measuring absorption as a function of that energy. Like a NIR-Vis-UV experiment applied to transition metal complexes, ligand-field-based information can be obtained from their X-ray absorption edge spectra [14–19], in particular, the splitting of the valence d orbitals and their covalent mixing with the ligand environment.

It is useful to consider the correlation between the classical visible spectroscopy photon experiment (i.e., 3d → 3d) and X-ray spectroscopies (metal K-edge, i.e., 1s → 3d and metal L-edge, i.e., 2p → 3d) taken at a synchrotron (Fig. 1). All three types of transitions are dominated by ligand-field orbital splittings, electron repulsion effects, and covalency and can be used to gain important insight into these properties in a diverse range of materials from solar cells to catalytic centers of metalloenzymes. Here, we systematically consider applications of ligand-field theory to X-ray absorption edge transitions. For both the metal K- and L-edge regions, we will first consider the relatively simple case of $3d^9$ Cu(II) with one hole (and the related $3d^{10}$ Cu(I)). We then extend the review to the more complex cases of Fe(II) and Fe(III) complexes with four and five electron holes, respectively.

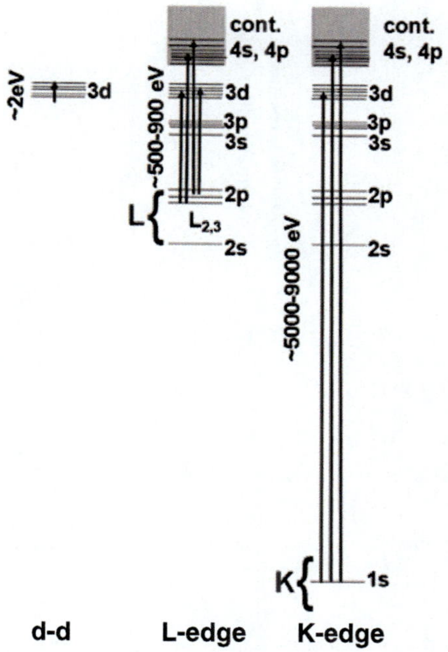


**Fig. 1** Energy level diagram comparing d → d transitions, metal L-edge and metal K-edge excitations for first transition series elements

## 2 Metal K-Edge XAS

X-ray absorption edge spectroscopies are divided on the basis of the shell from which the electron originates. The most widely used XAS spectral method involves the 1s or K-shell (Fig. 1, right). The spectra of $D_{4h}$ and $D_{2d}$ $[CuCl_4]^{2-}$ are typical of transition metal K-edge spectra and have one hole that simplifies the analysis. The main regions of the spectra are indicated in Fig. 2. For open shell transition metals, the pre-edge involves a weak 1s → 3d transition, while the edge contains the intense 1s → 4p transitions and excitation into the continuum. Finally, the near edge and extended X-ray absorption fine structure (EXAFS) regions are also indicated in Fig. 2. While the low-energy region of the spectrum is dominated by bound-state transitions, the near edge and EXAFS regions are dominated by multiple scattering and construction and destruction interferences of the ejected photoelectron that are sensitive to geometric structure and will not be further considered here; the reader is directed to the many reviews on the topic [20, 21].

### 2.1 *Copper Systems*

#### 2.1.1 Cu(I)

Among the most straightforward transition metal edge spectra to consider are those of Cu(I) complexes. Cu(I) has a filled $3d^{10}$ shell. Thus transitions arising from a 1s core

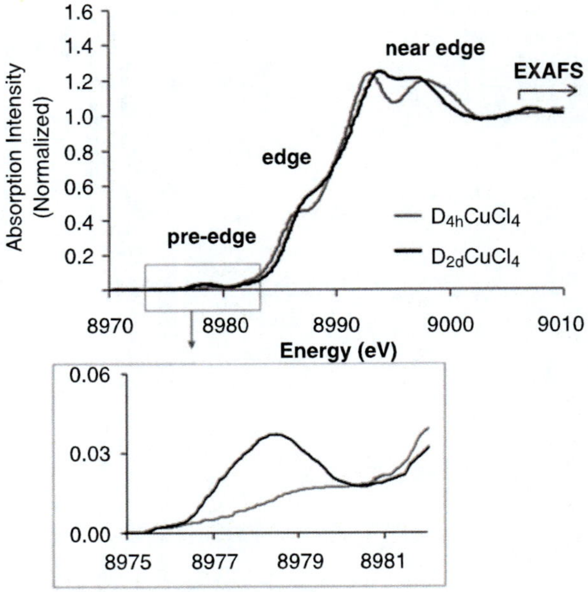


**Fig. 2** The basic features of a metal K-edge X-ray absorption spectrum highlighted on the spectra of two classical coordination compounds $D_{4h}$ $CuCl_4^{2-}$ (*pale line*) and $D_{2d}$ $CuCl_4^{2-}$ (*dark line*). The *inset* shows the pre-edge region is expanded at the bottom [22]

level can only involve 1s → 4s and 1s → 4p excitations, where only the transitions to the 4p orbitals are electric dipole allowed. Figure 3 shows representative members of the series of Cu(I) complexes with different coordination numbers: 2, 3, and 4. For each coordination environment, the ligand field has a different effect on the relative energies of the 4p orbitals. The change in their splitting pattern thus changes the shape of the Cu(I) K-edge. From Fig. 3A, the simplest case of two coordinate linear copper has a large repulsive interaction of the ligands along the *z*-axis shifting the $4p_z$ orbital to higher energy. This results in the intense low-energy feature in the XAS at 8,984 eV (the 1s → $4p_{x,y}$ transitions). The addition of a third ligand to the Cu(I) splits the degeneracy of the $4p_x$ and $4p_y$ levels and thus in 3 coordinate Cu(I) complexes the low-energy feature is split in energy. Finally, in four coordinate distorted tetrahedral Cu(I) complexes, all three 4p orbitals are equally raised in energy and thus one broad absorption feature is observed at energies >8,985 eV.

### 2.1.2 Cu(II)

Proceeding from the $3d^{10}$ electron configuration of Cu(I) to the $3d^9$ configuration of Cu(II) opens a hole in the $3d_{x^2-y^2}$ orbital (for square planar and tetragonal ligand

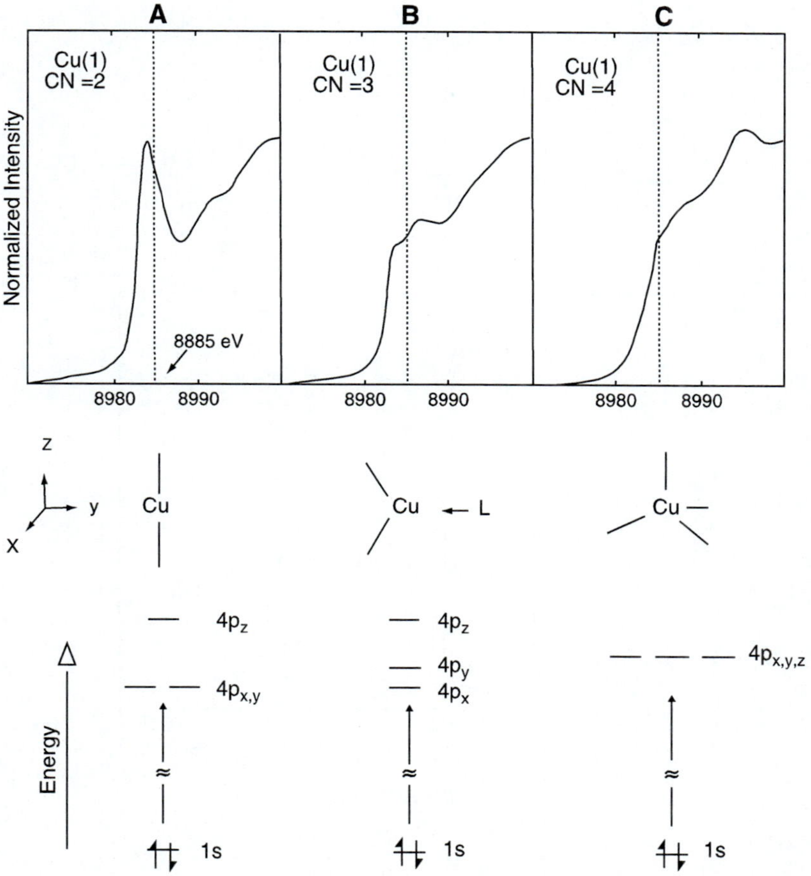


**Fig. 3** Pre-edge features of a series of Cu(I) complexes of varying geometry. (**a**) Linear two-coordinate, (**b**) three-coordinate, and (**c**) four-coordinate distorted tetrahedral. The resulting 4p orbital energy splitting by those ligand fields are shown at the bottom with the 1s → 4p transitions indicated. Altered from [23], with permission, Copyright American Chemical Society, 1993

fields), which generates a low-energy 1s → 3d transition at 8,979 eV, labeled "pre-edge" in Fig. 2. The 1s → 3d transition is electric dipole forbidden. However, this transition is observed even in centrosymmetric transition metal complexes as is the case for square planar $D_{4h}$ $[CuCl_4]^{2-}$. To determine the intensity origin of this transition, polarized XAS was performed [24]. The results of this experiment are shown in Fig. 4. Spectra were obtained with the polarization ($\vec{E}$) and propagation ($\vec{k}$) directions shown schematically: the $\vec{E}$ and $\vec{k}$ vectors are in the *xy* plane and the site was rotated about the $[CuCl_4^{2-}]$ *z*-axis ($C_4$) using oriented single crystals (Fig. 4a). As shown in Fig. 4b, the 8,979 eV "pre-edge" peak is polarized, and the intensity of this peak as a function of $\phi$ is plotted in Fig. 4c. The intensity maximized every 90° demonstrating that it is a quadrupole allowed transition.

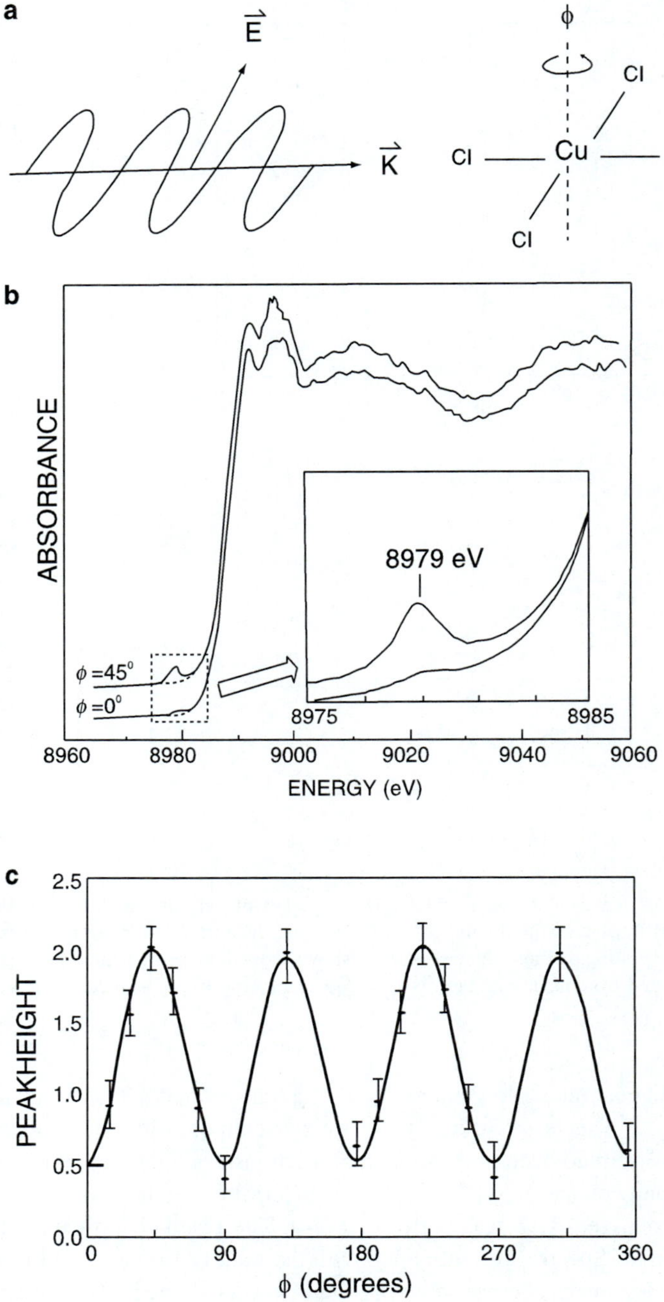


**Fig. 4** Single crystal XAS on $D_{4h}$ $CuCl_4^{2-}$. (**a**) Propagation and polarization direction of incident X-rays relative to $CuCl_4^{2-}$ single crystal which was rotated about its $C_4$ axis by an angle ϕ; (**b**) polarized K-edge XAS spectra of $D_{4h}$ $CuCl_4^{2-}$. *Inset* shows polarization of the pre-edge feature of 8,979 eV; (**c**) intensity of 8,979 eV peak plotted as a function of rotation angle $\phi$. ($\phi = 0$ starting along a Cu–Cl bond.) Figure adapted from [11] and [24]. Altered from [23], with permission, Copyright Elsevier, 1982

Note that while these are generally immeasurably weak in the NIR-Vis-UV spectrum of transition metal complexes, in the X-ray region, the wavelength of light is ~1.4 Å and thus higher terms in the multipole expansion for the interaction of radiation with matter can become sizeable. The electric quadrupole transition moment integral is given in (1), where $\vec{k}$ is the propagation direction of the light, $\vec{p}$ is the electron momentum operator, and $\vec{r}$ operates on the electron coordinates. From this equation and Fig. 4a, where $\vec{E}$ defines the $x$-axis and $\vec{k}$ the $y$-axis, the electric quadrupole moment transforms as $M(xy)$ and the transition moment is non-zero for a transition to the $\Psi_e = d_{x^2-y^2}$ half occupied orbital when the propagation direction $\vec{k}$ bisects adjacent Cl–Cu bonds, ($\phi = 45°$) in Fig. 4c.

$$I \propto \langle 1s \left| (\vec{E}\bullet\vec{p})(\vec{k}\bullet\vec{r}) \right| \Psi_e \rangle \tag{1}$$

In Fig. 2 (bottom), we show two isotropic XAS spectra of $[CuCl_4]^{2-}$ , those of $D_{4h}$ and $D_{2d}$ distorted complexes. It is clear that the pre-edge of the $D_{2d}$ distorted $[CuCl_4]^{2-}$ is 6× more intense. For Cu(II) in a non-centrosymmetric site, there can be 4p mixing into the $3d_{x^2-y^2}$ orbital via interaction with the ligands. This will contribute electric dipole character to the "1s → 3d" transition where "3d" now corresponds to a 3d/4p mixed ½ occupied d orbital as described below. It has been found that the electric dipole intensity for a pure 1s → 4p transition is ~100× the electric quadrupole intensity for a pure 1s → 3d transition [15, 22, 25], thus this large change of pre-edge intensity reflects only 4% 4p mixing.

#### 2.1.3 Ligand-Based Origin of the 4p Contribution to Metal K Pre-edge Intensity

As described above, the additional intensity in the Cu K pre-edge results from the mixing of some metal 4p character into the ½ occupied metal $3d_{x^2-y^2}$ orbital in a non-centrosymmetric ligand field [24]. This can reflect an electrostatic effect of the non-centrosymmetric ligand field which is very small [1, 2] and a dominant covalent effect [26]. The covalent contribution to this group theory allowed mixing can be visualized by examining the symmetry adapted linear combination (SALC) of Cl 3p orbitals which interacts with the Cu $3d_{x^2-y^2}$ orbital to form a σ-bond. In the case of $D_{4h}$ $CuCl_4{}^{2-}$, the complex is centrosymmetric and the Cl 3p SALC with the correct linear combination ($b_{1g}$) to form a sigma bond with the $d_{x^2-y^2}$ orbital is illustrated, in Fig. 5 (left). These orbitals cannot overlap with the Cu 4p orbitals in $D_{4h}$ ($a_{2u}$, $e_u$). When the symmetry is lowered to $D_{2d}$, this combination of Cl 3p orbitals which forms the σ bond with the $3d_{x^2-y^2}$ orbital can now additionally overlap the Cu($4p_z$) orbital ($D_{4h} \rightarrow D_{2d}$: $b_{1g} \rightarrow b_2$, $a_{2u} \rightarrow b_2$). This is shown in Fig. 5, right, and is expressed by the CI matrix below [26]. The Cl 3p-based ligand SALC is at lowest energy. The metal 3d and metal 4p orbitals are at higher energy, with the energy separation $\Delta_1$ and $\Delta_2$, respectively. $T_1$ is the overlap matrix element between the Cl 3p SALC and the metal $3d_{x^2-y^2}$ orbital and $T_2$ the interaction matrix

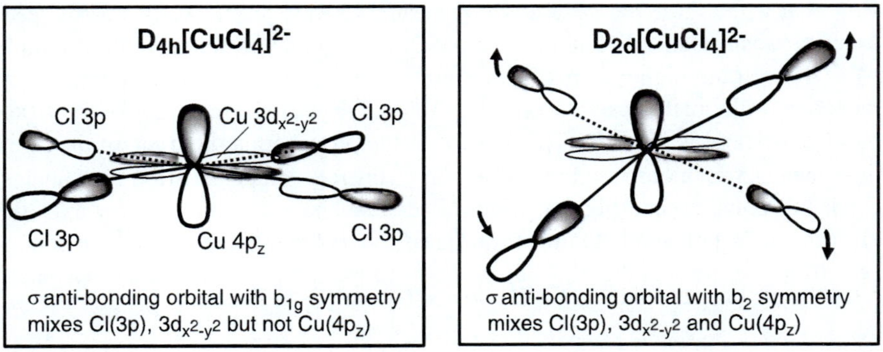


**Fig. 5** The symmetry adapted combination so the Cl 3p orbitals which interact with the Cu $d_{x^2-y^2}$ orbital in $D_{4h}$ and $D_{2d}$ distorted $[CuCl_4]^{2-}$. The Cu 4p and the Cu $3d_{x^2-y^2}$ orbitals cannot directly mix, however they do through overlap with the ligand σ orbitals in $D_{2d}$ symmetry

between the $Cl_{3p}$ SALC and the metal $4p_z$ orbital. The simultaneous overlap of Cl-based 3p with both the Cu-based $3d_{x^2-y^2}$ and the Cu-based $4p_z$ leads to a net mixing of Cu $4p_z$ character into the Cu $d_{x^2-y^2}$ orbital and electric dipole intensity in the XAS pre-edge spectrum. Solving the matrix for $D_{2d}$ $CuCl_4{}^{2-}$ gives a ½ occupied Cu d-based LUMO wave-function which is + 61% Cu $3d_{x^2-y^2}$ + 35% Cl 3p + 4% Cu 4p in character.

$$\begin{array}{l} \text{Cu 4p} \\ \text{Cu 3d} \\ \text{Cl 3p} \end{array} \begin{vmatrix} \Delta_2 - E & & T_2 \\ & \Delta_1 - E & T_1 \\ & & 0 - E \end{vmatrix} = 0 \qquad (2)$$

### 2.1.4 Ligand to Metal Shake-Down Transition in Cu(II)

The polarized spectrum of $D_{2d}$ $[CuCl_4]^{2-}$ (Fig. 6A) shows that the pre-edge peak at 8,779 eV is $z$ polarized consistent with the $4p_z$ mixing into the $d_{x^2-y^2}$ orbital due to the ligand field as described in Sect. 2.1.3. At higher energy, two additional $z$ polarized transitions are observed at 8,987 eV and 8,993 eV. One would expect the $1s \rightarrow 4p_z$ transition in this region; however, final state electronic relaxation splits this into two features, the $1s \rightarrow 4p_z$ and a $1s \rightarrow 4p_z$ + LMCT shakedown which is at lower energy than the main peak [23, 27].

In the main, Cu $1s \rightarrow 4p$, and the shakedown, Cu $1s \rightarrow 4p$ + LMCT, processes, excitation of a 1s core electron into the 4p orbital (an electric dipole –allowed transition; $z$-pol for $1s \rightarrow 4p_z$) creates a core hole, which results in an increased effective nuclear charge on the Cu felt by the valence d orbitals shifting these to deeper binding energies as shown schematically in Fig. 6B, $1 \rightarrow 2$. For Cu(II) complexes, the half occupied $3d_{x^2-y^2}$ orbital relaxes to an energy below that of the filled ligand valence orbitals. As a result of this relaxation, a lower energy

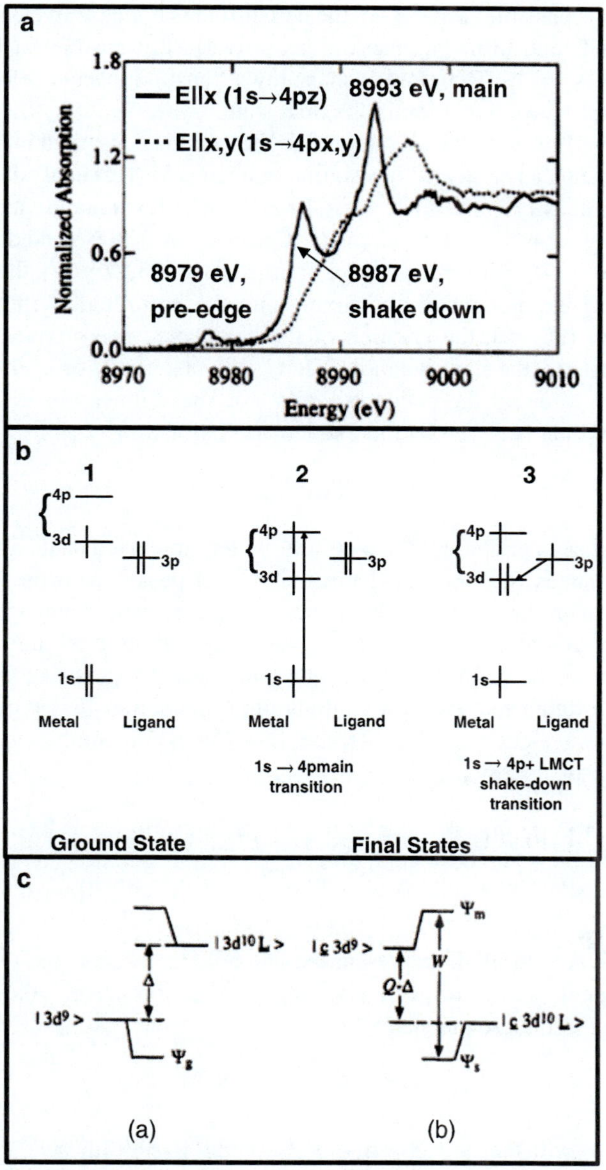


**Fig. 6** **A.** Polarized Cu K-edge spectra of $D_{2d}$ $CuCl_4^{2-}$. *z*-Polarized spectrum (*solid line*) and *x*,*y*-polarized spectrum (*dotted line*). The feature in the *z*-polarized spectrum ~8,987 eV is assigned as the 1s → 4p + LMCT shakedown transition, the feature at 8,993 eV the main transition. **B.** Schematic of 1s → 4p Cu K-edge transition and its associated shakedown transition. 1. Ground 2. The 1s → 4p final state with relaxation of valence levels due to the creation of a core hole. 3. The final state associated with the 1s → 4p + LMCT shakedown transition. **C.** Configuration interaction formalism for analysis of the Cu K-edge near-edge structure. (a) Ground-state wave-function as determined by the parameters *T* and *Δ*; (b) Final-state wave-functions. The lower-energy 1s → 4p + LMCT shakedown final state ($\Psi_s$) is separated in energy from the main 1s → 4p transition final state ($\Psi_m$) by the splitting *W*. Altered from [22], with permission, Copyright American Chemical Society, 1993

configuration is possible (Fig. 6B-3); the d manifold is filled by an electron from the ligand, i.e., a ligand to metal charge transfer (LMCT) transition at lower energy than the pure 1s → 4p. Therefore, a shakedown transition occurs below the energy of the main 1s → $4p_z$ transition in the final state.

The observed shakedown is analogous to the satellites observed in photoelectron spectroscopy and can be treated in a similar manner [25]. This analysis enables us to quantify the intensity of the 1s → 4p + LMCT shakedown feature as a percentage of the total 1s → 4p transition intensity. Comparison of the shakedown intensity to that of the 1s → 3d(+4p) pre-edge transition (at ~8,979 eV) then allows for quantitation of the amount of $4p_z$ mixing into the 3d orbital. Within the sudden approximation [28–30], the creation of the core hole occurs rapidly, before the electron adjusts to the new potential. Here, the intensity $I_i$ of a given transition corresponding to either the main (1s → 4p ) or shake down (1s → 4p + LMCT) final state can then be given by the Sudden approximation

$$I_i = |\langle \Psi_i(\underline{c}) \Psi_R(\underline{c}) \rangle|^2,$$

where $\underline{c}$ donates a core hole, $\Psi_i$ denotes the initial unrelaxed state, and $\Psi_R$ are the relaxed final states (the main and the shakedown peaks) with the core electron removed. This implies that only final states with the same symmetry as the main peak can contribute to shakedown intensity. Quantitation of this process and determination of the intensity ratio of the main peak to the shakedown peak can be obtained through the use of a configuration interaction (CI) type model. The ground-state wave-function $\Psi_G$ for the $D_{2d}$ $CuCl_4^{2-}$ complex is obtained by diagonalizing the energy matrix:

$$\begin{vmatrix} \langle \Psi(3d^9)|H|\Psi(3d^9)\rangle - E' & \langle \Psi(3d^9)|H|\Psi(3d^{10}\underline{L})\rangle \\ \langle \Psi(3d^9)|H|\Psi(3d^{10}\underline{L})\rangle & \langle \Psi(3d^{10}\underline{L})|H|\Psi(3d^{10}\underline{L})\rangle - E' \end{vmatrix} = 0 \quad (3)$$

where $\Psi(3d^9)$ and $\Psi(3d^{10}\underline{L})$ represent the one-hole metal state and the $3d^{10}$ metal/one-hole ligand state, respectively. Diagonalization gives the eigenvector corresponding to the ground state:

$$\Psi_g = \cos\theta|3d^9> - \sin\theta|3d^{10}\underline{L}> \quad (4)$$

$$\text{with} \tan 2\theta = 2T/\Delta, \quad T = \langle \Psi(3d^9)|H|\Psi(3d^{10}\underline{L})\rangle \quad (5)$$

$$\text{and}\ \Delta = <\Psi(3d^9)|H|\Psi(3d^9)> - <\Psi(3d^{10}\underline{L})|H|\Psi(3d^{10}\underline{L})> \quad (6)$$

$T$ is the interaction matrix element between the configurations contributing to the ground-state wave-function, and $\Delta$ is the energy difference between the two configurations, as shown in Fig. 6C-a.

The 1s → 4p transition gives the main ($\Psi_m$) and shakedown ($\Psi_s$) peaks. The interaction matrix for those excited states is constructed in an analogous fashion as for the ground state, with the addition of Q to the diagonal energy of the $\underline{c}3d^9$ configuration to account for the increase in the effective nuclear charge felt by the $3d^9$ state on the metal due to the core hole. The solutions for the excited-state wave-function are given by the expressions below, where $\underline{c}$ indicates a Cu 1s core hole and $\tan 2\theta' = 2T/(\Delta - Q)$; $0 < \theta' < 90°$.

$$\Psi_m = \sin\theta'|3d^9> + \cos\theta'|\underline{c}3d^{10}\underline{L}> \tag{7}$$

$$\Psi_s = \cos\theta'|\underline{c}3d^9> - \sin\theta'|\underline{c}3d^{10}\underline{L}> \tag{8}$$

The energy splitting W between the main and the shakedown peaks is given by

$$W = [(\Delta - Q)^2 + 4T^2]^{1/2} \tag{9}$$

The main-to-shakedown peak intensity ratio ($I_m/I_s$) is

$$\frac{I_m}{I_s} = \left(\frac{\sin\theta' \cos\theta - \cos\theta' \sin\theta}{\cos\theta' \cos\theta - \sin\theta' \sin\theta}\right) = \tan^2(\theta' - \theta) \tag{10}$$

Thus, the intensity ratio between the main and shakedown peaks is determined by the change in the wave-function $\theta' - \theta$ upon creation of a 1s core hole. PES satellite features give $\Delta = 0.88$ eV, $T = 1.5$ eV, $Q = 8.9$ eV, and $W = 8.2$ eV for the $D_{2d}$ $CuCl_4{}^{2-}$ complex [22]. Because $\Delta$ and $T$ are ground-state parameters, the values determined by PES can be used in the analysis of the XAS shakedown intensity in $D_{2d}$ $CuCl_4{}^{2-}$. The value of $Q$ for the 1s → 4p promotion is not known. As a first approximation, $Q$ is assumed to be the same as for the 2p ionization.

This analysis indicated that the shakedown peak at 8,987 eV in Fig. 6A is 60% of the total 1s → $4p_z$ intensity. This allowed for quantitation (given the pre-edge intensity) of the amount of $4p_z$ mixing into the 3d orbital, as ~4%. This was used to rule out the extensive $4p_z$ mixing that had been invoked for the small parallel hyperfine splitting observed in the EPR spectrum of $D_{2d}$ $CuCl_4$ [31].

## 2.2 *Iron Systems*

### 2.2.1 Multiplet Effects

In going from a $3d^9$ Cu(II) system to a transition metal complex with more than one hole, the analysis of the pre-edge becomes dependent on ligand-field and electron repulsion effects. Depending on the ligand-field strength and the electron configuration, these two effects can be comparable in energy, or one can dominate [15]. The impact of the open shell on high-spin Fe(II) and Fe(III) XAS spectra is outlined below.

For Fe(II), the high-spin $d^7$ free ion final states (the $(1s)^1$ core hole is spherical and thus does not contribute an additional orbital splitting to the states) are $^4F$

and $^4P$ which split in an octahedral ligand field into $^4A_{2g}$, $^4T_{1g}$, $^4T_{2g}$, and $^4T_{1g}$ states. These states are indicated on the weak field side of the $d^7$ Tanabe–Sugano diagram in Fig. 7 (left) [32]. Note that the energies of these states will differ from those given in the Tanabe–Sugano diagrams due to the presence of the core hole. However, the basic splitting patterns caused by the combination of electron repulsion and ligand-field effects remain the same as those given in the ligand-field correlation diagram.

For high-spin $3d^5$ Fe(III), the spin allowed free ion $3d^6$ final state is $^5D$ which splits into $^5T_{2g}$ and $^5E_g$ states separated by 10Dq in an octahedral ligand field. This is indicated on the $d^6$ Tanabe–Sugano diagram in Fig. 7, right. Thus, the ligand field of the high-spin Fe(III) pre-edge is more straightforward to interpret than that of high-spin Fe(II) complexes that exhibit additional multiplet effects due to $e^-$–$e^-$ repulsion.

The quadrupole allowed intensity for these final states is given at the top of the center inset to Fig. 7. In going to a tetrahedral ligand field, the energy splitting of the $3d^{n+1}$ final state changes. In addition, as described above for Cu, metal 4p mixing into the 3d orbitals becomes allowed. This mixing provides an additional electric dipole contribution to the intensities of the pre-edge features. From Fig. 7, center bottom, it is important to note that this electric dipole mixing can affect one final state relative to another thus, having a significant effect on the energy distribution of the intensity in a pre-edge. For example, for $T_d$ Fe(III), Fig. 7, the effect of 4p mixing only occurs for the $t_2$ orbitals and has no effect on the e set as the 4p have $t_2$ symmetry in $T_d$. Therefore, the 1s → 3d pre-edge features in Fe(II) and Fe(III) sites will, in general, be more intense for four or five coordinate non-centrosymmetric geometries, and the intensity distribution of the pre-edge shape is dependent on the specific geometry.

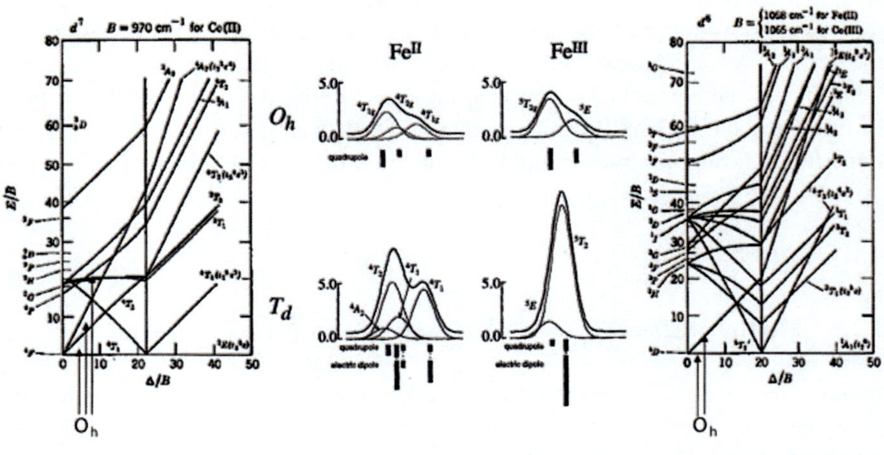


**Fig. 7** *Middle panel* – ferrous (*left*) and ferric (*right*) pre-edge features in the Fe K-edge including the relative contribution of (quadrupole and dipole) to the intensity at the bottom of each spectrum, upper $O_h$ and lower $T_d$. The states responsible for the splitting are indicated on the Tanabe–Sugano diagrams; $d^7$ *left*: final state of Fe(II), $d^6$ *right*: final state of Fe(III). Altered from [15], with permission, Copyright American Chemical Society, 1997

# 3 Metal L-Edge XAS

## 3.1 From K-Edge to L-Edge Absorption

Metal L-edge absorption features formally probe the transitions from the second ($2s^2 2p^6$) or L-shell, Fig. 1. The $L_1$ edge arises from excitation of electrons from the $2s^2$ sub-shell where both the $L_2$ and $L_3$ edges arise from the spin–orbital split 2p sub-shell. For first transition series metals, the $L_1$ edge is weak and offers little spectroscopic information that cannot be better obtained from the K-edge. Thus, when referring to the L-edge spectra of a first transition series metal complex, this usually indicates the $L_{2,3}$ edges.

The $2p \rightarrow 3d$ transition of the $L_{2,3}$ edges is electric dipole allowed producing $2p^5 3d^{n+1}$ final states. The $2p^5$ core configuration has an orbital angular momentum of $l = 1$ which will couple to the spin angular momentum of $s = 1/2$ to produce $J = 3/2$ and $J = 1/2$ final states with the $J = 3/2$ state, the $L_3$ feature, at lower energy and with approximately twice the intensity of the $J = 1/2$, $L_2$ peak in Fig. 8a. The $L_2$ edge is broader due to an additional Auger decay channel of the excited state which is absent for the $L_3$-edge [33]. In the absence of multiplet

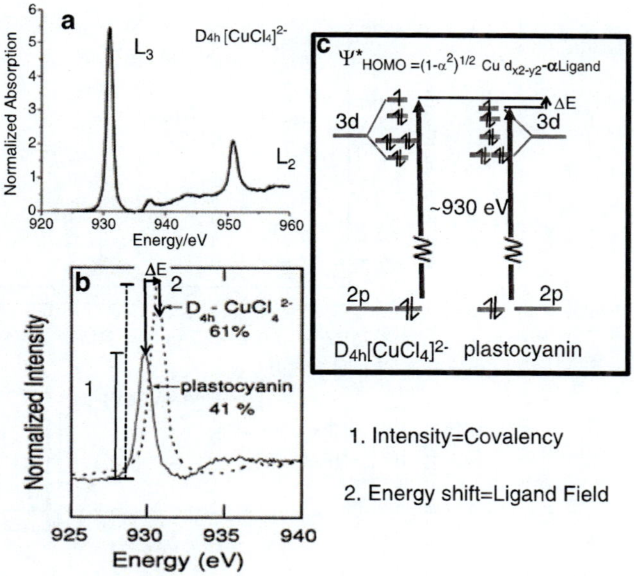


**Fig. 8** (**a**) The Cu L-edge spectrum of $[CuCl_4]^{2-}$ the $L_3$ and $L_2$ peaks are indicated; (**b**) Comparison of the $L_3$ of $[CuCl_4]^{2-}$ and plastocyanin. The two features of an L-edge spectrum are indicated. 1. Intensity which reflects covalency. 2. Energy shift which reflects ligand field. (**c**) Ligand-field splitting comparison of plastocyanin and $[CuCl_4]^{2-}$ indicating the origin of the intensity and energy difference. Altered from [17], with permission, Copyright American Chemical Society, 1997

effects, these two peaks would have the intensity ratio of 2:1 mentioned above (i.e., the ratio of degeneracies); however, multiplet effects redistribute intensity between the $L_2$ and $L_3$ edges, thus it is usually better to consider them together [17, 34, 35].

As both the $L_3$ and $L_2$ arise from transitions to the 3d orbitals, they are affected by ligand-field and multiplet effects similar to the K pre-edge spectra described above. However, as different selection rules govern the K (quadrupole) and L (dipole) pre-edges, these provide access to complementary types of spectroscopic information. The Fe K and $L_3$ pre-edges of $[Fe(tacn)_2]^{2+}$ and $[Fe(tacn)_2]^{3+}$ are compared in Fig. 9. While the two edges occur at different energy, the scale of both Fig. 9a, b has been set to 10 eV. From the comparison, it can be seen that sharper features are present in the $L_3$ pre-edge compared to the K pre-edge reflecting better energy resolution (1.0–1.2 eV for K-edge vs. 0.1–0.15 eV for an L-edge). There are two contributions to the better L-edge resolution. First, the resolution of a monochromator is better at lower energy, and second, the core hole lifetime broadening effects are less at the L-edge relative to the K-edge. However, it is also clear that the spread of the features in Fig. 9b (the $L_3$ edge) is larger than in Fig. 9a, (the K-edge). This is due to the addition of exchange interactions with the $^2P$ core hole, which will be described below.

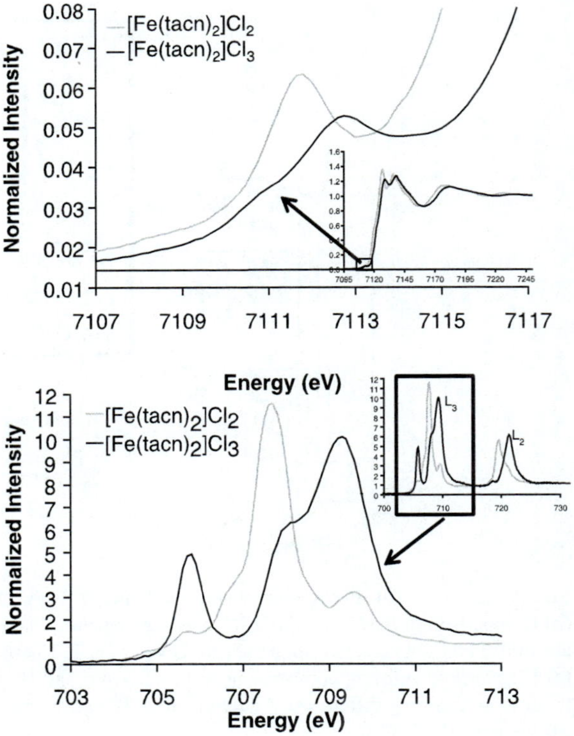


**Fig. 9** A comparison of the $L_3$ pre-edge and K pre-edge of $[Fe(tacn)_2]Cl_2$ and $[Fe(tacn)_2]Cl_3$. For both spectra, the *x*-axis is scaled to 10 eV to facilitate a comparison of spectral shape

## 3.2 Copper Systems

### 3.2.1 Intensity/Covalency Correlation

As for the metal K-edges above, it is important to first consider the simple $3d^9$ Cu(II) electron configuration. In this case, the L-edge involves a Cu 2p → 3d transition which is electric dipole allowed and produces a $2p^5 3d^{10}$ final state.

As the 2p orbital is localized on the metal and the 2p → 3d is electric dipole allowed, the intensity of this peak will reflect the amount of metal d character in the half occupied $\Psi$* molecular orbital of a Cu(II) complex as shown in Fig. 8c. This is a direct probe of covalency in transition metal complexes. Figure 8b compares the $L_3$ intensity of $D_{4h}$ $[CuCl_4]^{2-}$ and plastocyanin. The more covalent plastocyanin is reduced in intensity relative to $[CuCl_4]^{2-}$ reflecting the decreased metal and increased ligand character in the half occupied $\Psi$*. For $D_{4h}$ $CuCl_4{}^{2-}$ from the Cl K-edge (see [36]), there is 39% Cl 3p thus, 61% Cu d character in the ½ occupied HOMO; therefore, this intensity can be used to calibrate the L-edge intensity and determine the metal character, i.e., covalency, of the blue copper center and other highly covalent sites. For blue copper, this gives 39% Cu d character, the rest of the wave-function being delocalized into a thiolate S ligand orbital. This gives a large, covalent contribution to rapid, directional electron transfer by this metalloprotein site. As more holes are opened in the d shell for earlier transition metal complexes, this effect of the covalency of the d orbitals on the L-edge intensity is still present. This must be weighted by the number of unoccupied orbitals on the metal, and by the different covalencies of the different d orbitals (i.e., differential orbital covalency (DOC)). That is then distributed by electron repulsion over the many multiplet final states (see Sect. 3.3).

### 3.2.2 Ligand-Field Effects

In addition to the intensity differences described above, there is a clear difference in the energy of the $L_3$ peaks of plastocyanin and $[CuCl_4]^{2-}$; this is indicated by the arrow, $\Delta E$ in Fig. 8b. There are two potential contributions to this energy difference. First, the effective nuclear charge on the Cu is affected by the ligation and if this has a differential effect [19] on the energy of the 2p relative to the 3d orbital this would shift the transition energy. Second is the effect of the ligand field. It is found from an analysis of XPS data for Cu(II) complexes that a change of $Z_{\mathrm{eff}}$ does not have a substantially different energy shift on electrons in the 2p relative to the 3d orbitals, the energy shift of the 3d shell is about 70% that of the 2p [37]. Thus, the energy difference $\Delta E$ in $L_3$ edge energies for Cu(II) complexes is mostly due to differences in ligand-field strengths, as indicated in the orbital splitting patterns given in Fig. 8c [38]. For blue copper sites, the decrease by ~1.0 eV in $L_3$ edge energy relative to $D_{4h}$ $CuCl_4{}^{2-}$ in Fig. 8b is due to the decreased ligand field on the copper as the blue Cu site has three equatorial ligands destabilizing the energy of the $3d_{x^2-y^2}$ orbital relative to four in $D_{4h}$ $CuCl_4{}^{2-}$.

## 3.3 *Iron Systems*

### 3.3.1 Low-Spin Fe(III)

#### Multiplet Effects

To consider the effect of more than one hole in the d-orbitals, we first focus on the relatively simple case of low-spin Fe(III) with a $(t_{2g})^5(e_g)^0$ ground configuration, Fig. 10a. From this ground configuration, a 2p core $e^-$ will have transitions to both the $t_{2g}$ hole ($d_\pi$ orbital) and into the four holes in the $e_g$ set ($d_\sigma$ orbitals). This means that in the absence of electron repulsion and covalency effects, both K and L pre-edges would appear as two transitions split by 10Dq with an intensity ratio 1:4, K-edge quadrupole, L-edge dipole. However, as described above for the K-edges, this is generally not what is observed owing to multiplet effects that complicate the analysis of both the K pre-edge and the L pre-edge for metals with more than one valence hole. First, considering the $d^6$ excited configuration (the Tanabe–Sugano diagram given in Fig. 10d), the transition to the $t_{2g}$ hole results in an $^1A_1$ final state at lowest energy, while the transitions to the $e_g$ hole give $^3T_1$, $^3T_2$, $^1T_1$, and $^1T_2$ final states (Fig. 10c). This describes the final states of a K pre-edge [15]. However, for the L-edge, we must include the 2p spin–orbit and the 2p–3d $e^-$–$e^-$ repulsion in the $2p^53d^6$ final states, that are lifetime and instrument broadened into the L-edge spectrum shown in Fig. 10e. In addition, we must add the 3d final state spin–orbital coupling, Fig. 10f. The addition of this makes the low-energy $t_{2g}$ feature disappear at the $L_2$-edge and increase in intensity at the $L_3$ edge. This is discussed below. Note that all the lines in Fig. 10e, f reflect the ligand-field and electron repulsion split final states of a $(2p)^5(3d)^6$ core hole excited configuration.

All of the simulations given in Fig. 10a–f do not include covalency, only the combined effects of spin–orbit coupling, electron repulsion, and ligand-field splitting of the d orbitals. Covalency effects will be considered below. It is important to note that a ligand-field effect can have a significant effect on L-edge spectral shape. This is highlighted in Fig. 11, where the spectrum of a low-spin Fe(III)corrole is compared to that of a low-spin Fe(III)porphyrin. A significant change in the tetragonal splitting of the $a_1$ ($d_{z^2}$) and $b_1$ ($d_{x^2-y^2}$) orbitals in going from $D_{4h}$ porphyrin to $C_{2v}$ corrole has a significant effect on L-edge spectral shape.

#### Effects of 3d Spin–Orbit Coupling

As can be seen from the sequence of simulations shown in Fig. 10e–f, the addition of 3d spin–orbit coupling makes the low-energy $t_{2g}$ feature disappear from the $L_2$ and increase in intensity at the $L_3$ edge (comparison of Fig. 9f-g and Fig. 11a, $J = 1/2$, $\Gamma_7$ spectral simulation).

For low-spin Fe(III), the ground ${t_{2g}}^5$ configuration gives a $^2T_2$ ground state; the $T_2$ has effective orbital angular momentum which leads to spin–orbit coupling, this

**Fig. 10** Contributions to L-edge spectral shape: (**a**) Ligand field and electron configuration of low-spin Fe(III). (**b**) The two transitions with a ratio of 1:4 that would be obtained in the absence of spin–orbit coupling and electron repulsion. (**c**) The five states that arrive from the addition of $e^{-}$–$e^{-}$ repulsion. (**d**) $d^6$ Tanabe–Sugano diagram with the transitions indicated on the strong field (*right*) side. (**e**) Spectrum simulated with the effects of the metal p core hole (both spin–orbit and repulsion). (**f**) Adds 3d spin–orbit coupling. (**g**) Shows the σ and π density of states distributed across the final state spectrum

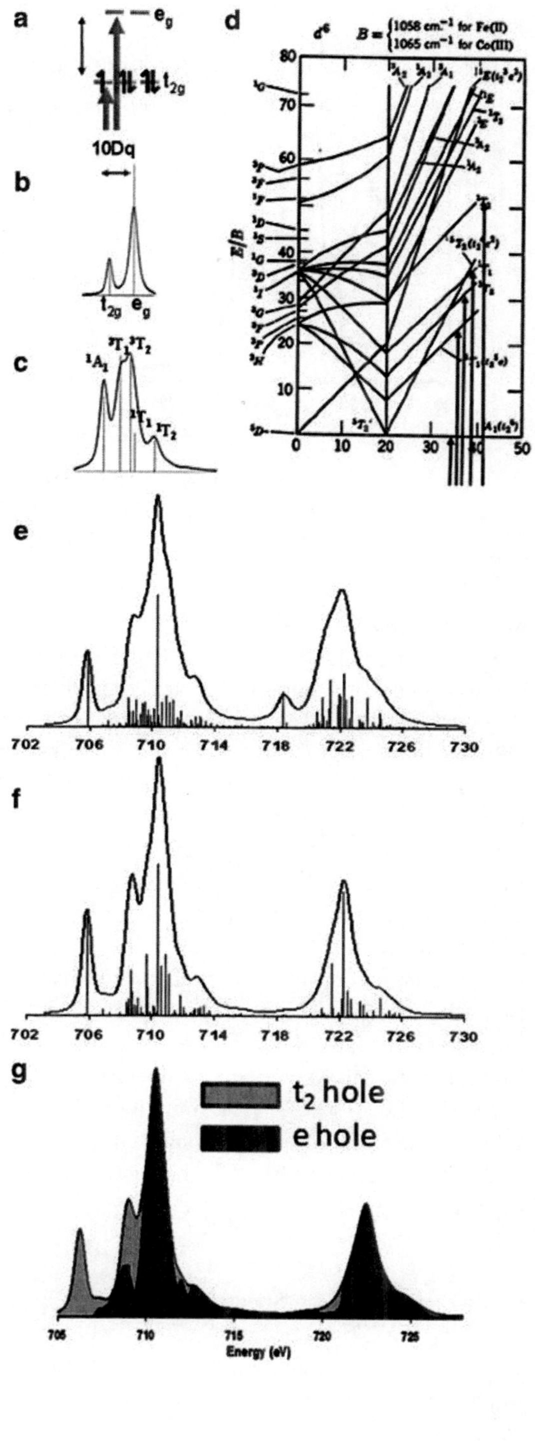

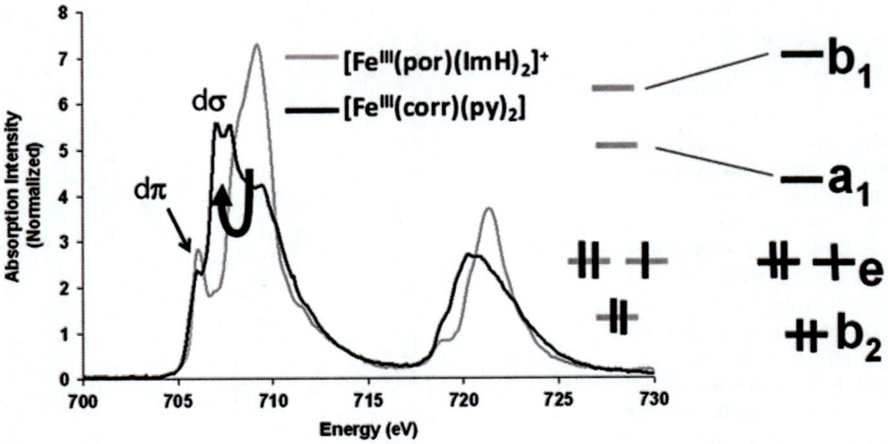


**Fig. 11** The effects of ligand-field splitting on the L-edge spectral shape of two low-spin Fe(III) compounds [39]. Altered from [39], with permission, Copyright American Chemical Society, 2009

gives ($S = 1/2 \times T_{2g}$) or $\Gamma_6 \otimes \Gamma_5 = \Gamma_7 + \Gamma_8$ ($E_2$ + G) of which $\Gamma_7$ ($J = 1/2$) is lowest in energy (Fig. 12b, left) [40]. The $2p^5t_{2g}{}^6$ excited configuration gives a $^2T_1$ excited state. This spin–orbit couples to give $\Gamma_6 \otimes \Gamma_4 = \Gamma_6 + \Gamma_8$, ($E_1$ + G). The "$t_{2g}$ peak" at the $L_3$ edge is $\Gamma_8$ and that at the $L_2$ edge is $\Gamma_6$. An electric dipole allowed transition ($T_1 = \Gamma_4$) from a $\Gamma_7$ spin–orbit ground state gives $\Gamma_7 \otimes \Gamma_4 = \Gamma_7 + \Gamma_8$. Thus the transition to the $\Gamma_6$ at the $L_2$ edge is not allowed and only the $L_3$ $t_{2g}$ peak ($\Gamma_8$) has electric dipole intensity. Alternatively, starting from the higher energy, $\Gamma_8$, spin–orbit component of the $^2T_2$ ground-state transitions ($\Gamma_8 \times \Gamma_4 = \Gamma_6 + \Gamma_7 + 2\Gamma_8$) to both the $L_2$ ($\Gamma_6$) and $L_3$ ($\Gamma_8$) pre-edges has intensity. These differences are clear from the multiplet simulations given in Fig. 12, where simulations for the $J = 3/2$ and $J = 1/2$ states are separated. The energy separating the two components of the ground state is given by $(3/2)\lambda$ or ~422 cm$^{-1}$. Thus there is no thermal population of the $\Gamma_8$ component of the ground state, and it cannot contribute intensity to the $L_2$ pre-edge feature. However, this energy separation is small and in real systems minor deviations from $O_h$ symmetry will mix these two states (Fig. 12b) [33, 35]. A simulation done in $D_{4h}$ symmetry (i.e., the dynamic Jahn–Teller effect for a $^2T_2$ ground state) with a separation between $d_{xy}$ and $d_{xz/yz}$ of ~700 cm$^{-1}$ is given in Fig. 12a, bottom, where the $t_{2g}$ peak is now present (arrow) at the low-energy side of the $L_2$ edge [19].

## Covalency Effects on Spectral Shape

Covalency (i.e., the amount of metal d character lost in bonding to the ligands) can impact spectral shape at the L-edge in several ways. These are described systematically below.

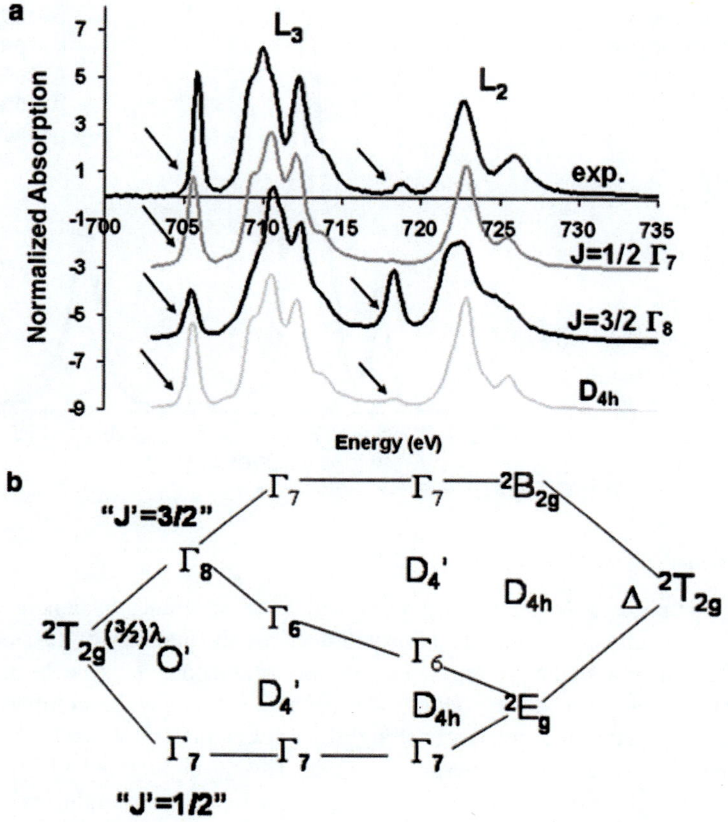


**Fig. 12** (**a**) Comparison of the experimental data (*top*) to spectral simulations, $O_h$ ligand field: $J = 1/2\ \Gamma_7$ (2nd from *top*), $J = 3/2$, $\Gamma_8$ (2$^{nd}$ from *bottom*) and in a $D_{4h}$ ligand field (*bottom*). *Arrows* indicate the "$t_{2g}$ feature," arising from the $2p^6t_{2g}{}^5 \rightarrow 2p^5t_{2g}{}^6$ transition. (**b**) Spin–orbit and low-symmetry ($D_{4h}$) effects on the $^2T_{2g}$ ground state of low spin $d^5$ complexes. *Left*, $^2T_{2g}$ spin orbit splits in the octahedral double group to give $\Gamma_7$ and $\Gamma_8$ states ($J' = 1/2$) and ($J' = 3/2$). These are then low-symmetry split in the $D_4'$ double group. To the *right*, the $^2T_{2g}$ state is split by tetragonal elongation ($\Delta$) giving $^2B_{2g}$ and $^2E_g$ states, which upon spin–orbit coupling give $\Gamma_7$, $\Gamma_6$, and $\Gamma_7$ states; the two limits are correlated. Reproduced with permission from [19], with permission, Copyright American Chemical Society, 2006

### *Nephelauxetic Effect*

As with ligand-field absorption spectra, the Nephelauxetic effect [41, 42] will also impact L-edge XAS data. For L-edge spectra, the effect of a reduction in inter-electron repulsion is also to reduce the free-ion state splitting and make the spectra somewhat more "orbital like." This is demonstrated by the sequence of simulated low-spin Fe(III) L-edge X-ray absorption spectra shown in Fig. 13, where the $e^- - e^-$ repulsion is reduced systematically from $\beta = 100\%$ to 60%. This starts from 80% of the Hartree–Fock calculated values of the Slater Condon Shortley parameters for $e^- - e^-$ repulsion.

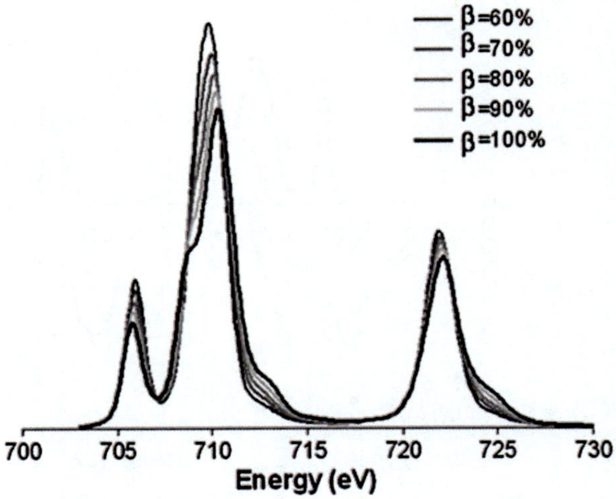


**Fig. 13** The effect of Nephelauxetic reduction on the spectrum of low-spin Fe(III)

*Differential Orbital Covalency*

For $O_h$ complexes, the $t_{2g}$ and $e_g$ holes will have different covalencies due to differences in $\pi$ and $\sigma$ bonding and this will decrease the intensity of the associated $2p \rightarrow 3d$ transition in the spectrum in a non-uniform manner. This has been termed differential orbital covalency (DOC) [18]. For low-spin Fe(III) complexes, the effect on the spectrum is relatively straightforward as the $t_2$ hole and e hole final states are well separated in energy, Fig. 10g. This is highlighted in Fig. 14 by comparison of the experimental spectrum of $[Fe(tacn)_2]^{3+}$, a low-spin Fe(III) complex with $\sigma$-only donor amine ligands to the calculated low-spin Fe(III) spectrum at the ionic limit, including the multiplet and ligand-field effects discussed above, but without including the effect of covalency, as well as a simulation with equal covalency in the d-orbitals (isotropic).

Clearly, the intensity of the transitions to the $e_g$ set is greatly reduced in the data relative to the simulation. As described in [18], we can include the covalency of the different symmetry sets of d orbitals through a valence bond configuration interaction model (VBCI). This allows ligand–metal charge transfer (LMCT) mixing into the d orbitals, which is varied to fit the experimental data [18]. We can then extract the resultant mixing using a projection method, (described below). From the fit in Fig. 14c, $[Fe(tacn)_2]Cl_3$ has, as expected, no $\pi$ covalency but its $d_\sigma$ orbitals have 30% ligand character.

*Back-Bonding*

In addition to $\sigma$ and $\pi$ donor ligands, another significant impact that covalency can have on spectral shape is through back-bonding. The effect of back-bonding on

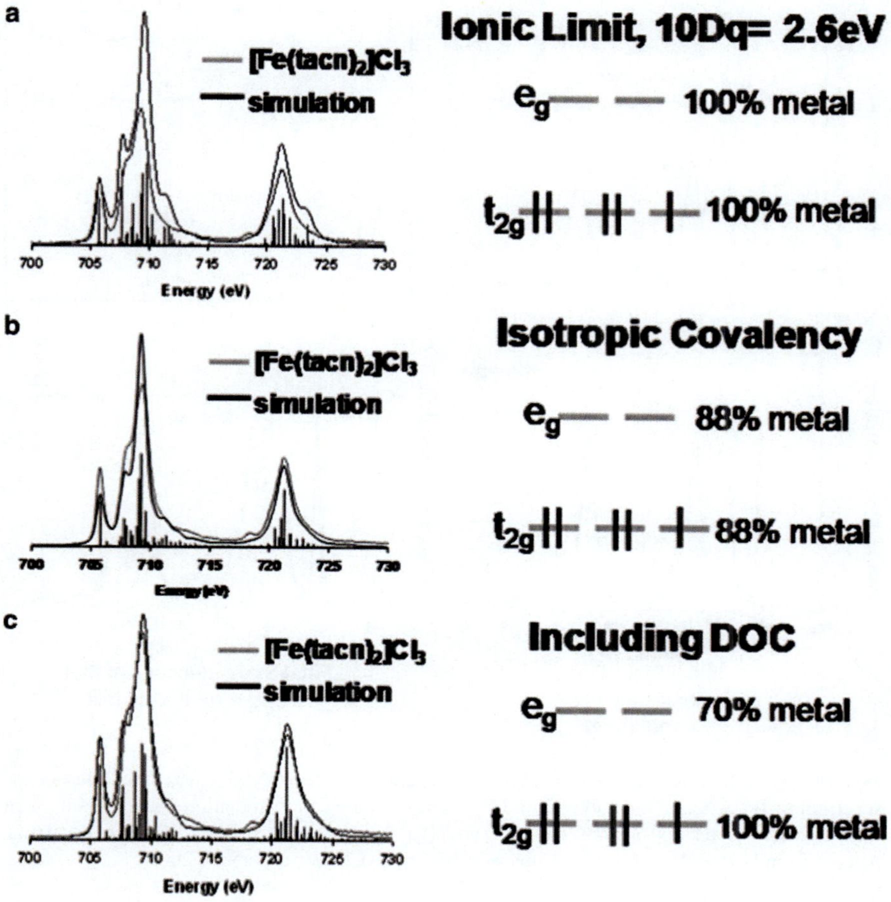


**Fig. 14** Comparison of ionic limit (**a**); isotropic (**b**) and differential orbital covalency simulation of $[Fe(tacn)_2]^{3+}$. Data are *gray*, simulations are *black*

L-edge spectral shape can be very pronounced as back-bonding adds additional transitions to the L-edge spectrum. Figure 15a compares the experimental L-edge spectrum of $[Fe(tacn)_2]^{2+}$ discussed above to that of $[Fe(CN)_6]^{3-}$. It is clear that in $[Fe(CN)_6]^{3-}$, an additional peak has appeared in the spectrum in Fig. 15a labeled as $\pi^*$, and $e_g$ intensities are greatly decreased relative to $[Fe(tacn)_2]Cl_2$ This reflects the Fe(2p) transition to the $CN^-$ $\pi^*$ orbital, which derives its intensity from Fe(3d) mixing due to back-bonding. Also, there is a final state mixing mechanism that redistributes intensity from the main $e_g$ peak into the $\pi^*$ orbitals [19]. These data could be fit by a model which mixed metal to ligand charge transfer (LMCT) to determine back-bonding as well as LMCT for $\sigma$ and $\pi$ donation.

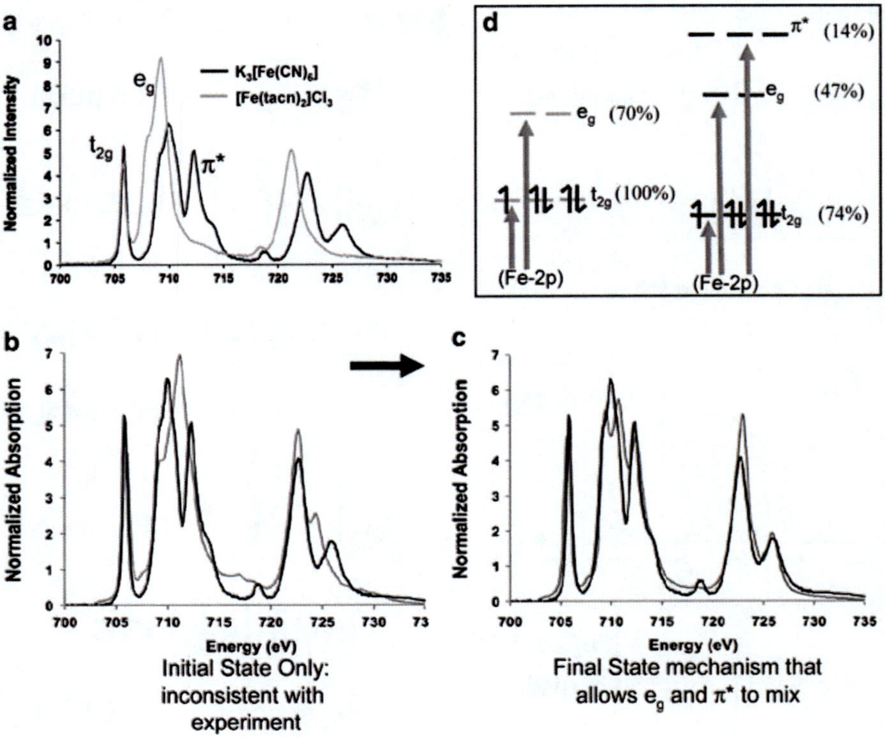


**Fig. 15** (**a**) Fe L-edge spectrum of $[Fe(tacn)_2]^{2+}$ relative to $[Fe(CN)_6]^{3-}$, (**b**) simulation of the spectrum of $[Fe(CN)_6]^{3-}$ including initial state back-bonding, (**c**) simulation also including final state effects, (**d**) results of simulations, $[Fe(CN)_6]^{3-}$ (*right*) compared to $[Fe(tacn)_2]^{2+}$ (*left*). Altered from [19] with permission, Copyright American Chemical Society, 2006

### *Quantifying Covalency Through a Ligand-Field Plus Charge Transfer Multiplet Model*

The above description is intended to give the reader qualitative insight into the spectral differences that can be identified and interpreted in terms of differential orbital covalency. These are done quantitatively with the aid of the ligand-field multiplet model using the software suite developed by R.D. Cowan and coworkers [43] and modified by B.T. Thole [34, 44, 45]. This suite of software programs can simulate both ligand-field limit spectra and spectra modified by covalency effects (through charge transfer mixing in a valence bond model).

In the ligand-field multiplet model, the ground state is a single $d^n$ electron configuration. This is shown in Fig. 10f for the $3d^5$ electron configuration. Each line represents a single transition and the number of transitions increases when, electron repulsion and spin–orbital coupling are included.

In the atomic limit (with no ligand field effect), this model is often written as:

$$H = H_{\text{ave}} + H_{\text{ls}} + H_{\text{ee}} \tag{11}$$

where,

$$H_{\text{ave}} = \sum_N \frac{p_i^2}{2m} + \sum_N \frac{-Ze^2}{r_i} \tag{12}$$

and $\Sigma_N$ is the sum over $N$ electrons $i$.

$$H_{\text{ls}} = \sum_N \xi(r_i) l_i . s_i \tag{13}$$

and

$$H_{\text{ee}} = \sum_{ij} \frac{e^2}{r_{ij}} \tag{14}$$

where $ij$ represents all electron pairs.

In these expressions, $H_{\text{ave}}$ is the average energy of the configuration and contains the kinetic term and the interactions of the electrons with the nucleus. In comparing a ground to an excited configuration, this essentially determines the weighted energy of a transition. Within these simulations, this is empirically estimated by comparison to the experimental data. This weighted average energy is then perturbed by both $H_{\text{ee}}$, which gives the electron–electron interaction in terms of the Slater–Condon-Shortly parameters $F_{\text{i}}$ (ground and excited states) and $G_{\text{i}}$ (excited state only), and $H_{\text{SO}}$ which is the spin–orbit coupling that has the two components (i.e., within the p and d manifolds). The systematic effects of adding the different components $H_{\text{ee}}$ and $H_{\text{so}}$ to $H_{\text{ave}}$ are shown by the series of simulations given in Fig. 10a–f.

However, in many spectra such as the $[Fe(tacn)_2]^{3+}$ spectrum shown in Fig. 14, clear and identifiable parts of the system are inadequately described by this ligand-field model. In such cases, we can model the differential orbital covalency (DOC) contribution to spectral shape by incorporating a valence bond configuration interaction model (VBCI) which couples two or more configurations together. In the case of donor bonding, the charge transfer multiplet model adds a $d^{n+1}\underline{L}$ configuration to the single ground $d^n$ configuration of the ligand-field model. The lowest energy $d^n$ configuration is set at an energy $\Delta_{\text{GS}}$ above the lowest energy $d^n$ configuration, the two states are coupled by a configuration interaction represented by the covalent mixing term, $T$

$$T = < 3d^n|h|3d^{n+1}\underline{L} > \tag{15}$$

where $h$ is the molecular Hamiltonian and $T$ is proportion to the metal–ligand overlap. In the case of a two configuration system (eqn 16), its solutions are

$$\begin{vmatrix} 0 - E & T \\ T & \Delta_{GS} - E \end{vmatrix} = 0 \tag{16}$$

$$\Psi_{GS,B} = a_1|3d^n> +b_1|3d^{n+1}\underline{L}> \tag{17}$$

$$\Psi_{GS,AB} = b_1|3d^n> -a_1|3d^{n+1}\underline{L}> \tag{18}$$

Similarly, the bonding ($\Psi_{ES,B}$) and anti-bonding ($\Psi_{ES,AB}$) change transfer solutions for the XAS excited states may be written as:

$$\begin{vmatrix} 0 - E' & T \\ T & \Delta_{ES} - E' \end{vmatrix} = 0 \tag{19}$$

$$\Psi_{ES,B} = a_2|\underline{c}3d^{n+1}> +b_2|\underline{c}3d^{n+2}\underline{L}> \tag{20}$$

$$\Psi_{ES,AB} = b_2|\underline{c}3d^{n+1}> -a_2|\underline{c}3d^{n+2}\underline{L}> \tag{21}$$

where the charge transfer configuration in the excited state is at an energy $\Delta_{ES} = (\Delta_{GS} + U - Q)$, where U is the 3d–3d electron interaction (the "Hubbard U") and Q is the 2p–3d repulsion, above the ground configuration. The estimated energy difference between $\Delta$ and $\Delta_{ES}$ is usually ~2 eV. This has significant implications on the spectra of Cu(III) compounds which are discussed further in [38]. The interaction parameter $T$ is usually (but not always) [46] approximated as equivalent between the ground and excited states [18, 47]. These models can be expanded to include back-bonding [19, 46] with the addition of CI with an MLCT state, $3d^{n-1}(L^-)$, in a three state configuration and additional π and σ donor (i.e., LMCT) contributions in highly covalent systems [48].

The simulations within the ligand-field charge transfer multiplet model do not lend themselves to interpretation in terms of orbitals because they naturally mix many states as part of the simulation. This is highlighted in Fig. 10, where it can be clearly seen that many final states (vertical lines in Fig. 10e, f) contribute to the spectrum. To interpret the data in an MO description, the parameters used to simulate the spectra can be projected via a dummy transition onto strong-field components to determine the covalency contribution to each orbital.

$$\begin{aligned}\Psi_{GS} = {} & \alpha_1|t_{2g}^{n}e_g^0> + \alpha_2|t_{2g}^{n-1}e_g^1> + \alpha_3|t_{2g}^{n-2}e_g^2> + \alpha_4|t_{2g}^{n-3}e_g^3> \\ & + \alpha_5|t_{2g}^{n-4}e_g^4> + \alpha_6|t_{2g}^{n+1}e_g^0\underline{L}> + \alpha_7|t_{2g}^{n}e_g^1\underline{L}> + \alpha_8|t_{2g}^{n-1}e_g^2\underline{L}> \\ & + \alpha_9|t_{2g}^{n-2}e_g^3\underline{L}> + \alpha_{10}|t_{2g}^{n-3}e_g^4\text{-}\underline{L}>, \text{where} \sum_i \alpha_i^2 = 1.\end{aligned} \tag{22}$$

Within the projection methodology, the ground state ($\Psi_{GS}$) is comprised of a linear combination of contributions from the main and charge transfer configurations. The latter reflect the covalency that can be projected onto a hypothetical state in which the multiplet effects have been removed (i.e., with no $e^-$–$e^-$ repulsion) leaving only the ligand-field splittings. The covalency of the individual $t_{2g}$ (π) or $e_g$ (σ) orbital sets is found from the $\alpha_i^2$ of the $t_{2g}^{n+1}e_g^{m}\underline{L}$ and $t_{2g}^{n}e_g^{m+1}\underline{L}$ configurations corresponding to the LMCT from the $t_{2g}^{n}e_g^{m}$ ground state into the metal $t_{2g}$ and $e_g$ orbitals, respectively. The intensity of the dummy transition is distributed among the strong-field $3d^n$ and $3d^{n+1}$ configurations according to their contributions to the ground-state wave-function. For example, for a low-spin octahedral Fe(III) complex, the intensities will be non-zero for the $t_{2g}^5$, the $t_{2g}^5e_g^1$ (σ covalency), and the $t_{2g}^6$ (π covalency) configurations. The relative contributions of these configurations can then be weighted to get the differential orbital covalency. Further details of the projection method used can be found in [18].

### 3.3.2 High-Spin Fe(III)

Above, we considered low-spin Fe(III), where the ligand-field is large relative to $e^- - e^-$ repulsion and clear features of the spectrum can be identified with "orbital characteristics." In going to high-spin Fe(III) complexes, the ligand field is decreased and the multiplet effects can thus be comparable. This complicates the analysis of the L-edge as states as both the $d_\pi$ and $d_\sigma$ orbital character are mixed by the $e^- - e^-$ repulsion to produce σ and π intensity distributions (i.e., π and σ density of states over the multiplet final states). Thus, the $d_\pi$ and $d_\sigma$ states are distributed across the L-edge spectrum. As described above for the metal K-edge, $e^- - e^-$ repulsion in the weak field (left-hand) side of the Tanabe–Sugano diagram predicts that two peaks split by 10Dq with a 3:2 intensity ratio. However, as for low-spin Fe(III) complexes, inclusion of p–d electron repulsion and spin–orbit coupling leads to the L-edge spectra like those shown in Fig. 16 for a series of Fe(III) simulations all high-spin but with 10Dq systematically varied. Multiplet lines (i.e., final states) contribute to each spectral feature and each of these states contains both $d_\sigma$ and $d_\pi$ character. The relative contributions of π and σ bonding to the L-edge XAS spectrum of high-spin Fe(III) can still be sorted out. From Fig. 16a, the low-energy feature of the L-edge multiplet spectrum becomes more pronounced as 10Dq increases. In Fig. 16b, we have projected out the density of states (DOS) associated with the $d_\pi$ and $d_\sigma$ contributions to the L-edge spectral shape. While they are more mixed than for the low-spin Fe(III) case (compare to Fig. 10g), the contributions are still significantly different, with the $d_\pi$ character more strongly distributed over the lower energy region. Thus the spectral shape for high-spin Fe(III) complexes is still sensitive to DOC and can be used to probe σ and π bonding in complexes.

The spectra shown in Fig. 16 have been used to probe the bonding in high-spin Fe(III) complexes related to the siderophores. The L-edge spectra of three siderophore model compounds $[Fe(ox)_3]^{3-}$, $[Fe(cat)_3]^{3-}$, $[Fe(aha)_3]$, and the

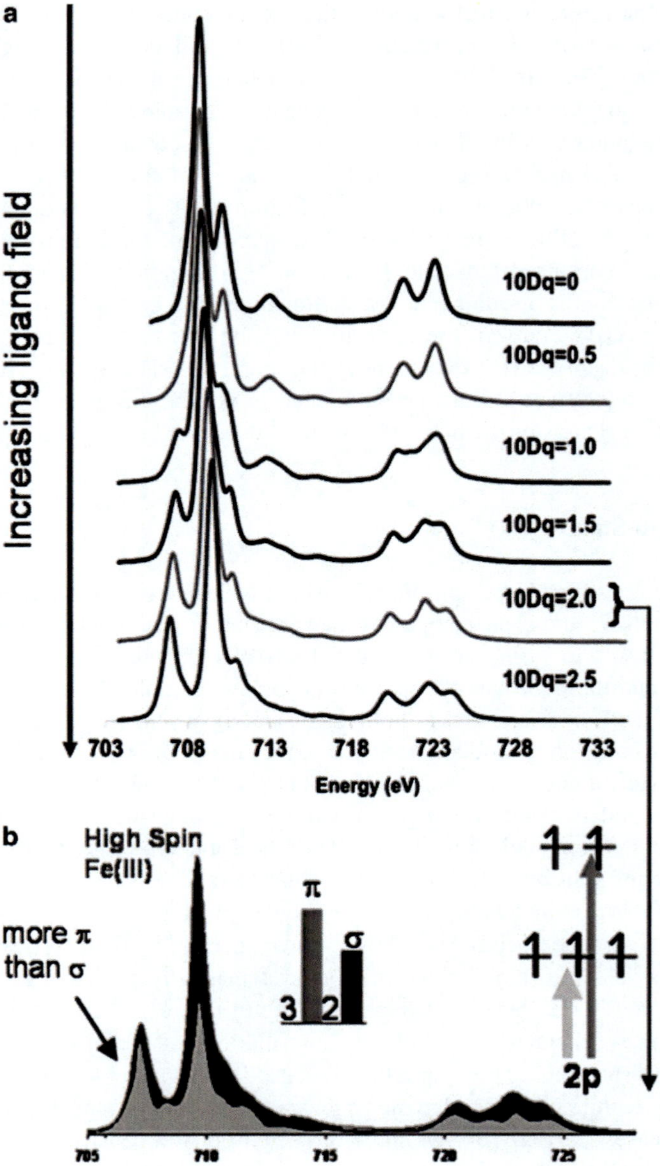


**Fig. 16** (**a**) The effect of ligand field on the L-edge spectra of high-spin Fe(III). Top spectrum has 10Dq = 0.0 eV, bottom spectrum has 10Dq = 2.5. (**b**) σ and π density of state distributions over an $O_h$ Fe(III) high-spin L-edge (10Dq = 1.8 eV). Altered from [52] with permission, Copyright American Chemical Society, 2010

reference compounds $[Fe(ida)_2]^-$ are given in Fig. 17. These data have been used to quantify the differential orbital covalencies of the catecholate and hydroxamate complexes relative to oxalate. These were important model compounds in

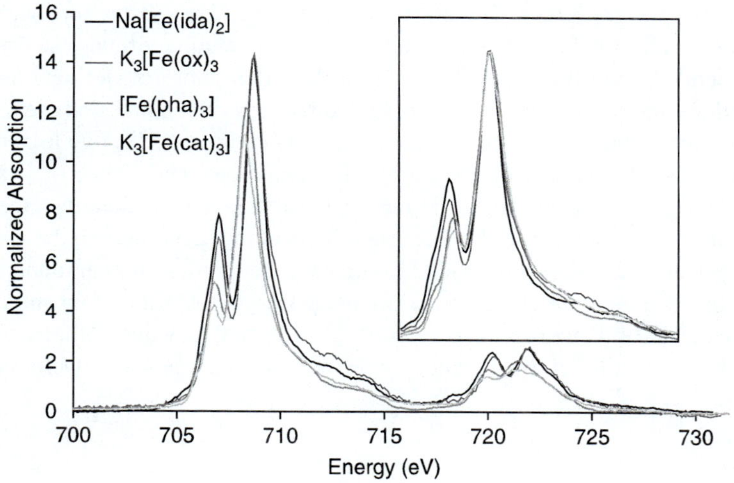


**Fig. 17** Fe L-edge absorption spectra for the compounds: $Na[Fe(ida)_2]$, $K_3[Fe(ox)_3]$, $[Fe(pha)_3]$ and $K_3[Fe(cat)_3]$. Inset shows the four spectra shifted in energy to superpose and normalized to the same intensity on the main peak. Altered from [52] with permission, Copyright American Chemical Society, 2010

determining the electronic structural origin of the increased stability constants of the siderophores, whose binding groups are most typically catecholate, hydroxamate, and carboxylate. High binding constants have both enthalpic (covalent and ionic contributions to bonding as well as solvent effects) and entropic contributions. The stability constant for $[Fe(cat)_3]^{3-}$ ($cat^{2-}$ = catecholate) is $10^{44.9}$ [49], that of the hexadentate catecholate-based enterobactin is $10^{49}$, that of tris-hydroxamate compound $[Fe(aha)_3]$ ($aha^-$ = acetohydroxamate) is $10^{28.33}$ [50], and that of the hexadentate tris-hydroxamate complex ferrioxamine B is $10^{30.6}$. Thus, both the catecholate and hydroxamate bidentate ligands make a significant contribution to the overall binding constants of the siderophores. For comparison, the stability constant of $[Fe(ox)_3]^{3-}$ is $10^{18}$ [51]. The bonding contributions to these differences in stability constants have been difficult to quantify experimentally. We see from the spectra in Fig. 17 that in going from the reference $[Fe(ox)_3]^{3-}$ complex to the $[Fe(cat)_3]^{3-}$ complex, both the $d_\pi$ and $d_\sigma$ regions of the $[Fe(cat)_3]^{3-}$ spectrum have lower intensity, therefore, increased covalency. Based on the simulations of these L-edge spectra, we were able to quantify the electronic structure contributions to the high stability constants in terms of $\sigma$ and $\pi$ covalent bonding from experiment. As the coefficients of the oxygen in the oxalate and catecholate free ligand valence orbitals are similar (see [52]), we can estimate the relative energy contributions to bonding from the experimental $\sigma$ and $\pi$ covalencies. In a molecular orbital description, bonding energy is proportional to $(H_{M\text{–}L})^2/\Delta$, where $H_{M\text{–}L}$ is the resonance integral (proportional to metal–ligand overlap) and $\Delta$ is the energy separation of the orbitals before bonding; the coefficient squared of ligand character in the metal d-orbital (i.e., covalency) is proportional to $(H_{M\text{–}L}/\Delta)^2$. Therefore, the bonding energy is proportional to the covalency scaled by the energy differences of the interacting orbitals (i.e., energy $\alpha$ (coefficient)$^2\Delta$). On the basis of the fits to the siderophore model data and DFT calculations, the coefficients of mixing for oxalate and catecholate complexes could be determined, and related to the different binding energies of the two ligands. This analysis showed that 20% of the energy difference between $[Fe(ox)_3]^{3-}$ and $[Fe(cat)_3]^{3-}$ ($\Delta G = 35$ kcal mol$^{-1}$) is due to the increase in $\pi$ bonding; the larger contribution is in fact due to the increase in $\sigma$ bonding, even though the $\sigma$- and $\pi$-orbital covalencies in the $[Fe(cat)_3]^{3-}$ complex increase by similar amounts. This is due to the $\sigma$ donor orbital having a greater energy separation from the metal d orbitals than the $\pi$ donor orbitals (see [52]). Thus, a small covalent molecular orbital coefficient can still reflect a strong bond between energetically well separated orbitals.

## 4 Concluding Comments

Carl Ballhausen taught us how to interpret ground and low-lying excited state spectral features in terms of ligand-field and covalency effects. With the availability and development of synchrotron radiation, this has been extended to the X-ray region where ligand-field and covalency effects impact metal K- and L-edges. The L-edge intensity and its distribution over the multiplets (i.e., DOC) are particularly powerful in experimentally determining the bonding in highly covalent systems where ligand spectral features can obscure the use of more traditional methods.

Since the L-edge is done in ultra-high vacuum, its application to interpret catalytic and bioinorganic systems can be limited. However, one can still accomplish this using a K-edge (an experiment that can be done on a frozen solution), to obtain an L-edge like spectrum via resonant inelastic X-ray scattering (RIXS). This involves a 1s $\rightarrow$ 3d transition followed by a 2p $\rightarrow$ 1s K$\alpha$ photon emission to fill the 1s hole and, therefore, access the same $1s^2 2p^5 3d^{n+1}$ final state as in L-edge XAS. This method is currently under development.

**Acknowledgments** E.I.S. would like to acknowledge NIH GM 040392 and NSF CHE 0948211 grants for funding this research. R.K.H. would like to thank Monash University for a beam-line research fellowship.

## References

1. Ballhausen CJ (1962) Introduction to ligand field theory. McGraw-Hill, New York
2. Ballhausen CJ (1979) Molecular electronic structures of transition metal complexes. McGraw-Hill, New York
3. Ballhausen CJ (1979) J Chem Ed 56:215–218
4. Ballhausen CJ, Gray HB (1971) ACS monograph 168 (Coord Chem v1), 3–83

5. Cotton FA, Ballhausen CJ (1956) J Chem Phys 25:617–619
6. Liehr AD, Ballhausen CJ (1957) Phys Rev 106:1161–1163
7. Liehr AD, Ballhausen CJ (1959) Ann Phys 2:134–155
8. Gray HB, Malstroem BG (1983) Comments Inorg Chem 2:2024–2028
9. Holm RH, Kennepohl P, Solomon EI (1996) Chem Rev 96:2239–2314
10. Solomon EI, Hare JW, Gray HB (1976) Proc Natl Acad Sci 73:1389–1393
11. Solomon EI, Hanson MA (1999) In: Solomon EI, Lever ABP (eds) Inorganic electronic structure and spectroscopy, vol 2. Wiley, New York
12. Cramer SP, Hodgson KO (1979) Prog Inorg Chem 25:1–39
13. Brown GS, Doniach S (1980) Synch Radiat Res 353–385
14. Glaser T, Hedman B, Hodgson KO, Solomon EI (2000) Acc Chem Res 33:859–868
15. Westre TE, Kennepohl P, DeWitt JG, Hedman B, Hodgson KO, Solomon EI (1997) J Am Chem Soc 119:6297–6314
16. DeBeer George S, Metz M, Szilagyi RK, Wang H, Cramer SP, Lu Y, Tolman WB, Hedman B, Hodgson KO, Solomon EI (2001) J Am Chem Soc 123:5757–5767
17. George SJ, Lowery MD, Solomon EI, Cramer SP (1993) J Am Chem Soc 115:2968–2969
18. Wasinger EC, deGroot FMF, Hedman B, Hodgson KO, Solomon EI (2003) J Am Chem Soc 125:12894–12906
19. Hocking RK, Wasinger EC, deGroot FMF, Hodgson KO, Hedman B, Solomon EI (2006) J Am Chem Soc 128:10442–10451
20. Koningsberger DC, Prins R (1988) X-ray absorption: principles, applications, techniques of EXAFS, SEXAFS and XANES. Wiley, New York
21. Zhang HH, Hedman B, Hodgson KO (1999). In: Solomon EI, Lever B (eds). Inorganic electron structure and spectroscopy, John Wiley and Sons, USA 1:315–554
22. Shadle SE, Penner-Hahn J, Schuger HJ, Hedman B, Hodgson KO, Solomon EI (1993) J Am Chem Soc 115:767–776
23. Kau L-S, Spira-Solomon DJ, Penner-Hahn J, Hodgson KO, Solomon EI (1987) J Am Chem Soc 109:6433–6442
24. Hahn JE, Scott RA, Hodgson KO, Doniach S, Desjardins SR, Solomon EI (1982) Chem Phys Lett 88:585–588
25. Gerwith AA, Cohen SL, Schugar HJ, Solomon EI (1987) Inorg Chem 26:7
26. DeBeer George S, Brant P, Solomon EI (2005) J Am Chem Soc 127:667–674
27. Smith TA, Penner-Hahn JE, Berding MA, Doniach S, Hodgson KO (1985) J Am Chem Soc 107:5945–5955
28. Manne R, Aberg T (1970) Chem Phys Lett 7:282–284
29. Larsson S (1977) Phys Scr 16:378–380
30. Haas C, Sawatzky GA (1981) Phys Rev B 23:4369–4380
31. Shadle SE, Hedman B, Hodgson KO, Solomon EI (1994) Inorg Chem 33:4235–4255
32. Sugano S, Tanabe Y (1970) Multiplets of transition-metal ions in crystals. Academic, New York
33. deGroot FMF, Hu ZW, Lopez MF, Kaindi G, Guillot F, Tronc M (1994) J Chem Phys 101:6570–6576
34. Thole BT, van der Laan G (1988) Phys Rev 38B:3158–3170
35. deGroot FMF (2005) Coord Chem Rev 249:31–63
36. Shadle SE, Hedman B, Hodgson KO, Solomon EI (1995) J Am Chem Soc 117:2259–2272
37. Kennepohl P, Solomon EI (2003) Inorg Chem 42:689–695
38. Sarangi R, Nermeen A, Fujisawa K, Tolman WB, Hedmand B, Hodgson KO, Solomon EI (2006) J Am Chem Soc 128:8286–8296
39. Hocking RK, Debeer George S, Gross Z, Walker FA, Hodgson KO, Hedman B, Solomon EI (2009) Inorg Chem 48:1678–1688
40. Figgis BN, Hitchman MA (2000) Ligand field theory and its applications. Wiley-VCH, New York
41. Jorgenson CK (1962) Prog Inorg Chem 4:73–124

42. Jorgenson CK (1971) Modern aspects of ligand field theory. North-Holland Publishing Company, Amsterdam, London
43. Cowan RD (1981) The theory of atomic structure and spectra. University of California Press, Berkeley
44. Thole BT, van der Laan G, Fuggle JC, Sawatzky GA, Karanatak RC, Esteva J-M (1985) Phys Rev B 32:5107–5118
45. van der Laan G, Thole BT, Sawatzky GA (1987) Phys Rev B 37:6587–6589
46. Hocking RK, Wasinger EC, Yan Y, DeGroot FMF, Walker FA, Hodgson KO, Hedman B, Solomon EI (2007) J Am Chem Soc 129:113–125
47. deGroot F (2001) Chem Rev 101:1779–1808
48. De Groot F, Kotani A (2008) Core level spectroscopy of solids. CRC Press, Taylor & Francis Group, 6000 Broken Sound Parkway NW
49. Dari-Abu K, Barclay SJ, Riley PE, Raymond KN (1983) Inorg Chem 22:3085–3089
50. Gerrett TM, Miller PW, Raymond KN (1989) Inorg Chem 28:128–133
51. Deneux M, Meiller R, Benoit RL (1968) Can J Chem 46:1383
52. Hocking RK, Debeer George S, Raymond KN, Hodgson KO, Hedman B, Solomon EI (2010) J Am Chem Soc 132:4006–4015

Struct Bond (2012) 142: 185–212
DOI: 10.1007/430_2011_58

Published online: 15 November 2011

# Time-Resolved X-Ray Diffraction: The Dynamics of the Chemical Bond

**Klaus B. Møller and Niels E. Henriksen**

**Abstract** We review the basic theoretical formulation for pulsed X-ray scattering on nonstationary molecular states. Relevant time scales are discussed for coherent as well as incoherent X-ray pulses. The general formalism is applied to a nonstationary diatomic molecule in order to highlight the relation between the signal and the time-dependent quantum distribution of internuclear positions. Finally, a few experimental results are briefly discussed.

**Keywords** Diatomic molecule · Inversion · Nonstationary states · Pump probe · X-ray scattering

## Contents



K.B. Møller (✉) and N.E. Henriksen (✉)
Department of Chemistry, Building 207, Technical University of Denmark, 2800 Kgs. Lyngby, Denmark
e-mail: klaus.moller@kemi.dtu.dk; neh@kemi.dtu.dk

# 1 Introduction

This review is dedicated to the memory of Carl Johan Ballhausen, our scientific grandfather so to speak. We have both received our formal training in quantum chemistry from Jens Peder Dahl, a former graduate student of Ballhausen, and one of us (NEH) was personally introduced to Carl Johan Ballhausen around 1980 – at that time already the grand old man of quantum chemistry in Denmark. Towards the end of his scientific career – after his seminal contributions to the theory of chemical bonding – Carl Ballhausen considered various nonstationary time-dependent problems in molecular quantum mechanics. That is, topics which are related to this review and to the dynamics of the chemical bond.

Bond making and bond breaking are key events in chemistry. Until recently, direct experimental observation of these events – as they unfolds – was not possible. This was mainly due to the underlying ultrafast dynamics on the femtosecond timescale; a timescale which already could be inferred from early theoretical works in molecular reaction dynamics (see, e.g., [1]).

The first successful experimental observation of the dynamics of the chemical bond was accomplished in the 1980s [2]. The clocking of such an ultrafast event requires an almost instantaneous initiation of the dynamics which can be accomplished by a femtosecond laser pulse. After this flash of electromagnetic radiation – the pump pulse – the dynamics of the induced photochemical reaction is probed via another ultrashort interaction with a probe pulse. This is repeated at various time delays between the pump and probe pulse and a series of "snapshots" are obtained which together constitute a "molecular movie" of the dynamics.

The coherent molecular motion initiated by the pump pulse can survive for a long time in (low pressure) gasses. In solution, incoherent motion will set in when the solutes start to interact with the solvent shell. Probing of molecular motion in the latter case, after the early-time coherent dynamics, will therefore represent an ensemble average which can contain a variety of different motions.

Early experiments in this new field of femtosecond chemistry took the form of time-resolved spectroscopy since the probing involved absorption or emission spectroscopy. Theoretical interpretation of the spectroscopic data is clearly required in order to obtained the desired information, i.e., snapshots of the time-dependent distribution of atomic positions. To that end, extensive quantum chemical calculations of energies of excited electronic states are needed, which even today can be cumbersome for larger molecular systems. Soon after the first successful experiments using time-resolved spectroscopy, there was, therefore, efforts to use alternative probing techniques like diffraction. The advantage is that a simpler and more direct connection between the diffraction signals and molecular structure is available.

Thus, time-resolved electron diffraction has evolved into a powerful technique [3], which so far allows for a time resolution of a few picoseconds. Such probing on the picosecond timescale via time-resolved electron diffraction has revealed the structures of many short-lived intermediates, $A_1 \rightarrow A_2 \rightarrow A_3 \rightarrow \ldots$, e.g., in reaction mechanisms of organic chemistry.

An alternative track which has been pursued recently is time-resolved X-ray diffraction. An important goal is to improve the time resolution such that direct observation of the dynamics of the chemical bond is possible, corresponding to the observation of the time-dependent distribution of atomic positions. Pulsed X-rays are, e.g., obtained from synchrotron radiation or plasma sources. The temporal duration of these pulses is currently in the range of ~100 ps–100 fs [4]. Very recently, the first free-electron laser has produced short ~100–10 fs coherent and highly intense bursts of X-rays [4–6].

Conventional X-ray diffraction is carried out on crystalline samples and sharp diffraction peaks are observed due to the underlying periodic crystal structure. With the high intensity X-rays obtained from synchrotron radiation or free-electron lasers it is now possible to extract the scattering/diffraction from solute molecules in dilute solutions. The diffraction pattern from isolated molecules is compared to crystalline samples weak and diffuse. In this review, we do not distinguish between X-ray diffraction and X-ray scattering since both phenomena have the same underlying theoretical framework.

The focus of this review is on the underlying theoretical concepts which are required in order to extract the dynamics of the chemical bond from time-resolved X-ray diffraction. In Sect. 2, a review of the theory of time-resolved X-ray diffraction is given. Section 3 presents an application of the theory to the dynamics of the diatomic molecule including numerical results for the NaI molecule. Section 4, gives a short review of some experimental results of time-resolved X-ray diffraction and Sect. 5 gives finally an outlook.

## 2 Review of the Theory of Time-Resolved X-Ray Diffraction

In order to set the stage for the discussion of time-resolved X-ray diffraction on nonstationary molecular states, we start with a short summary of some key results of conventional X-ray diffraction on static molecular structures.

An incident photon with wave vector $\mathbf{k}_0$, where $|\mathbf{k}_0| = 2\pi/\lambda$, is scattered on the molecular framework. Typically, 10 keV X-rays are used corresponding to a wave length $\lambda$ of about 1 Å. The short wave length comparable to molecular dimensions implies, in particular, that the often used electric-dipole approximation for the interaction with the electromagnetic field breaks down. The wave vector of the scattered photon is $\mathbf{k}_s$ and $\mathbf{q} = \mathbf{k}_0 - \mathbf{k}_s$ is the scattering vector where for elastic scattering $|\mathbf{k}_s| = |\mathbf{k}_0|$ and $|\mathbf{q}| = 2|\mathbf{k}_0|\sin(\theta/2)$, where $\theta$ is the angle between $\mathbf{k}_0$ and $\mathbf{k}_s$.

The differential scattering cross-section $d\sigma/d\Omega$, which is the scattering rate per solid angle $d\Omega$ per incident radiation flux in the continuous wave limit for the incident X-ray field, can for elastic scattering on a static molecular structure specified by the nuclear coordinates $\mathbf{R}$ be written as [7]

$$\frac{d\sigma}{d\Omega}(\mathbf{q}) = \left| \int \psi_e^*(\mathbf{r}_1 \ldots \mathbf{r}_n; \mathbf{R}) \hat{L} \psi_e(\mathbf{r}_1 \ldots \mathbf{r}_n; \mathbf{R})\, d\mathbf{r}_1 \ldots d\mathbf{r}_n \right|^2 \tag{1}$$

The cross-section is here (and in the following) expressed in units of the scattering of a free electron, i.e., the Thomson cross-section, $\psi_e$ is the electronic wave function for $n$ electrons, and $\hat{L}$ is the scattering operator

$$\hat{L} = \sum_{i=1}^{n} e^{i\mathbf{q}\cdot\mathbf{r}_i}. \tag{2}$$

Using that electronic wave functions only changes sign when interchanging particles, this can also be written as

$$\begin{aligned} \frac{d\sigma}{d\Omega}(\mathbf{q}) &= \left| \int \psi_e^*(\mathbf{r}_1 \ldots \mathbf{r}_n; \mathbf{R}) \sum_{i=1}^{n} e^{i\mathbf{q}\cdot\mathbf{r}_i} \psi_e(\mathbf{r}_1 \ldots \mathbf{r}_n; \mathbf{R}) d\mathbf{r}_1 \ldots d\mathbf{r}_n \right|^2 \\ &= \left| \sum_{i=1}^{n} \int \psi_e^*(\mathbf{r}_1 \ldots \mathbf{r}_n; \mathbf{R})\, e^{i\mathbf{q}\cdot\mathbf{r}_1} \psi_e(\mathbf{r}_1 \ldots \mathbf{r}_n; \mathbf{R}) d\mathbf{r}_1 \ldots d\mathbf{r}_n \right|^2 \\ &= \left| \int \rho_e(\mathbf{r}; \mathbf{R}) e^{i\mathbf{q}\cdot\mathbf{r}} d\mathbf{r} \right|^2 \\ &\equiv |F(\mathbf{R}, \mathbf{q})|^2 \end{aligned} \tag{3}$$

where in the last integral, the electron density is obtained by integrating over all electron coordinates except one, i.e.,

$$\rho_e(\mathbf{r}; \mathbf{R}) = n \int |\psi_e(\mathbf{r}, \mathbf{r}_2, \ldots, \mathbf{r}_n; \mathbf{R})|^2 d\mathbf{r}_2 \ldots d\mathbf{r}_n \tag{4}$$

Clearly, the observed signal is related to a Fourier transform of the electron density, also denoted as the molecular form factor/scattering amplitude $F(\mathbf{R}, \mathbf{q})$.

Direct access to structure determination, i.e., atomic positions is obtained via the independent atom model (IAM, also denoted as the isolated-spherical-atom model). Here, the electron density is written as a sum of free atomic densities

$$\rho_e(\mathbf{r}; \mathbf{R}) = \sum_J \rho_J(\mathbf{r} - \mathbf{R}_J) \tag{5}$$

where $\rho_J$ is the electron density for the ground state of the $J$th atom and $\mathbf{R}_J$ is the position of the nucleus in the $J$th atom. This approximation is in general considered to be quite adequate for structure determination [7] although it clearly ignores the effect of chemical bonding. When (5) is introduced into the molecular form factor of (3), we obtain after a change of variables

$$F(\mathbf{R},\mathbf{q}) = \int \rho_e(\mathbf{r};\mathbf{R})e^{i\mathbf{q}\cdot\mathbf{r}}d\mathbf{r} = \sum_J f_J(q)e^{i\mathbf{q}\cdot\mathbf{R}_J} \tag{6}$$

where

$$f_J(q) = \int \rho_J(\mathbf{r})e^{i\mathbf{q}\cdot\mathbf{r}}d\mathbf{r} \tag{7}$$

is the well-known atomic form factor also known as the scattering factor of the $J$th atom, which only depends on the magnitude $q = |\mathbf{q}|$ of the scattering vector. From a fit of the experimentally determined scattering cross-section at different scattering angles, it is then possible to determine the unknown atomic positions $\mathbf{R}_J$.

The theory of time-resolved X-ray diffraction has been discussed in various papers [8–21]. In seminal work by Wilson and coworkers [8], it was suggested to use transient X-ray diffraction to follow the dynamics of chemical reactions in the liquid phase. This idea was explored by calculating transient X-ray diffraction signals based on a molecular dynamics simulation of the photodissociation of $I_2$ in hexane. The signals were calculated using the conventional X-ray diffraction theory on static molecular structures, as outlined above. That is, the X-ray probe pulse was assumed to be ultrashort corresponding to instantaneous probing, whereas a configuration space average was included in line with a standard molecular dynamics approach. The finite duration of the pulses was in subsequent works taken into account in a heuristic way (see, e.g., [13]) simply by integrating the transient results over the time envelope of the pulses. Similarly, the distribution of atomic positions was taken into account by averaging the conventional expressions for static molecular structures over the nuclear degrees of freedom.

Furthermore, an interesting observation was made [13] the scattering intensity from a set of coherently excited electronic states, e.g., the electronic ground state and an excited electronic state, is in general not given by the incoherent sum of the scattering intensities from the individual electronic states but by a coherent sum which involves cross-terms.

Inspired by previous works and, in particular, by the work in [14], the main aim of our recent work [19–21] has been to develop the theory of time-resolved X-ray diffraction from first principles.

## 2.1 *Quantum Theory of Pulsed X-Ray Scattering*

In this section, we review the derivation of the scattering signal from a pulsed X-ray field on a time-dependent molecular system. The key quantity of interest is the differential signal strength $dS/d\Omega$, which we define as the total number of scattered photons arriving at the detector per solid angle. Various derivations of $dS/d\Omega$ exist in the literature based on perturbation or response theory [14, 15, 19]. In the perturbative treatment [14, 19], we consider the material state $|\Psi(t)\rangle$ created by the pump laser pulse. The perturbation on this state due to the interaction with the X-ray pulse is given by,

$$|\Psi^{(1)}(t)\rangle = \frac{1}{i\hbar}\int_0^t dt' e^{-i\hat{H}_M(t-t')/\hbar}\hat{H}_{int,M}(t')|\Psi(t')\rangle \tag{8}$$

where $\hat{H}_M$ is the Hamilton operator for the free molecular system and $\hat{H}_{int,M}$ is the perturbation caused by interaction with the X-ray field. The differential signal will be proportional to the magnitude of $|\Psi^{(1)}(t)\rangle$. Following the treatment of absorption/emission by Sakurai [22], $\hat{H}_{int,M}$ can be constructed as the matrix element of the full interaction Hamiltonian in the field states corresponding to the scattering event of interest.

### 2.1.1 The Quantized Electromagnetic Field

In the case of high-energy X-ray scattering, the Hamiltonian describing the interaction between the material system and the radiation field takes the form [9, 10, 23],

$$\hat{H}_{int} = \sum_j \frac{q_j^2 \hat{A}^2(\mathbf{r}_j, t)}{2m_j} \tag{9}$$

where $j$ runs over all the charged particles, of charge $q_j$ and mass $m_j$, in the material system. In this expression, $\hat{\mathbf{A}}(\mathbf{r},t) = \hat{\mathbf{A}}^{(+)}(\boldsymbol{r},t) + \hat{\mathbf{A}}^{(-)}(\boldsymbol{r},t)$ is the (Heisenberg) operator for the A field of the radiation with positive component ($\omega_{\mathbf{k}} = ck$),

$$\hat{\mathbf{A}}^{(+)}(\mathbf{r},t) = \sum_u \sum_{\mathbf{k}} \boldsymbol{\varepsilon}_u \sqrt{\frac{\hbar}{2\varepsilon_0 V \omega_{\mathbf{k}}}} \hat{a}_{u\mathbf{k}} e^{i(\mathbf{k}\cdot\mathbf{r}-\omega_{\mathbf{k}}t)} \tag{10}$$

and $\hat{A}^{(-)} = [\hat{A}^{(+)}]^{\dagger}$. Here, $\hat{a}_{u\mathbf{k}}$ is the photon annihilation operator in the second quantization description of the radiation field and $\boldsymbol{\varepsilon}_u$ is the polarization vector. Due to the high charge–mass-ratio of electrons compared to nuclei, it is customary to limit the sum in (9) to electrons, i.e.,

$$\hat{H}_{int} = \frac{e^2}{2m_e}\sum_{i=1}^{n} \hat{A}^2(\mathbf{r}_i, t). \tag{11}$$

Now, a single scattering event involves, in a manner of speaking, the scattering of an incident photon with wave vector $\mathbf{k}_0$ and polarization $u$ into a photon with wave vector $\mathbf{k}_s$ with polarization $v$. Quantum mechanically, single photons can be described by the so-called number state given by $|u\mathbf{k}\rangle = \hat{a}^{\dagger}_{u\mathbf{k}}|vac\rangle$ and $|v\mathbf{k}\rangle = \hat{a}^{\dagger}_{v\mathbf{k}}|vac\rangle$, respectively. Using these states, only the two cross-terms of $\hat{A}^2(\mathbf{r},t)$ give nonzero contributions:

$$\langle v\mathbf{k}_s|\hat{A}^2(\mathbf{r},t)|u\mathbf{k}_0\rangle = 2\mathbf{A}^*_{v\mathbf{k}_s}(\mathbf{r},t)\cdot\mathbf{A}_{u\mathbf{k}_0}(\mathbf{r},t) \tag{12}$$

where

$$\mathbf{A}_{uk}(\mathbf{r},t) = \boldsymbol{\varepsilon}_u A_{\mathbf{k}} e^{i(\mathbf{k}\cdot\mathbf{r}-\omega_{\mathbf{k}}t)} \tag{13}$$

where $A_{\mathbf{k}} = \sqrt{\hbar/2\varepsilon_0 V\omega_{\mathbf{k}}}$ This expression resembles a (complex) classical monochromatic field. Equation (12) can, therefore, be viewed as a semiclassical expressed for the perturbation on the material system. In the time-independent first-order perturbation theory treatment of X-ray scattering [9, 10, 23], the above-mentioned differential cross-section for elastic scattering is obtained from the matrix element $\langle v\mathbf{k}_s, \psi_e|\hat{H}_{int}|u\mathbf{k}_0, \psi_e\rangle$.

The field states can be characterized by their photon number intensity, which is defined by

$$I(\mathbf{r},t) = \frac{2\varepsilon_0 c}{\hbar\omega_{\mathbf{k}}}\left\langle \hat{\mathbf{E}}^{(-)}(\mathbf{r},t)\cdot\hat{\mathbf{E}}^{(+)}(\mathbf{r},t)\right\rangle \tag{14}$$

where $\hat{\mathbf{E}}^{(\pm)} = -\partial\hat{\mathbf{A}}^{(\pm)}/\partial t$ is the positive (negative) component of the electric field operator [24–26]. For the number states, the photon number intensity equals $c/V$, that is, the number states represent in this sense a continuous wave, despite the fact that the expectation value of the electric field is zero.

To describe an X-ray pulse, we need a wave-packet description of the incident X-ray field. The simplest wave-packet is the coherent polarized one-photon, multimode wave-packet [24, 25]

$$\left|\psi_{u\mathbf{k}_0}\right\rangle = \sum_k c_{k-k_0}|u\mathbf{k}\rangle \tag{15}$$

where $c_k$ defines a distribution which is peaked around zero with a width $\Delta k$. This wave-packet consists of field eigenstates, each containing one photon, with wave vectors all pointing in the same direction (different from the direction of $\mathbf{k}_s$) with varying length around the average length $k_0$. The wave-packet $\left|\psi_{u\mathbf{k}_0}\right\rangle$ is a one-photon state because it is an eigenstate of the photon number operator, $\hat{N} = \sum_u \sum_{\mathbf{k}} \hat{a}^{\dagger}_{u\mathbf{k}}\hat{a}_{u\mathbf{k}}$ with the eigenvalue 1, while the average number of photons in the $k$th mode equals $\left|c_{k-k_0}\right|^2 / \sum_k |c_k|^2$. Normalizing by the *average* photon energy, $\hbar\omega_{\mathbf{k}_0}$, the photon number intensity for this state becomes

$$I(\mathbf{r},t) = \frac{2\varepsilon_0 c}{\hbar\omega_{\mathbf{k}_0}}\langle\hat{\mathbf{E}}^{(-)}(\mathbf{r},t)\hat{\mathbf{E}}^{(+)}(\mathbf{r},t)\rangle = \frac{c}{V}|h(\mathbf{r},t)|^2. \tag{16}$$

where $h(\mathbf{r},t) = \sum_k \sqrt{(k+k_0)/k_0} c_k e^{i(\mathbf{k}\cdot\mathbf{r}-\omega_\mathbf{k}t)} \sim \sum_k c_k e^{i(\mathbf{k}\cdot\mathbf{r}-\omega_\mathbf{k}t)}$, if we assume that the width $\Delta k$ is small compared with the average value $k_0$. Hence, the photon number intensity is proportional to the square magnitude of the Fourier transform of $c_k$, which means that the intensity is a pulse with duration $\Delta t \sim 1/c\Delta k$ at a given $\mathbf{r}$ or, equally, a pulse with a spatial extension in the range $\Delta r \sim 1/\Delta k$ in $r$ at a given time. Thus, coherent X-ray pulse durations ranging from 0.02 fs [4], which results in $\Delta r$ ranging from 60 Å, originate from a $\Delta k$ smaller than 0.02 Å$^{-1}$. This $\Delta k$ is (reasonably) small compared to a $k_0$ on the order of 1 Å$^{-1}$ for X-rays with energies in the tens of keV regime, which justifies the above-mentioned assumption.

Since our goal is to calculate the differential scattering signal, we can use any representation of the scattered field as long as we in the calculation of the differential scattering signal add all contributions to the signal in the solid angle $d\Omega$. We, therefore, take the scattered field state to be $|v\mathbf{k}_s\rangle$ as in the usual treatment.

Using these states, again only the two cross-terms of $\hat{A}^2(\mathbf{r}, t)$ give nonzero contributions:

$$\left\langle v\mathbf{k}_s \middle| \hat{A}^2(\mathbf{r},t) \middle| \psi_{u\mathbf{k}_0} \right\rangle = 2\mathbf{A}^*_{v\mathbf{k}_s}(\mathbf{r},t) \cdot \mathbf{A}_{\psi;u\mathbf{k}_0}(\mathbf{r},t) \tag{17}$$

where

$$\mathbf{A}_{\psi;u\mathbf{k}}(\mathbf{r},t) = h(\mathbf{r},t)\mathbf{A}_{u\mathbf{k}}(\mathbf{r},t) = \boldsymbol{\varepsilon}_u h(\mathbf{r},t) A_\mathbf{k} e^{i(\mathbf{k}\cdot\mathbf{r}-\omega_\mathbf{k}t)} \tag{18}$$

This expression resembles a (complex) classical pulsed field with a carrier frequency $\omega_\mathbf{k}$.

Despite the fact that the matrix element of $\hat{A}^2(\mathbf{r}, t)$ resembles the product of two classical fields, neither the number state nor the wave-packet themselves have classical analogs in the sense that these states are eigenstates of the $\hat{A}$ operator. A state that fulfills this criteria is a highly excited multimode coherent state [27],

$$\left|\psi_{u\mathbf{k}_0}\right\rangle = \prod_k \left|\alpha_{k-k_0} u\mathbf{k}\right\rangle \tag{19}$$

where a single-mode coherent states is a wave-packet given by [26–28]

$$\left|\alpha_\mathbf{k} u\mathbf{k}\right\rangle = e^{-|\alpha_\mathbf{k}|^2/2} \sum_n \frac{\left(\alpha_\mathbf{k} \hat{a}^\dagger{}_{u\mathbf{k}}\right)^n}{n!} |vac\rangle. \tag{20}$$

Hence, $\left|\psi_{u\mathbf{k}_0}\right\rangle$ consists of single-mode coherent states with wave vectors all pointing in the same direction (different from the direction of $\mathbf{k}_s$) but of varying length around the average length $k_0$. This state is not eigenstates of the photon number operator $\hat{N} = \sum_u \sum_\mathbf{k} \hat{a}^\dagger_{u\mathbf{k}} \hat{a}_{u\mathbf{k}}$ but rather an eigenstate of the annihilation operator, $\hat{a}_{u\mathbf{k}} \left|\psi_{u\mathbf{k}_0}\right\rangle = \alpha_{k-k_0} \left|\psi_{u\mathbf{k}_0}\right\rangle$. Hence, the $k$th mode in (19) contains an *average* of $|\alpha_{k-k_0}|^2$ photons, and by analogy with the treatment in Sect. 2.1, $\alpha_k$ in (19) defines a distribution which is peaked around zero with the width $\Delta k$, where $\Delta k \ll \mathbf{k}_0$. The state $|\psi_{u\mathbf{k}_0}\rangle$ is an eigenstate of the operator $\hat{\mathrm{A}}^{(+)}(\mathbf{r}, t)$ with the eigenvalue $A_{\mathbf{k}_0,u}(\mathbf{r}, t)$ [see (18)], where $h(\mathbf{r}, t) \sim \sum_k \alpha_k e^{i(\mathbf{k}\cdot\mathbf{r}-\omega_\mathbf{k} t)}$. By a similar analysis, $I(\mathbf{r}, t) = (c/V)|h(\mathbf{r}, t)|^2$. The relative uncertainty in the electric field amplitude diminishes with increasing mean photon number, and the average field variation resembles that of a pulsed classical field with the carrier wave defined by $\mathbf{k}_0$. For an, in this sense "classical-like," incident X-ray wave-packet of highly excited coherent states, the field undergoes no noticeable change in the average photon number in the direction of $\mathbf{k}_0$ by scattering one photon into the direction $\mathbf{k}_s$ (different from the direction of $\mathbf{k}_0$). Hence, if we use the final state $|\psi_{u\mathbf{k}_0}\rangle|v\mathbf{k}_s\rangle$, the matrix element of $\hat{A}^2(\mathbf{r}, t)$ takes the exact same form as in (17) and (18).

To summarize, using either the one-photon or the coherent-state wave-packet, the matrix element of $\hat{A}^2(\mathbf{r}, t)$ resembles the product of two classical fields representing the incoming pulse and the outgoing plane wave. Assuming that the $\mathbf{r}$-dependence of the pulse envelope can be neglected, which is a very reasonable assumption for a nanometer sized (or smaller) system considering the extent of an ultrashort pulse, as discussed above, the perturbation can be written as

$$\hat{H}_{int,M} = \frac{e^2}{m_\mathrm{e}} \boldsymbol{\varepsilon}_u \cdot \boldsymbol{\varepsilon}_v A_{\mathbf{k}_0} A_{\mathbf{k}_s} h(t) \hat{L} e^{i(\omega_{\mathbf{k}_s} \; \omega_{\mathbf{k}_0})}. \tag{21}$$

### 2.1.2 The Scattering Signal

The differential scattering signal, i.e., the number of scattered photons per solid angle can be written as,

$$\frac{dS}{d\Omega} = \int d\omega_{\mathbf{k}_s} \frac{d^2S}{\mathrm{d}\Omega d\omega_{\mathbf{k}_s}}. \tag{22}$$

Here

$$\frac{d^2S}{d\Omega d\omega_{\mathbf{k}_s}} = \rho(\omega_{\mathbf{k}_s}) \lim_{t\to\infty} \langle \Psi^{(1)}(t) | \Psi^{(1)}(t) \rangle \tag{23}$$

where

$$\rho(\omega_{\mathbf{k}_s}) = \frac{\omega_{\mathbf{k}_s}^2 V}{(2\pi)^3 c^3} \tag{24}$$

is the frequency density of scattered states $|v\mathbf{k}_s\rangle$ that contribute to the signal per solid angle [23]. Using the results of the previous section, we obtain (in units of the classical Thomson cross-section)

$$\frac{d^2S}{d\Omega d\omega_{\mathbf{k}_s}} = \frac{1}{2\pi}\frac{\omega_{\mathbf{k}_s}}{\omega_{\mathbf{k}_0}}\int_0^{\infty} d\tau I(\tau)\int_{-\infty}^{\infty} d\delta C_p(\delta) e^{-i\omega_{\mathbf{k}_s}\delta}$$
$$\times \left\langle \Psi\left(\tau+\frac{\delta}{2}\right)\left|\hat{L}^{\dagger}\hat{U}\left(\tau+\frac{\delta}{2},\tau-\frac{\delta}{2}\right)\hat{L}\right|\Psi\left(\tau-\frac{\delta}{2}\right)\right\rangle \tag{25}$$

where again we assume that all particles in the material system experience the same pulse envelope. $I(t)$ is the X-ray pulse photon number intensity and $C_p(\delta)$ is the (normalized) X-ray pulse time-correlation or coherence function, which is related to the (normalized) frequency density-of-states spectrum of the incident X-ray pulse through a Fourier transform [26], $F_p(\omega) = (2\pi)^{-1}\int_{-\infty}^{\infty} dt C_p(t)e^{-i\omega t}$. This is a central result of our analysis. Obviously, for the fully coherent X-ray pulse that led to this expression, the coherence time is essentially the same as the pulse duration. However, for an incoherent X-ray pulse, consisting of many coherent subpulses, the coherence time is independent of (i.e., much shorter than) the pulse duration. The expression for the signal arising from an incident incoherent X-ray pulse is the same as (25).

Expressions similar to (25) have previously been reported in the literature [14, 15], where a perturbation on the form given in (17) was either simply postulated [14] or derived from a classical description of the incident X-ray pulse [15]. Here, (25) has been derived from first principles, i.e., the quantum theory of light–matter interactions, with a quantum mechanical description of the incident X-ray pulse.

Further analysis of the general expression can be performed either in the time or the energy domain [14, 19, 20] and various simplifications can be employed depending on the specifics of the incoming X-ray field and the energy/time scales of the molecular system.

First, we consider the situation where the coherence time, i.e., the time during which $C_p(t)$ is substantially different from zero, is so short (sub-fs) that we can ignore all of the dynamics in the material system during this time. This is a relevant limit for X-rays from *incoherent* synchrotron radiation and laser plasma sources [4]. The time-dependent material system is created by the pump UV–laser pulse and it, therefore, contains amplitude on the electronic ground state as well as on one (or several) excited electronic state(s). Adopting a density-operator description, we expand the state of the material system in the adiabatic electronic basis,

$$\hat{\rho} = \sum_{mn} \hat{\rho}_{mn}|n\rangle\langle m| \tag{26}$$

where the operators $\hat{\rho}_{mn}$ act only on the nuclear degrees of freedom, $\mathbf{R}$. Then,

$$\frac{dS}{d\Omega} = \int_0^{\infty} d\omega_{\mathbf{k}_s}\frac{\omega_{\mathbf{k}_s}}{\omega_{\mathbf{k}_0}} F_p(\omega_{\mathbf{k}_s})\int_0^{\infty} d\tau I(\tau)\sum_{mn}\int d\mathbf{R}\rho_{mn}(\mathbf{R},\tau)s_{mn}(\mathbf{R}) \tag{27}$$

where $s_{mn}(\mathbf{R}) = \langle m|\hat{L}^{\dagger}\hat{L}|n\rangle$ and $\rho_{mn}(\mathbf{R},\tau) = \langle \mathbf{R}|\hat{\rho}_{mn}|\mathbf{R}\rangle$. The terms $\rho_{nn}(\mathbf{R},\tau) = \rho_n(\mathbf{R},\tau)$ are nuclear densities and define the probability density of finding a nuclear configuration **R** in the electronic state $|n\rangle$. The nondiagonal terms $\rho_{mn}(\mathbf{R}, \tau)$ are the nuclear coherences describing the inferences between the different electronic states which will oscillate with a frequency of roughly the inverse electronic energy spacing. If the time-resolution of the experiment, given by X-ray pulse length defined by $I$ and the jitter between pump and probe pulses, permits it these may show up as "beating" patters in the diffraction pattern. However, this is most often not the case for *incoherent* pulses from synchrotron and laser plasma sources and, in addition, arguments can be given that $s_{mn}(\mathbf{R})$ are either zero due to symmetry or, in any case, much smaller than $s_{nn}(\mathbf{R})$ [20]. Hence, the off-diagonal terms in (27) will be dropped in the following. We note that for an incoherent sum of coherent sub-pulses, both the spectral widths of each of these subpulse as well as the possible variation in their carrier frequency contribute to the width of $F_p$.

Obviously, the details in the time-profile, $I$, and the frequency spectrum, $F_p$, of the incident X-pulse, depend on the experimental setup. However, if the duration of the pulse is either sufficiently short or sufficiently long compared to the time scale of the nuclear dynamics, $I$ may be replaced by either a delta function or a constant on the nuclear time scale. Likewise, if the width of $F_p$ can be neglected (known as the "static approximation"), we can obtain simplified expressions for the differential scattering signal. However, as pointed out earlier, the frequency widths of X-ray pulses obtained from, e.g., synchrotron radiation are typically on the order of percent of the carrier frequency. Hence, in order to simulate the finer details of the experimental signal, the actual frequency distribution of the incident X-ray pulse must be taken into account [29].

Having dropped the off-diagonal terms in (27), only the terms $s_{nn}$ remain. These terms can be separated into an elastic and a purely inelastic terms,

$$s_{nn} = \left|\langle n|\hat{L}|n\rangle\right|^2 + \sum_{n\neq m} \left|\langle m|\hat{L}|n\rangle\right|^2. \tag{28}$$

The inelastic contributions to the signal are only important for light atoms [30] and, in addition, relatively insensitive to the atomic arrangement [20]. Hence, in a practical experiment where the nuclear dynamics is elucidated by making difference diffraction images; one with the pump laser turned on minus one with the pump laser turned off, the inelastic terms can be ignored.

Collecting all of the above, we arrive at the following expression for the difference signal obtained from either a synchrotron or plasma source,

$$\Delta\frac{dS}{d\Omega} = \int_0^\infty d\omega_{\mathbf{k}_s} \frac{\omega_{\mathbf{k}_s}}{\omega_{\mathbf{k}_0}} F_p(\omega_{\mathbf{k}_s}) \int_0^\infty d\tau I(\tau) \sum_n \int d\mathbf{R} \left|\langle n|\hat{L}|n\rangle\right|^2 \Delta\rho_n(\mathbf{R}, \tau). \tag{29}$$

This expression looks just like the continuous-wave differential cross-section for a fixed nuclear geometry in a single electronic state (1), weighted by the probability distribution of nuclear geometries in the populated electronic states, summed up over the duration of the incoming X-ray pulse and, finally, weighted by the frequencies of the X-ray pulse. Equation (29) resembles the general expression given in [16], which is derived from a purely classical treatment of the X-ray field and its interaction with the material system followed by quantization of the latter. This is consistent with the fact that the classical analysis presented in [16] discards the coherence of the X-ray field by considering each scattering event to be independent and instantaneous (see also [23]), which is reasonable with the above-mentioned X-ray sources [4].

Now, future X-ray sources (XFELs) may have coherence times comparable to or even longer than the time scale of electronic motion (but still relatively short compared to the time scale of nuclear dynamics) [4, 5]. The analysis in the time-domain must then be redone for this situation [19]. However, this situation is more conveniently analyzed in the energy domain because in this case the frequency spread of the incoming X-ray pulse will be narrower due to the longer coherence time and the static approximation can safely be applied. Doing so, we arrive at the same expression as (29), albeit without the integral over frequencies [20],

$$\Delta \frac{dS}{d\Omega} = \int_0^\infty d\tau I(\tau) \sum_n \int d\mathbf{R} \left| \langle n | \hat{L} | n \rangle \right|^2_{k_s = k_0} \Delta\rho_n(\mathbf{R}, \tau). \tag{30}$$

Finally, in order to simplify the relation between the differential scattering signal and the nuclear motion, we adopt as in conventional X-ray diffraction theory, the independent atom model (IAM) for the electrons [12–14, 17]. In the IAM, according to (6)

$$\langle \hat{L} \rangle = F(\mathbf{R}, \mathbf{q}) \approx \sum_J f_J e^{i\mathbf{q}\cdot\mathbf{R}_J} \tag{31}$$

where the sum runs over all the atoms (nuclei), $f_J$ is the atomic scattering factor, $\boldsymbol{R}_J$ is the position of the atom, and $\mathbf{q} = \mathbf{k}_0 - \mathbf{k}_s$ is the scattering vector.

The signal then becomes,

$$\Delta \frac{dS}{d\Omega} = \int_0^\infty d\omega_{\mathbf{k}_s} \frac{\omega_{\mathbf{k}_s}}{\omega_{\mathbf{k}_0}} F_p(\omega_{\mathbf{k}_s}) \int_0^\infty d\tau I(\tau) \sum_n \int d\mathbf{R} |F(\mathbf{R}, \mathbf{q})|^2 \, \Delta\rho_n(\mathbf{R}, \tau) \tag{32}$$

where $F_p(\omega_{\mathbf{k}_s}) \approx \delta(\omega_{\mathbf{k}_s} - \omega_{\mathbf{k}_0})$ in the static approximation, i.e., if the coherence time of the incoming X-ray pulse is on the order of femtoseconds.

### 2.1.3 Discussion of Final Formulas

Equation (32) is our final result for the difference-diffraction signal. A convenient decomposition and interpretation of this result is given in the following. Within the static approximation, we write the diffraction signal as follows

$$\frac{dS}{d\Omega} = \int_0^\infty d\tau I(\tau) \frac{dS_{\text{inst}}}{d\Omega}(\tau) = \int \tilde{\rho}(\mathbf{R}; t_p) |F(\mathbf{R}, \mathbf{q})|^2 d\mathbf{R}. \tag{33}$$

Thus, the signal can be interpreted as an instantaneous scattering signal integrated over the X-ray probe pulse or, alternatively, as a conventional X-ray diffraction signal weighted by $\tilde{\rho}(\mathbf{R}; t_p)$, which is the time-averaged density of nuclear geometries given as a convolution of the instantaneous distribution of atomic positions $\rho(\mathbf{R}, \tau)$ with the intensity profile $I$ of the X-ray pulse

$$\tilde{\rho}(\mathbf{R}; t_p) = \int_0^\infty d\tau I(\tau) \rho(\mathbf{R}, \tau) \tag{34}$$

where the parameter $t_p$ indicates that the probe pulse is centered at $\tau = t_p$ corresponding to the pump–probe delay time and $\rho(\mathbf{R}, \tau) = \sum_n \rho_n(\mathbf{R}, \tau)$ is the nuclear density created by the pump pulse. This density is distributed over two or more electronic states.

Thus, (33) has a close resemblance to the conventional expression in (3). Equation (33) takes the distribution of atomic positions into account as well as the finite duration of the X-ray probe pulse. Clearly, if the temporal duration of the X-ray pulse is short compared to the molecular dynamics, the diffraction signal corresponds to diffraction from the instantaneous distribution of atomic positions, $\rho(\mathbf{R}, t_p)$, at time $t_p$. Structure determination of short-lived intermediates can be accomplished with "long" probe pulses, i.e., long on the timescale of atomic motion but short compared to the lifetime of the population. This limit corresponds to current experiments where fast kinetics is studied by X-ray pulses from synchrotrons.

The diffraction signal, after the pump pulse is fired, contains contributions from molecules in the electronic ground state as well as contributions from excited electronic states. Typically, the vast majority of the molecules remains in the electronic ground state, and in order to amplify the interesting structural dynamics in the excited states, a diffraction-difference method is employed. Thus, one records a difference signal between, respectively, pump pulse "on" and pump pulse "off," i.e.,

$$\Delta \frac{dS}{d\Omega} = \left(\frac{dS}{d\Omega}\right)_{\text{on}} - \left(\frac{dS}{d\Omega}\right)_{\text{off}} = \int [\tilde{\rho}_{\text{on}}(\mathbf{R}; t_p) - \tilde{\rho}_{\text{off}}(\mathbf{R}; t_p)] |F(\mathbf{R}, \mathbf{q})|^2 d\mathbf{R} \tag{35}$$

where we again used that the electronic density and hence $|F(\mathbf{R}, \mathbf{q})|^2$ is approximately unchanged after laser excitation. This equation is in agreement with (32) within the static approximation.

## 3 Application to Dynamics of a Diatomic Molecule

It is instructive to consider the time-resolved X-ray scattering for a nonstationary diatomic molecule. Our aim here is to present a simplified analysis which highlights the relation between the key features in the dynamics and the time-dependent diffraction signals. We assume in the following that the nuclear density created by the pump pulse, $\rho_{\mathrm{on}}(\mathbf{R}, t)$, is distributed over the electronic ground state and an excited electronic state, that is, $\rho_{\mathrm{on}}(\mathbf{R}, t) = \rho_{\mathrm{gr}}(\mathbf{R}, t) + \rho_{\mathrm{ex}}(\mathbf{R}, t)$.

Since the interaction potential of the diatomic molecule is spherical symmetric, it is natural to consider the Hamiltonian in polar coordinates. Thus, the nuclear Hamiltonian for the relative motion takes the well-known form,

$$\hat{H} = -\frac{\hbar^2}{2\mu}\left(\frac{\partial^2}{\partial R^2} + \frac{2}{R}\frac{\partial}{\partial R} - \frac{\hat{L}^2}{\hbar^2 R^2}\right) + V(R) \tag{36}$$

where $\mu$ is the reduced mass, $R$ is the internuclear distance, $\hat{L}$ is the angular momentum operator of the relative motion, and $V(R)$ is the electronic energy. Rotational invariance implies $[\hat{H}, \hat{L}] = 0$, and the angular momentum associated with the relative motion is a constant of motion. We consider the dynamics of a state with a definite angular momentum, and write the radial part of the wave function in the form $u_l(R, t)/R$. We note that

$$\hat{H}\left(\frac{u_l(R,t)}{R}Y_{lm}(\theta,\phi)\right) = \frac{Y_{lm}(\theta,\phi)}{R}\hat{H}(l)u_l(R,t) \tag{37}$$

where $Y_{lm}$ are the spherical harmonics eigenstates associated with the angular momentum and $\hat{H}(l)$ is equivalent to a one-dimensional Hamiltonian given by

$$\hat{H}(l) = -\frac{\hbar^2}{2\mu}\frac{\partial^2}{\partial R^2} + V_l(R) \tag{38}$$

with the effective potential

$$V_l(R) = V(R) + \frac{\hbar^2 l(l+1)}{2\mu R^2} \tag{39}$$

where the angular momentum quantum number $l = 0, 1, \cdots$. Using (37), the time evolution of a state which is in a stationary rotational state, can then be written in the form

$$\mathrm{e}^{-i\hat{H}t/\hbar}\left(\frac{u_l(R)}{R}Y_{lm}(\theta,\phi)\right) = \frac{Y_{lm}(\theta,\phi)}{R}e^{-i\hat{H}(l)t/\hbar}u_l(R) \tag{40}$$

Thus, the radial motion takes place in the effective one-dimensional potential, $V_l(R)$. For a strongly repulsive potential, the motion is to a good approximation independent of $l$.

In connection with the evaluation of the diffraction signal, we note that the scattering amplitude, (6), for a diatomic molecule with bond vector $\mathbf{R} = \mathbf{R}_1 - \mathbf{R}_2$, takes the form

$$\begin{aligned} |F(\mathbf{R},\mathbf{q})|^2 &= \sum_{\alpha,\beta}^{2} f_\alpha(q)^* f_\beta(q) e^{i\mathbf{q}\cdot(\mathbf{R}_\alpha - \mathbf{R}_\beta)} \\ &= |f_1(q)|^2 + |f_2(q)|^2 + 2 f_1(q) f_2(q) \mathrm{Re}\{e^{i\mathbf{q}\cdot\mathbf{R}}\} \end{aligned} \tag{41}$$

using that the atomic scattering/form factors are real valued. The complex exponential in (41) can be expanded in the form [31]

$$e^{i\mathbf{q}\cdot\mathbf{R}} = 4\pi \sum_{l=0}^{\infty} \sum_{m=-l}^{m=l} i^l j_l(qR) Y^*_{lm}(\alpha, \delta) Y_{lm}(\theta, \phi) \tag{42}$$

where $(q,\alpha, \delta)$ specifies the magnitude and direction of the $\mathbf{q}$ vector, $(\theta, \phi)$ gives the direction of the bond vector $\mathbf{R}$, and $j_l(qR)$ is a spherical Bessel function. The spherical Bessel function is related to the Riccati–Bessel function $\hat{j}_l(qR)$ (eigenfunction of the Hamiltonian in (38) for $V(R) = 0$), $j_l(qR) = \hat{j}_l(qR)/(qR)$.

## 3.1 Ground-State Signal

First, we consider the scattering signal before the arrival of the pump pulse when the molecule is in the stationary vibrational–rotational ground state. The stationarity implies that the time-averaged density in (34) is identical to the stationary density, $\tilde{\rho}_{\mathrm{off}}(\mathbf{R}; t_p) = \rho_{\mathrm{off}}(\mathbf{R}) = \rho_{\mathrm{gr}}(\mathbf{R})$. Within a harmonic approximation to the vibrational ground state, $u_0(R)$ can to a good approximation be represented by a Gaussian, i.e.,

$$\rho_{\mathrm{off}}(\mathbf{R}) = u_0^2(R)/4\pi R^2 \sim [2\pi(\Delta R)_0^2]^{-1/2} \exp\left[-\frac{(R-R_e)^2}{2(\Delta R)_0^2}\right]/4\pi R^2 \tag{43}$$

where $(\Delta R)_0^2 = \hbar/(2m\omega)$. The scattering signal in (33) becomes,

$$\begin{aligned} \left(\frac{dS}{d\Omega}\right)_{\mathrm{off}} &= \int \rho_{\mathrm{off}}(\mathbf{R}) |F(\mathbf{R},\mathbf{q})|^2 d\mathbf{R} \\ &= |f_1(q)|^2 + |f_2(q)|^2 + 2 f_1(q) f_2(q) \mathrm{Re}\left\{\int d\mathbf{R} e^{i\mathbf{q}\cdot\mathbf{R}} \rho_{\mathrm{off}}(\mathbf{R})\right\} \\ &= |f_1(q)|^2 + |f_2(q)|^2 + 2 f_1(q) f_2(q) \int_0^\infty dR \frac{\sin(qR)}{qR} \rho_{\mathrm{off}}(\mathbf{R}) \end{aligned} \tag{44}$$

since in this case, where the initial state is spherical symmetric, only the $l = m = 0$ term in (42) contributes to the integral due to the orthonormality of the spherical harmonics,

$$\langle Y_{l'm'}|Y_{lm}\rangle = \int_0^{\pi}\int_0^{2\pi} Y^*_{l'm'}(\theta,\phi)Y_{lm}(\theta,\phi)\sin\theta d\theta d\phi = \delta_{ll'}\delta_{mm'} \tag{45}$$

and $j_0(qR) = \sin(qR)/qR$. Typically $(\Delta R)_0/R_e$ corresponds to a few percent and $(\Delta R)_0$ is largest for molecules with small masses and small force constants. In the classical limit $(\Delta R)_0^2 \rightarrow 0$, the integrand in (44) is nonzero only for $R = R_e$, and

$$\left(\frac{dS}{d\Omega}\right)_{\text{off}} = |f_1(q)|^2 + |f_2(q)|^2 + 2f_1(q)f_2(q)\frac{\sin(qR_e)}{qR_e} \tag{46}$$

which is just the classical Debye formula for a spherical symmetric molecule.

## 3.2 Excited-State and Difference-Density Signal

Next, we consider the scattering signal when the molecule is in a nonstationary excited state obtained by excitation out of the initial stationary vibrational–rotational ground state.

The interaction with the pump pulse is described within the electric-dipole approximation,

$$\hat{H}_{\text{int}} = -\boldsymbol{\mu}(R)\cdot\mathbf{E}_L(t) \tag{47}$$

We consider an electronic transition in a molecule, from the electronic ground state "gr" to an excited state "ex." Within first-order perturbation theory for the interaction with an electromagnetic field [see (8)], the state vector associated with the nuclear motion in the excited electronic state "ex" is given by (at times, $t$, when the laser pulse has vanished) [32, 33]

$$\chi_{\text{ex}}(\mathbf{R},t) = \frac{1}{i\hbar}\int_{-\infty}^{\infty} dt' \mathrm{e}^{-i\varepsilon_0 t'/\hbar}E_L(t')\phi(\mathbf{R},t-t') \tag{48}$$

where $\phi(\mathbf{R},t-t') = \langle\mathbf{R}|\exp[-i\hat{H}_{\text{ex}}(t-t')/\hbar]|\phi\rangle$, $E_L(t)$ is the (pump) laser field, and

$$|\phi\rangle = \mu_{12}|\chi_{\text{gr}}\rangle \tag{49}$$

is the Franck–Condon wave-packet. In this expression, $|\chi_{\text{gr}}\rangle$ is the initial stationary nuclear state in the electronic ground state with energy $\varepsilon_0$, and $\mu_{12}$ is the projection of the electronic transition-dipole moment on the polarization vector of the electric field, which we define as the $z$-axis. Thus, the excited state wave function can be thought of as a coherent superposition of Franck–Condon wave packets promoted to the upper state at times $t'$ with different weighting factors (given by $E_L(t')$) and phases. At time $t$, each of these wave packets in the superposition has evolved for a time $t - t'$.

We note in passing that the angles in (42) were defined with respect to a $z$-axis and hence with respect to the polarization vector of the electric field.

We specialize now to a delta pulse $E_L(t') = E_L\delta(t')$ on the nuclear timescale and a parallel transition where the transition-dipole moment is parallel to the bond vector $\mathbf{R}$. Then,

$$\chi_{\mathrm{ex}}(R,t) = \frac{1}{i\hbar}\mathrm{e}^{-i\hat{H}_{\mathrm{ex}}t/\hbar}\left(E_L|\mu_{12}|\frac{u_l(R)}{R}\cos\theta Y_{lm}(\theta,\phi)\right) \tag{50}$$

Assume $l = 0$, then $\cos\theta Y_{00} = \sqrt{1/3}\,Y_{10}$ and according to (40)

$$\chi_{\mathrm{ex}}(\mathbf{R},t) = \frac{E_L|\mu_{12}|}{i\hbar R}\sqrt{1/3}\,Y_{10}\,e^{-i\hat{H}_{\mathrm{ex}}(l=1)t/\hbar}u_0(R) \tag{51}$$

where we have assumed that the transition-dipole moment $\mu_{21}$ is constant, i.e., independent of the internuclear distance (the Condon approximation) and where the spherical harmonic $Y_{10} = \sqrt{3/(4\pi)}\cos\theta$. The time-dependent probability density associated with the atomic positions is then

$$\rho_{\mathrm{ex}}(\mathbf{R},t) = \frac{E_L^2|\mu_{12}|^2}{3\hbar^2}Y_{10}^2\,|\hat{U}(t)u_0(R)|^2/R^2 \tag{52}$$

with the time-evolution operator $\hat{U}(t) = \exp(-\mathrm{i}\hat{H}_{\mathrm{ex}}(l=1)t/\hbar)$. Thus, the pump laser creates a time-dependent anisotropic distribution of atomic positions.

The scattering signal in (33), for an ultrashort X-ray probe pulse centered at time $t_{\mathrm{p}}$, becomes

$$\begin{aligned}\left(\frac{dS_{\mathrm{inst}}}{\mathrm{d}\Omega}\right)_{\mathrm{on}} &= \int \rho_{ex}(\mathbf{R},t_p)|F(\mathbf{R},\mathbf{q})|^2 d\mathbf{R}\\ &= A\left(|f_1(q)|^2 + |f_2(q)|^2\right) + 2f_1(q)f_2(q)\mathrm{Re}\left\{\int d\mathbf{R}e^{i\mathbf{q}\cdot\mathbf{R}}\rho_{\mathrm{ex}}(\mathbf{R},t_p)\right\}\end{aligned} \tag{53}$$

where $A = E_L^2|\mu_{12}|/3\hbar^2$. In order to integrate over the angles $\theta$ and $\phi$, we use the following relation

$$Y_{10}^2 = (2Y_{20}/\sqrt{5} + Y_{00})/2\sqrt{\pi} \tag{54}$$

and (42). Then, from the orthonormality of the spherical harmonics in (45), we obtain

$$\int_0^{\pi}\int_0^{2\pi} e^{i\mathbf{q}\cdot\mathbf{R}}Y_{10}^2(\theta,\phi)\sin\theta d\theta d\phi = j_0(qR) - j_2(qR)(3\cos^2\alpha - 1) \qquad (55)$$

and the signal can be written in the form,

$$\left(\frac{dS_{\text{inst}}}{d\Omega}\right)_{\text{on}} = A\left(|f_1(q)|^2 + |f_2(q)|^2\right) + 2Af_1(q)f_2(q)$$
$$\times\left\{\int_0^{\infty} dR\, j_0(qR)|\hat{U}(t_p)u_0(R)|^2 - (3\cos^2\alpha - 1)\right.$$
$$\left.\times\int_0^{\infty} dR\, j_2(qR)|\hat{U}(t_p)u_0(R)|^2\right\} \qquad (56)$$

with Bessel functions $j_0(x) = \sin(x)/x$ and $j_2(x) = (3/x^3 - 1/x)\sin x - (3/x^2)\cos x$. We observe that the signal for the nonstationary excited state consists of a term which is independent of the internuclear distance and a term which depends on the time-dependent distribution of internuclear positions. The latter term contains an isotropic distribution in $\mathbf{q}$ and, due to the nonisotropic distribution of the molecules prior to X-ray scattering, a term which depends on $q$ as well as on the angle $\alpha$ between the polarization vector of the laser pulse and the $\mathbf{q}$ vector.

We note that when the effect of the pump pulse is described within first-order perturbation theory, the density in the electronic ground state is unchanged and $\Delta\tilde{\rho}=\tilde{\rho}_{\text{on}}(\mathbf{R};t_p) - \tilde{\rho}_{\text{off}}(\mathbf{R};t_p) = \tilde{\rho}_{\text{ex}}(\mathbf{R};t_p)$, that is, (56) describes also the difference signal in (35) within first-order perturbation theory. To second order, $\tilde{\rho}_{\text{gr}} \neq \tilde{\rho}_{\text{off}}$ and the difference signal contains signatures of the modified ground-state distribution, i.e., the ground-state "hole" created by the pump pulse, see Sect. 3.4.

The change in the signal as a function of time $t_p$ is due to the change in the average internuclear distance as well as the spreading of the wave-packet $|\hat{U}(t_p)u_0(R)|^2$. In the limit of an ultrashort $\delta$ pump – the Franck–Condon wave-packet is localized at early times; however, the momentum uncertainty of the initial state gives rise to fast spreading of the wave-packet. Longer pulses produces broader wave-packets with slower spreading. Thus, at any given time, the signal is due to scattering from a distribution of structures. The nonnegligible width of the wave-packet implies that the classical concept of a molecular structure is ill defined. Note that the wave-packet dynamics encoded in the signal is identical for the isotropic and the anisotropic term.

Above, we have considered the time-resolved X-ray scattering signal assuming that the initial state is the vibrational–rotational ground state. At any given time, the signal is due to scattering from a distribution of structures. This distribution is temperature dependent and the signal is obtained as a thermal average over the initial states. The signal depends in particular on the initial probability density of interatomic distances. Thus, at temperature $T$, the width of the dynamical non-equilibrium structures will be broadened by a time-independent factor [19].

### 3.3 Inversion

A key point is how to extract the dynamics from the experimental diffraction signal. For a diatomic molecule, an exact inversion procedure is available [14, 21]. First, we note that (56) can be written in the form

$$\left(\frac{dS_{\text{inst}}}{d\Omega}\right)_{\text{on}} = S_0(q, t_p) - P_2(\cos\alpha)S_2(q, t_p) \tag{57}$$

where $P_2(x) = (3x^2 - 1)/2$ is a Legendre polynomial and $S_0$ and $S_2$ are identified by comparison to (56), i.e.,

$$S_2(q, t_p) = 4Af_1(q)f_2(q)\int_0^\infty dR\, j_2(qR)\left|\hat{U}(t_p)u_0(R)\right|^2 \tag{58}$$

The two functions $S_0(q, t_p)$ and $S_2(q, t_p)$ can be obtained from the experimental diffraction signal. For a fixed delay time $t_p$ and a fixed length of the scattering vector $q$, corresponding to a circle on the detector plate, each pixel on this circle can be assigned a value of the signal and a value of $P_2(\cos\alpha)$. A plot of the signal versus $P_2(\cos\alpha)$ gives according to (57) a straight line with intercept $S_0(q, t_p)$ and slope $-S_2(q, t_p)$. This procedure is then repeated for a range of $q$ values.

Alternatively, the isotropic signal $S_0(q, t_p)$ can be obtained by "magic angle detection" [34]. That is, when the angle $\alpha$ between the incoming X-ray beam and the laser polarization axis, fulfills $\cos\alpha = 1/\sqrt{3}$, the last term in (57) disappears.

$S_2(q, t_p)$ (or $S_0(q, t_p)$) is now transformed according to

$$\begin{aligned}\int_0^\infty dq\, q^2 j_2(qR)\frac{S_2(q, t_p)}{4Af_1(q)f_2(q)} &= \int_0^\infty dR' \left|\hat{U}(t_p)u_0(R')\right|^2 \int_0^\infty dq\, q^2 j_2(qR) j_2(qR') \\ &= \frac{\pi}{2}\left|\hat{U}(t_p)u_0(R)\right|^2/R^2\end{aligned} \tag{59}$$

where the following relation between spherical Bessel function was used

$$\int_0^\infty dq\, q^2 j_n(qr)\, j_n(qr') = \frac{\pi}{2r^2}\delta(r - r'). \tag{60}$$

Thus, the time-dependent radial density $\left|\hat{U}(t_p)\, u_0(R)\right|^2/R^2$ of the laser excited density, (52) is recovered from the transformation in (59).

Beyond the first-order impulsive description of the difference signal, given above, a time-averaged difference density $\Delta\tilde{\rho} = \tilde{\rho}_{on}(\mathbf{R};\, t_p) - \tilde{\rho}_{\text{off}}(\mathbf{R};\, t_p)$ is obtained from the transformation. The effect of the finite duration of the X-ray pulse is a convolution with the instantaneous signal according to (34). A deconvolution of the X-ray pulse is in principle possible, provided the intensity profile is known and that the pump–probe delay time $t_{\text{p}}$ is well defined.

For polyatomic molecules approximate inversion schemes have been suggested [3, 21]. Alternatively, theoretical modeling of experiments and comparison to experimental diffraction data, typically via an iterative loop, can provide the desired information concerning the time-dependent distribution of atomic positions.

## 3.4 Illustration: X-Ray Diffraction on Laser Excited NaI

In order to illustrate the general theory outlined in the last sections, we review the dynamics of NaI excited by a UV photon, all details can be found in [21]. NaI is an extensively studied molecule in femtochemistry.

The molecule was initially placed in the electronic, vibrational, and rotational ground state. It was subsequently excited by a laser with a wavelength of 328 nm, with a $\sin^2$-shape envelope with 10 fs FWHM, and an intensity of $10^{11}$ W/cm$^2$. The wave-packet propagation was carried out by employing a Gauss–Legendre DVR in $\theta$, a plane-wave expansion for the radial wave functions, and the split operator method for the time evolution. For simplicity, the propagation was carried out nonperturbatively; however, the laser intensity was chosen such that weak-field effects dominate.

The relevant potential energy curves and nuclear dynamics of NaI after photo-excitation are shown in Fig. 1. The calculations were performed using the program package `WavePacket` [35].

We note that the valence electron in NaI undergoes a substantial reorganization – corresponding to electron transfer between Na and I – whenever the internuclear distance passes the avoided crossing [36]. This, however, does not seriously invalidate the independent atom model used to represent the molecular electron density, since the vast majority of the electrons can be considered as core electrons associated with a specific atom.

Figure 2 shows the nuclear difference density for NaI. We observe the oscillating excited state wave-packet and the ground state "hole," which shows minor vibrations. The first-order description given in Sect. 3.2 cannot describe the formation of the ground state "hole," but a detailed description to second-order in the ground state can be found in [21].

For all X-ray diffraction results, we assumed an X-ray beam of 8 keV photon energy with a $\sin^2$-pulse shape, and a half-width of 100 fs, which should roughly correspond to the situation at the Linear Coherent Light Source (LCLS) including timing jitter [6].

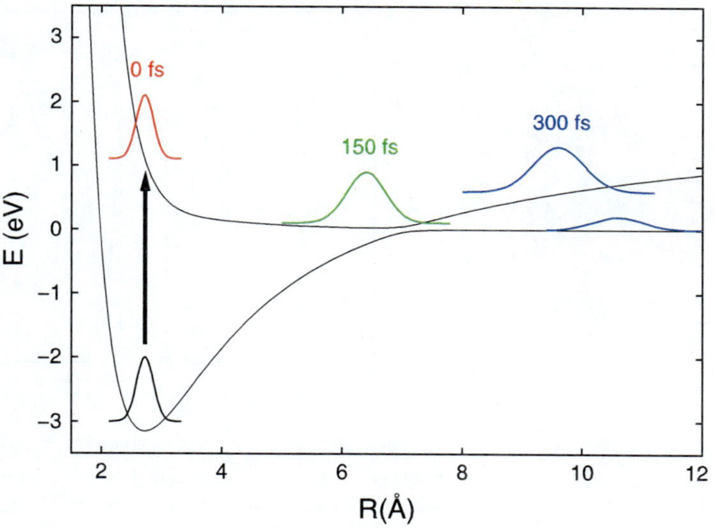


**Fig. 1** Potential energy curves and sketch of the relevant nuclear wave-packet motion of NaI. After an ultrashort excitation, NaI is vertically excited from the $X^1\Sigma^+$ ground state to the $A^1\Sigma^+$ first excited state, which exhibits a shallow minimum. The nuclei then vibrate on the excited state surface with bond distances ranging from approximately 3–12 Å. Through the avoided crossing around 7 Å, parts of the wave-packet can make a transition to the electronic ground state, and dissociate. Adapted from [21]

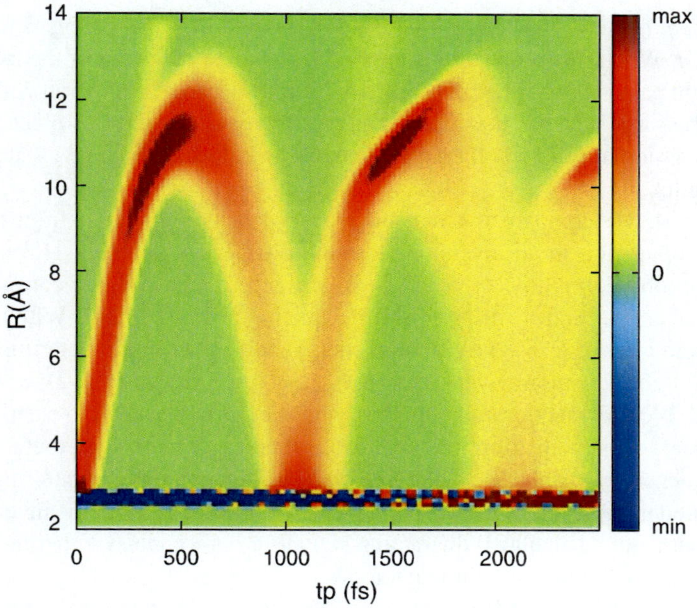


**Fig. 2** Nuclear vibrational difference density (in arbitrary units) after laser excitation weighted by $R^2$, as a function of the pump–probe delay time $t_p$ and internuclear distance $R$. Adapted from [21]

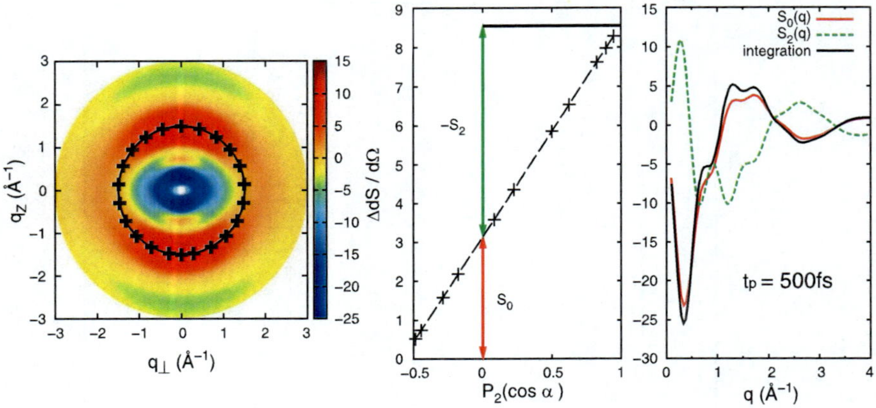


**Fig. 3** Left: calculated two-dimensional difference diffraction pattern $\Delta dS/d\Omega$ in arbitrary units for NaI at a pump-probe delay time of 500 fs. $q_z$ and $q_\perp$ give the length of the component of **q** parallel and perpendicular to the laser polarization axis, which was set perpendicular to the wave vector of the X-ray beam. *Middle*: plot of $\Delta dS/d\Omega$ as a function of $P_2(\cos\alpha)$ ($q = 1.5$ Å$^{-1}$, points correspond to those of the left image). Right: isotropic and anisotropic curves $S_n(q, t_p)$ encoded in the two-dimensional pattern. Also shown is the result from an azimuthal integration over the detector surface. Adapted from [21]

A typical diffraction pattern is shown in Fig. 3. As detailed in Sect. 3.2, this rather complex anisotropic pattern is built up from two diffraction curves $S_0(q, t_p)$ and $S_2(q, t_p)$ which encode the internal dynamics. In Sect. 3.3, we described how to obtain $S_0(q, t_p)$ and $S_2(q, t_p)$, as illustrated in the middle panel of Fig. 3 for a fixed value of $q$. When this procedure is applied for a range of $q$ values, the two curves in the right panel are obtained. The isotropic curve $S_0$ can be compared with various approximations, such as an azimuthal integration over the detector plane (see [34] for a discussion of this and other approximation schemes), which gives surprisingly good results.

In Fig. 4, we compare the real difference densities obtained from the wave packet propagation to an inversion from the diffraction data using (59).We used a pump–probe delay time of 1,650 fs which according to Fig. 2 corresponds to a stretched excited state distribution with a maximum around 12 Å. With the LCLS beam parameters of $E = 8$ keV photon energy, and a maximum scattering angle of $\theta_{\text{max}} = 60°$, only scattering vectors up to $q_{\text{max}} = 4\pi E/(hc)\sin(\theta_{\text{max}}/2) \approx 4$ Å$^{-1}$ are collected. We find that this range of scattering vectors is too small to clearly resolve the nuclear density distribution via (59), whereas perfect agreement with the original density distribution is obtained with a larger range of scattering vectors as indicated in the figure. To that end, it must also be recalled that the inversion was carried out from a calculated diffraction signal. In reality, noisy experimental data must be used as input for the inversion.

Above we only studied excitation of NaI out of the rotational ground state. In practice, one usually starts with a Boltzmann distribution of initial rotational

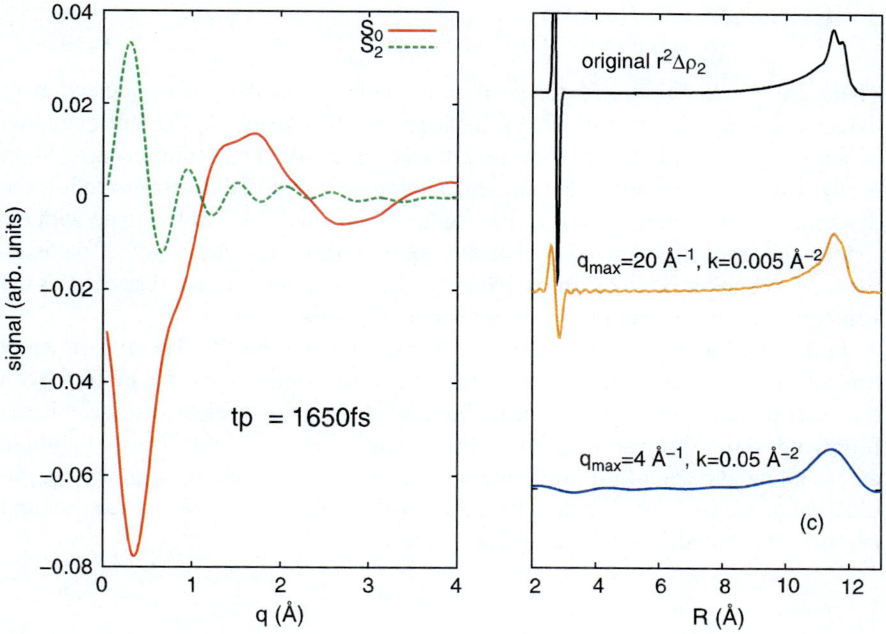


**Fig. 4** Inversion of the diffraction signal at a pump–probe delay time of 1,650 fs. Left panel: difference diffraction curves for the isotropic and anisotropic signal; Right panel: comparison between the original difference density and the inversion via (59). The upper curve is the original difference density, the other curves are reconstructions with different values of $q_{\max}$ and the damping constant $k$ (see [21] for details). Adapted from [21]

states. When the molecule absorbs a photon in the rotational ground state, the resulting excited state wave function consist of a single rotational quantum state with $l = 1$; the density $\rho_{\text{ex}}$ is always preferentially oriented towards the laser polarization axis [see (52)]. This is a special case; for a nonzero initial rotational quantum number $l_0$, the excited state wave function is composed of contributions with $l_0 + 1$ and $l_0 - 1$. As a consequence, the density also includes time-dependent rotational interference terms, and the anisotropic density can show amplitude changes that can be interpreted as rotation of the excited molecule. The rotational interference terms for ensemble members with different initial angular momentum oscillate with different time scales. When performing the ensemble average some time after the excitation, the different anisotropic contributions typically cancel. In effect, the diffraction pattern becomes almost isotropic on a timescale of few picoseconds with alignment revivals at certain times.

Very recently an anisotropic scattering pattern from photo-excited myoglobin was observed, including its temporal change from 100 ps to 1 μs [37]. The relatively slow rotational motion of myoglobin made this observation possible with the ~100 ps time resolution available at a synchrotron source.

# 4 Experimental Results

There are two interesting regimes of time evolution in the probing/detection of dynamical nonequilibrium structures. In the regime of "dynamics," the time evolution of atomic positions is detected on its intrinsic timescale, i.e., femtoseconds. Short X-ray pulses – on the timescale of atomic motion – are required in order to follow the dynamics of the chemical bond. In the regime of "kinetics," which has to do with the time evolution of populations – and in the context of time-resolved X-ray diffraction – the time evolution in an ensemble average of different interatomic distances or the structural determination of short-lived chemical species is considered.

In this section, we discuss briefly a few experimental results. The experimental time resolution has so far not allowed for direct observation of atomic motion (with the exception of a few experiments in the solid phase). However, as discussed in the following, short X-ray probe pulses allow for the detection of the time evolution in an ensemble average of different interatomic distances associated with the recombination of atoms following the interaction with a solvent cage or for the structural determination of short-lived chemical species.

## 4.1 *Transient Species from Recombination of Photodissociated $I_2$*

The photodissociation and recombination of $I_2$ in solution have been studied for many decades as a prototype of a simple chemical reaction in solution. As mentioned earlier, transient X-ray diffraction signals for the photodissociation of $I_2$ in hexane were calculated from molecular dynamics simulation back in 1986, in order to explore the potential of time-resolved X-ray scattering [8].

The first step towards an experimental realization of this idea was taken about 10 years ago (see [38] and references therein). This is clearly very demanding experiments. The signal from the solute is, e.g., very small compared to the signal from the solvent. The experiment was performed at the time-resolved beamline ID09B at the European Synchrotron Radiation Facility (ESRF). The experimental setup includes a pulsed synchrotron source, a chopper that selects single X-ray pulses, and a femtosecond pump pulse that runs in phase with the chopper. The duration of the X-ray probe pulse is about 100 ps.

The photodissociation of $I_2$ dissolved in $CCl_4$ was studied (at a concentration of $29 \times 10^{-3}$ mol/L). Although the process might look quite simple, it is actually quite complex. Several excited electronic states of $I_2$ participates in the process and after the dissociation, the solvent cage may trap the atoms leading to geminate recombination or the atoms may escape from the solvent cage and recombine with other partners. With the time resolution available at a pulsed synchrotron source (~100 ps), it is not possible to directly follow atomic motion, i.e., the dynamics of the chemical bond. Furthermore, it is expected that incoherent motion will set in when the solutes start to interact with the solvent cage. Probing of molecular motion in the latter case, after the early-time coherent dynamics, will therefore represent an ensemble average which can contain a variety of different motions. In this context, it can be anticipated that the time-convolution in (34) leads to an ensemble average of atomic positions, i.e., $\tilde{\rho}_{\text{average}}(\mathbf{R};\, t_p)$.

After 100 ps, the anisotropy introduced by the pump laser has disappeared due to the interaction with the solvent. Thus, with the isotropy of the liquid system before and after laser excitation, the contribution to the signal from the solute can be described by an equation similar to (44), with the quantum distribution of internuclear positions replaced by $\tilde{\rho}_{\text{average}}(\mathbf{R};\, t_p)$, i.e., the time-dependent I–I atom–atom distribution function.

The energy released by the excited iodine molecules heats the solvent. Thus, the solvent structure is also responding to the dynamics of the chemical reaction. This gives an additional contribution to the total difference diffraction signal [38, 39].

From the analysis of the experimental scattering data it was, in particular, possible to extract the time-dependent I–I atom–atom distribution function.

## 4.2 *Structure of an Excited Diplatinum Complex*

The next example concerns the determination of bonding distances in an electronically excited molecule in solution [40]. The structure of the excited $^3A_{2u}$ state of tetrakis-$\mu$-pyrophosphitodiplatinate(II), $Pt_2(P_2O_5H_2)_4^{4-}$, in aqueous solution was investigated on a timescale from 100 ps to 1 μs after optical pumping.

With a pump excitation wavelength of 267 nm and pulse duration of 2 ps, the molecule is excited via the $5d\sigma^* \rightarrow 6p\sigma$ transition [41]. Internal conversion within a few picoseconds leads to the triplet state $^3A_{2u}$ with a lifetime of 9.8 $\mu$s and phosphorescence wavelength of 515 nm. The excess energy in the internal conversion is transferred to the solvent on a timescale significantly shorter than 100 ps. This high-temperature configuration of the solvent leads to solvent expansion on the nanosecond timescale.

The experiment was performed at the time-resolved beamline ID09B at the ESRF (as described above). The X-ray probe pulse is 100 ps long, and sampling of scattering data was obtained from a large number of excitation cycles with a repetition rate of pump/probe cycles close to 1,000 Hz.

The fitting procedure was based on a calculation of the scattering intensity with the interatomic distances in the excited diplatinum complex as fitting parameters. Furthermore, the bulk-solvent contribution to the difference scattering signal is expressed in terms of the temperature change and the density change of the solvent. The response of the difference scattering signal to these parameters was determined in independent experiments where water was heated by short infrared pulses. The temperature and density changes were then treated as fitting parameters. Again, after 100 ps, or later, the anisotropy introduced by the pump laser has disappeared due to the interaction with the solvent. Thus, the scattering from the excited diplatinum complex was calculated via the Debye formula for an isotropic distribution of molecules (see (46) for the diatomic case).

After a careful analysis of the quality of the fit, for the 100 ps delay as well as the 1 $\mu$s delay, the Pt–Pt distance of the metastable triplet state was found to be 2.74 Å – which is 0.24 Å shorter than in the ground state.

## 5 Outlook

In the near future, experiments which allow for direct observation of coherent atomic motion in (photo-)chemical reactions, will be carried out [6]. Thus, real-time "filming" of atomic motions via time-resolved X-ray diffraction should become reality.

As emphasized in this review, time-resolved X-ray diffraction of dynamical nonequilibrium structures involve, at any given time, the signal from a distribution of structures, as described by quantum mechanical wave-packets in the case of pure states. Thus, the interpretation of experimental data is nontrivial. For isolated molecules, direct inversion of diffraction data is possible in some cases as illustrated in this review. This approach might be useful also for reactions in the liquid phase – for the short-time dynamics before interaction with the solvent plays an important role.

The high intensity of X-rays generated by free-electron lasers might also pave the way for direct observations of chemical reactions in the gas phase.

**Acknowledgments** This work was supported by The Danish National Science Foundation. We thank Dr. Ulf Lorenz for performing all the numerical calculations of this work.

## References

1. Hirschfelder J, Eyring H, Topley B (1936) J Chem Phys 4:170
2. Zewail AH (2000) J Phys Chem A 104:5660
3. Srinivasan R, Lobastov VA, Ruan C-Y, Zewail AH (2003) Helv Chim Acta 86:1763
4. Tschentscher T (2004) Chem Phys 299:271
5. Gaffney KJ, Chapman HN (2007) Science 316:144
6. Emma P et al (2010) Nat Photon 4:641
7. Coppens P (1992) Annu Rev Phys Chem 43:663
8. Bergsma JP, Coladonato MH, Edelsten PM, Kahn JD, Wilson KR, Fredkin DR (1986) J Chem Phys 84:6151
9. Ma H, Lin SH, Rentzepis P (1992) J Appl Phys 72:2174
10. Chao CH, Lin SH, Liu WK (1997) In: Helliwell JR, Rentzepis PM (eds) Time-resolved diffraction. Oxford University Press, Oxford, pp 260–283
11. Ben-Nun T, Martinez P, Weber R, Wilson KR (1996) Chem Phys Lett 262

12. Barty CPJ, Ben-Nun M, Guo T, Ráksi F, Rose-Petruck C, Squier J, Wilson KR, Yakovlev VV, Weber P, Jiang Z, Ikhlef A (1997) In: Helliwell JR, Rentzepis PM (eds) Time-resolved diffraction. Oxford University Press, Oxford, pp 44–70
13. Ben-Nun M, Cao J, Wilson KR (1997) J Phys Chem A 101
14. Cao J, Wilson KR (1998) J Phys Chem A 102:9523
15. Tanaka S, Chernyak V, Mukamel S (2001) Phys Rev A 63:063405
16. Bratos S, Mirloup F, Vuilleumier R, Wulff M (2002) J Chem Phys 116:10615
17. Rozgonyi T, Sauerbrey R, Feurer T (2005) J Appl Phys 97:013537
18. Debnarova A, Techert S, Schmatz S (2006) J Chem Phys 125:224101
19. Henriksen NE, Møller KB (2008) J Phys Chem B 112:558
20. Lorenz U, Møller KB, Henriksen NE (2010) Phys Rev A 81:023422
21. Lorenz U, Møller KB, Henriksen NE (2010) New J Phys 12:113022
22. Sakurai JJ (1967) Advanced quantum mechanics. Addison-Wesley, Reading
23. Als-Nielsen J, McMorrow D (2001) Elements of modern X-ray physics. Wiley, Chichester
24. Cohen-Tannoudji C, Dupont-Roc J, Grynberg G (1989) Photons and atoms. Wiley, New York
25. Cohen-Tannoudji C, Dupont-Roc J, Grynberg G (1992) Atom-photon interactions. Wiley, New York
26. Loudon R (1983) The quantum theory of light, 2nd edn. Oxford University Press, Oxford
27. Shore BW (1990) The theory of coherent atomic excitations, vol 1. Wiley, New York
28. Møller KB, Jørgensen TG, Dahl JP (1996) Phys Rev A 54:5378
29. Ihee H, Lorenc M, Kim TK, Kong Q, Cammarata M, Lee JH, Bratos S, Wulff M (2005) Science 309:1223
30. Lorenz U, Møller KB, Henriksen NE (2010) Phys Rev A 82:069901
31. Taylor JR (1972) Scattering theory. Wiley, New York
32. Henriksen NE (1995) Adv Chem Phys 91:433
33. Cao J, Wilson KR (1997) J Chem Phys 106:5062
34. Baskin JS, Zewail AH (2006) Chem Phys Chem 7:1562
35. Lorenz U, Schmidt B (2010) Wavepacket 4.7.2: a program package for quantum-mechanical wavepacket propagation and time-dependent spectroscopy. http://wavepacket.sourceforge.net, 2010
36. Grønager M, Henriksen NE (1998) J Chem Phys 109:4335
37. Kim J, Kim KH, Kim JG, Kim TW, Kim Y, Ihee H (2011) J Phys Chem Lett 2:350
38. Wulff M, Bratos S, Plech A, Vuilleumier R, Mirloup F, Lorenc M, Kong Q, Ihee H (2006) J Chem Phys 124:034501
39. Cammarata M, Lorenc M, Kim TK, Lee JH, Kong QY, Pontecorvo E, Lo Russo M, Schiro G, Cupane A, Wulff M, Ihee H (2006) J Chem Phys 124:124504
40. Christiansen M, Haldrup K, Bechgaard K, Feidenhans R (2009) J Am Chem Soc 131:502
41. Smith DC, Gray HB (1990) Coord Chem Rev 100:169

# Index